A POLITICAL ECONOMY OF FOREST RESOURCE USE

To the loving memory of the late Dr. Shafique Ahmed Khan,
who used to tell me: "forestry is more about people than about forests";
and
To Mustari Khan, who has an amazing capacity to spark the light in me whenever I am faced with darkness.

A Political Economy of Forest Resource Use

Case studies of social forestry in Bangladesh

NIAZ AHMED KHAN
University of Wales, Swansea, UK
and
University of Chittagong, Bangladesh

Studies in Green Research

Routledge
Taylor & Francis Group
LONDON AND NEW YORK

First published 1998 by Ashgate Publishing

Reissued 2018 by Routledge
2 Park Square, Milton Park, Abingdon, Oxon, OX14 4RN
605 Third Avenue, New York, NY 10017

Routledge is an imprint of the Taylor & Francis Group, an informa business

A Library of Congress record exists under LC control number: 98071400

ISBN 13: 978-1-138-60838-2 (hbk)
ISBN 13: 978-0-429-46452-2 (ebk)

Contents

List of Tables

List of Figures

Foreword

"Social forestry" has been an uncertain concept. Many practitioners accepted its early promise uncritically; and have equally uncritically accepted the subsequent shortcomings revealed in its approach. There are now many evaluations of failed and successful social forestry schemes; and a good number of programmatic statements about social forestry's use, limits and potential. Few people have seriously examined "social forestry" as an instance of rural development planning or its specific historical and social conditions. The importance of this present book is that Dr. Niaz Khan has tackled these wider intellectual issues of salience, historical context and ambiguity head-on, usually with verve and insight and often detailed documentation. In doing so, he illuminates important issues for Bangladesh rural studies and for our general understanding of the political economy of forest resource use in south Asia and beyond.

Dr. Khan shows how government interventions on behalf of forest conservation and pro-people rural development forestry are justified by high-sounding ideals of public service. The elevated ideals help structure the discourse within which quotidian calculations of technical forestry and the routine public administration of forest and village land take place. The "social forestry" discourse is powerful, provides a platform for discussions with foreign donors about increased forest department funding, yet also disguises or submerges the lessons being learned in the field. It is a vivid example of the development of institutional ignorance about poor people's real needs in which both national and international partners are complicit. The practices are too deeply seated to describe them as simply uncritical or even as overtly cynical. They involve, in his analysis, a complex mixture of institutional pathos, the pressures of political economy, predictable institutional maintenance and everyday, legal-rational administration.

The core of the book lies in an account of the extensive commercialization of forest products seen in historical perspective and the consequent increasing alienation of the surrounding villagers from their forest resources. In the gap between the ideals of "forestry for the people" and the rural population's concrete, sometimes embittered, disengagement from forest management lie the ambiguous rhetoric and practices concerning "social forestry" schemes. The same gap between aim and practice also provides the occasion for Dr. Khan to demonstrate his skills as a social analyst. He bridges the "micro" and "macro" by his concept of an evolving and complex political economic "frame" for social forestry scheme planning and management. "Alienation" is the concept he draws from Marx's social theory to characterize the rural folk's legal and perceived relation to the surrounding forest. It was, or it "is", "their" forest but many centuries of commercialization and government administration have separated them from its ownership; and from use of its resources.

The author's energy and dedication as a researcher is simply demonstrated in the book. It somehow manages to combine the scholarly development of concepts, a comparative analysis of "social forestry" elsewhere in south Asia, a detailed analysis of local social process in "social forestry" schemes, a very useful evaluation of scheme benefits and other outcomes, and an account of the regional and local history of forest products trade and administration. It is also based on fieldwork in two areas and an intimate knowledge of the government machinery. To have researched and woven a thread between all these elements was, I know from my close association with Dr. Khan's studies, very demanding; and it results in an impressive achievement. I am also impressed that he did not stop at the point he finished this work, content with his already considerable output. The very large number of journal articles published during his period as a research scholar while at the University of Wales Swansea signal an important contribution to social science in Bangladesh; and the arrival of an author of great industry with a thirst for investigation and high academic achievement.

Professor Alan Rew,
Professor of Development Policy and Planning,
and Director of the Centre for Development Studies,
University of Wales, Swansea.

Acknowledgements

"Praise be to God, the Cherisher and Sustainer of the Worlds" (Quran, Chapter 1:2) who has given me the ability to complete this difficult assignment, as He promised, "verily, with every difficulty there is a relief" (Quran, Chapter 94:6).

I have accumulated an unusually large number of debts for this work and my feeling at this time of acknowledgement is overwhelming. Many remain unnamed for the sake of brevity alone. However, I do hope that they will accept my sincere thanks, because they know who they are.

This book originated in the form of a doctoral thesis at the University of Wales, Swansea. I express my continuing debts to Professor Alan Rew, my supervisor, for half a decade of questioning and challenging we have shared which shaped my ideas.

My doctoral study in the UK was funded by a Commonwealth Scholarship under the Association of Commonwealth Universities, London. The University of Chittagong granted me study-leave and the Centre for Development Studies, Swansea provided a conducive working environment.

Some of my colleagues in Swansea and Chittagong have read parts of the original thesis and made valuable suggestions. My thanks to Chris Gerry, Richard Montogomary, Samiul Hasan, Tofail Ahmed, Prodromos Panayiotopoulos, Catherine Locke, Habibur Rahman, and Mike Shepperdson. My special appreciation is for Helen Hintjens for her good humour and readiness to help whenever I needed.

My deep gratitude is due to Afzalur Rahman and Abul Farah Saleh for their invaluable support and patience during the field-work.

I also extend my sincere thanks for the help I enjoyed from many government officials, especially the forest officials. Without their support, it would be impossible for me to access many historical and classified materials. I feel somewhat embarrassed at this point, because this book contains many references of the failure and limitations of public policies and practices. Such criticism may be considered inappropriate on the part of a researcher who has benefitted enormously from the assistance of public officials. Yet I decided to be probing and truthful of the realities. I avoided diplomacy while writing this book and tried to depict a candid and authentic picture of what I had observed in the field. My assessment of the actions of the government in general and the forest department in particular, in no way diminishes my personal debt to the concerned officials, especially to Mozammel Hossain, Shah Ali Imam, Mohammad Shafi, Fazlul Hoq, Munshi A. Islam, M. Mohiuddin, Zillur Rahman, Jahangir H. Chowdhury, Junaid K. Chowdhury, and Altaf Hossain.

I have benefitted much from conversations with the scientists in Bangladesh Forest Research Institute, Bangladesh Agricultural Research Council and Institute of Forestry including Farid U. Ahmed, Khairul Alam, Kamal Hossain, M. Mohiuddin, Rukunuddin Ahmed, Millat Mostafa, Mir M. Hasan, M. A. Sattar, A.B.M. Salehuddin, A.H. Majumdar, S. S. Islam, Jamil A. Chowdhury, Harun-or-Rashid, Waziullah, Mobashwar Ahmed, and Khairul Alam.

I also wish to thank Professor Chris Garforth of the University of Reading for many helpful comments on the original thesis and to Sarah Markham, Rachel Hedges, Pauline Beavers and others at Ashgate Publishing Limited for their help whilst preparing the manuscript for the bock.

I lived among the farmers and villagers in Betagi, Pomora and Chandra for a while, and shared some of their pains and sorrows. On numerous occasions, I had meals with them, sat in tea-stalls for hours together, and followed them through farms and fields. Nothing can possibly repay the gesture of goodwill they have shown to me; I feel a strange sense of kinship towards them.

Finally, my sincerest gratitude to my family members and my in-laws in Chittagong whose encouragement has always sustained me. I have dedicated this work to the memory of my father, the late Dr. Shafique Ahmed Khan and to my wife, Mustari, for reasons explained elsewhere. Although I owe a great deal to many individuals and institutions, the remaining errors and lack of judgement are my own responsibility.

Niaz. A. Khan
University of Wales, Swansea.

1 Setting the Scene

Forestry in Bangladesh: Growing Concerns

In 1987, a national Task Force while evaluating Bangladesh forestry, warned that unless some "decisive actions" were taken in the forestry sector, "the situation would soon deteriorate from crisis to catastrophe" (Task Force, 1987, p.24). The evidence in this book suggests that the prediction rings true for present circumstances. The recent Forestry Master Plan (FMP) of the country notes (FMP, 1992a, pp.17-18):

> The history of forestry in Bangladesh is one of continuous depletion in terms of area and quality of the forest resource....In 28 out of the 64 Districts of Bangladesh, there are no natural forests left....The percentage of land under the classification of "forest" is presently about 14 percent of the total land area; area under actual tree cover is about 8 to 9 percent ; and that of closed forest only 6.4 percent....The forest on State lands have been subjected to illicit cutting, shifting cultivation and encroachment, fire, grazing and other authopogenic influences. There are, thus, designated forest areas with low stock density and/or no tree cover at all....There has been an overall depletion in forest growing stock in all natural forests, and those of forest plantations are far below their potential.

Bangladesh currently possesses "an alarmingly small and shrinking productive forest base" (Task Force, 1987, p.24). An estimated 2.53 million hectare (m ha.) of the total land of the country is designated "forest land", out of which 2.25m ha. is government owned and 0.27m ha. is privately controlled homestead forest land. Out of 2.25m ha. of government forest land, 1.53m ha. is the national forest under the purview of the forest department; and the rest which is virtually barren, lies under the control of Local Government Councils in the Chittagong Hill Tracts (Table 1.1). The growing stock in all major forests of the country is drastically declining. The growing stock of Sundarban (the country's largest mangrove forest tract) has depleted from 717 million cubic feet (m cft.) in 1960 to 542m cft. in 1985; which implies a decline of 35% over 25 years (Chaffey et al., 1985). Similarly, there has been an estimated 17% decline in the resource base in the forests of Chittagong Hill Tracts over last 21 years (Bhuiyan, 1994, p.26). Most parts of the Inland *Sal* forest have been reduced to virtually shrub lands with only 25% tree cover (Table 1.2). Signs of desertification have emerged in northern

Bangladesh where only 2% area has effective tree cover (Bhuiyan, 1994, p.26).

Between 1971 and 1980, deforestation occurred at an annual rate of 8,000 ha. (Lai and Khan, 1990; Association of Development Agencies in Bangladesh (ADAB), 1992; Task Force, 1987). Referring to the Food and Agriculture Organisation's (FAO) recent estimates the Forestry Master Plan has suggested that between 1981 and 1990 "the annual loss of forests" has been 37,600 ha.; and has predicted, with great concern, that "if this rate is not contained, the forest of Bangladesh will not last through the period of [next] 20 years" (FMP 1992a, p.19). The area under forest cover has declined from 16% to 8% over a period of last 30 years (Rahman 1992, p.4). In 1990, the forest area with a 10% or more crown-cover was estimated to be only 769,000 ha. (FMP 1992a, p.19).

Table 1.1 Distribution of forest area by legal and ownership classes (area in million hectares)

Category	Area
State Lands:	0.53
Under the control of FD	1.37
* Reserved Forest	0.01
* Acquired Forest	0.02
* Vested Forest	0.04
* Protested Forest	0.09
* Transferred *Khas* (Revenue) Land	
Under control of District Administration:	0.72
* Unclassed State Forest	0.72
Private Lands:	
* Homestead/Village Forest	0.27
Others:	
* Community forest plantations of various kinds	0.01
Total:	2.53

Source: FMP, 1992, p.18.

Table 1.2 Estimated tree-cover in major forests of Bangladesh

Forest Type	Location	Approx. Extent (ha.)	Estimated Tree-cover
Sundarban Mangroves	Southeast Coast	600,000	75%
Mixed Dipterocarp Hill Forests	Chittagong and Sylhet Hills	600,000	50%
Sal Forests	Central Plains	125,000	25%
Homestead Forests	Throughout	270,000	80%

Source: Bangladesh Agricultural Research Council (BARC), 1991, p.5.

Most of the deforestation has occurred in the government forests (Asian Development Bank (ADB), 1992), underlining the inadequacy of the bureaucratic-custodian approach to forest management. Only about 60% of the forest department controlled forest is planted and the rest is either encroached or barren (Government of Bangladesh (GOB), 1990, p.VE-1; also see Task Force, 1987, p.24). The government forests meet only about 15% requirement of timber and fuel wood in the country (Forest Resources Management Project (FRMP), 1992, p.2). There has been, as the Task Force noted, "an increasing consensus that forestry sector policy was ineffective in maintaining the sector's contribution to the national economy" (Task Force, 1987, p.28). The forest department administered the forests under its jurisdiction through "Working Plans". These departmental manuals offered little room for public involvement in forestry. Recently, one major review and evaluation study on Bangladesh forestry also reached the conclusion that, there has been "a clear indication of the very poor condition of the forests under State control" (FMP, 1992, p.19; also see Byron, 1984; Task Force, 1987).

Many analysts argue that there is "very limited scope to increase and develop the forests under traditional forest management" by the government (Task Force, 1987, p.7; also see Magno and Ali, n.d., p.1). On the other hand, the village homestead forests have become the main source of wood supply in the country (Byron, 1984; Task Force, 1987; Douglas, 1982; GOB, 1990). These forests supply 85% of all wood requirements (including 90% of all fuel wood and 80% of timber) in Bangladesh (Douglas, 1982). On the contrary, target production from the government forests was further lowered because of a "drastic decline" in the growing stock on all major state forests in recent years (GOB, 1990, p.VE-2). Consequently, there has been mounting pressure on the village forests. One estimate, for example, suggests that total homestead forest resource has declined by more than 10% between 1977 and 1985 (Task Force, 1987, p.22).

The past history of forest use in Bangladesh, as I will elaborate in the coming chapters, demonstrates the processes of progressive commercialization of forest use, appropriation of forest land and revenue maximization by the state, and the predominance of commercial-industrial interests in forestry. The most significant social impacts of these trends in commercialisation and the consequent change over from community use of forests to state control, had been the systematic alienation of local communities from forests, and the progressive diminution of their traditional rights. The forest department with their paramilitary orientation had been trying to police "their" territories from local people with an uncompromising zeal. In the process, there has been a prodigious growth of conflict between the government and local communities. The Bangladesh government has consistently blamed population increase and local people for the destruction of forests (see Chapter 5). In fact, for many Third World governments in general, the natural reflex has been to blame local poverty and population growth for deforestation and other forestry-related problems (see Shepherd, 1992; Fernandes and Kulkarni, 1983; Barraclough and Ghimire, 1990; Shiva et al., 1987). But, as the Society for Promotion of Waste Lands Development (SPWD, 1992, p.39) convincingly argues:

> that [the people] have played a part in deforestation is not in dispute. But to blame people is a convenient diversion. [The] prescription, which has followed, has been counter productive. The solution should have been to devise policies to meet people's needs for forest products such as fodder...and fuelwood. Instead, it has been to demand and impose stringent laws, with military style controls, compulsory castration of bulls, banning people from entering forests, and curtailing the rights and concession of forest dwellers. These measures have their roots in the bureaucratic psyche which believes in the power of the stick and in rewards to those who wield it.

Similarly, in the general context of the developing countries Barraclough and Ghimire maintain that "the facile explanations that deforestation is primarily caused by poverty and population growth...are tautological" (1990, p.13). They view these problems as the "symptoms of [an] unequal exploitative development" process; and argue that "the symptoms" must not be "confused with causes of deforestation" (1990, p.13). These "exploitative styles of development" sponsored by many Third World governments, largely, lie at the root of major forestry problems; and they can assume different forms (see Chapters 2 and 5). For example, as Shepherd (1992a, p.8) argues, in many instances,

> demographic increase in a particular area is as frequently the result of squeezing a politically weak people out of good land into more marginal land as it is the result of a population explosion. Such evictions are frequently associated with the expansion of [state-sponsored] commercial agriculture or commercial forestry, driving the poor up hills...or into forests.

Thus, while conceding that local needs of the villagers caused some deforestation, evidence from this book and recent literature suggests that massive exploitation by "the commercial-industrial interests" (Fernandes, 1983, p.1) and the varied "development interventions" by governments have been the prime contributor to deforestation and other problems in the forestry sector in many countries of south Asia (Chapters 2 and 5).

In fact, the situation in the Bangladesh forestry sector was farther exacerbated by the growth-led economic interventions of the government (in line with other states of south Asia) in the name of "planned development". These interventions included such processes as the unplanned growth of forest-based industries without due regard to the validity of resource base; large scale eviction and displacement of local communities; neglect of traditional values and indigenous knowledge; and the intrusion of various exploitative elements into local territories. All these forces acting together have brought about grave consequences in the forestry sector including the massive deforestation; the failure of state efforts in forest conservation and regeneration; the widening mistrust between state foresters and local people; and the adverse impacts on the surrounding environment, especially on soil and water resources. The government was struggling to replenish the dwindling forests; but without popular support and participation, the state efforts were mostly doomed to failure (see Chapters 2 and 5). The government (GOB, 1992, pp.18 and 26) itself, commenting on the general state of forestry, notes with great concern that:

> Development of forestry in Bangladesh, specially the state owned forests is beset with many problems and constraints....The traditional custodian approach of forest management practices have been found to be inadequate to cope with the problems....Depletion of forests is perhaps the most serious environment problem faced by Bangladesh today. Availability of wood for industries and housing, biomass for fuel, different forest and non-forest products....Increase of frequency and magnitude of flash flooding,...run-off, erosion and reduction of water

absorption and consequent lowering of water tables, draught, increased salination, loss of top soil are amongst many deleterious effects of deforestation that can be noticed in Bangladesh.

Social Forestry as a Panacea

One widely promoted solution to these trends in deforestation and alienation has been Social Forestry (hereafter referred to as "SF"). Since the early 1980s, Bangladesh forestry has witnessed the vigorous march of SF programmes, apparently in an attempt to redress the process of public alienation, and to make room for wider participation of local people in forest use and management. A note of clarification regarding the terms "SF" and "local people" seems necessary here, considering their frequent use throughout the book. The term SF defies any universal definition. In this study, it is used as an umbrella term for such public initiatives which aim at ensuring "active participation by the rural people in planning, implementation and benefit-sharing of tree-growing schemes" (Task Force, 1987, p.1). SF is viewed within the broader framework of rural development in Bangladesh (as well as in other south Asian countries); and it has become one of the most dominant strategies for both rural development and forest management (Rahman, 1991; 1992; Alim, 1988; Khan, 1996). It includes public afforestation programmes in marginal and degraded state forest lands, the expansion and extension of forestry for local community development, the "strip plantations" (along highways, railways, embankments), and the "woodlot plantations" on state or community lands on "benefit-sharing" arrangements (for the common components of public SF programmes in Bangladesh, see Byron, 1984; Task Force, 1987; Alim, 1988; also see Chapter 3 for the south Asian perspective). The philosophy, stated goals and activities of SF, as commonly observed in south Asia, will be elaborated in chapter 3. The term "local people" is used rather flexibly in this study. It has been broadly used in different contexts to include the poor and disadvantaged sections of rural communities, forest-based populations (including tribal groups), and the common (generally poor) households neighbouring the forests. It also includes the "indigenous people", who are defined as the "descendants of inhabitants of a country prior to its colonization, whose rights are not still recognised...[they] live in a particular territory with their own social cultural characteristics" (Gray, n.d., p.7).

There has been unprecedented enthusiasm for SF in Bangladesh in line with other south Asian states. It was seen almost "as a panacea for all forestry related economic and environmental ills in the rural area" (Banerjee, n.d., p.25). The following excerpts from some studies are typical representations of the common trend of great appreciation for and expectation from SF projects:

> [SF is] the means of rural uplift [and] the final expression of the people's involvement in the .. development and exploitation of forests for the local communities' own benefit and benefit of the nation. (Maleskar, 1983, p.137)
>
> [SF is] the most practicable approach to the solution of wood deficit and environmental problems and rehabilitation of the depleted homestead and degraded forests. (Bhuiyan, n.d., p.ii)
>
> [SF is] the prime possible solution to the reverse the current trends in the forestry [sector]. (Ahmed and Bhuiyan, 1993, p.1)
>
> [SF is] the new and sustainable frontier to expand tree cover and the resource base, generate employment and income, ensure active participation of women, youth and the unemployed specially in the rural sector. (GOB, 1992, p.27)
>
> There can be hardly any other scheme of development which can benefit such a large number of people living below the grinding poverty line. (Tiwari, 1983, p.64, commenting on a prescribed SF model.)
>
> [SF] is a social change directed equally to foresters to cast off their traditional role from guardians of the forest to agents of development; and to people themselves in rural areas to shift from individualistic to a cooperative state of mind. (Rahman, 1994, p.95)

It has been unequivocally announced that SF "has supplanted traditional forest management as the top priority of the forestry sector in Bangladesh" (Task Force, 1987, p.1). SF programmes comprise the major part of the forest department's activities. The following table summarizes the activities under the public SF programmes in Bangladesh during the initial years (1980-1991).

Table 1.3 Activities under SF programmes (1980-1991)

Strip plantations (in Kilo Metre)	6032
Fuelwood plantation (in ha.)	8412
Agroforestry (in ha.)	1040
Number of seedlings distributed (in million)	37.87
Village afforestation (in number)	2806

Source: Hussain, 1992, p. 15.

The stated goals of SF cover a wide array of social, economic and environmental factors. The stated social and economic goals can include: the participation of women, youth and other disadvantaged sections of the society; the creation of people's organisation; reorientation of the values and practices of state foresters; the creation of a participatory environment for people and government officials within which to cooperate in meeting the needs of local people; increased living standard and social status of the participating farmers; creation of more avenues for employment, savings and investment; and fostering social equity (see Chapters 3 and 6). On the environmental side, the goals have included: control of erosion; checking desertification and natural calamities; and bringing denuded landscape under tree-cover. Thus, the assigned tasks of SF are somewhat all-encompassing and the coverage is gigantic by any standard. Hopes and aspirations about SF have been equally high-blown.

An Emergent Scepticism?

Although SF still remains the most dominant pursuit in the forestry sector, there have been some signs of disenchantment. Amidst years of obsession with SF, from the late 1980s, a few studies on SF projects in different locations of south Asia started to hint that the goals seemed too ambitious, the political-economic context in which SF projects were operating was not conducive, and the ambitious purposes of SF remained mostly unrealised (see Chapter 3). Reporting on India's experience with SF, Sen and Das (1987, pp.4-5), for example, argue that "the present practice of planning and management of [SF] forestry in the states hardly fulfils the conditions of participatory programmes". In the same vein, Khan (1994, p.343), from a review of selected south Asian SF projects, concludes that "albeit the widespread popularity and advocacy, the role of SF in ensuring participation and in empowering the local populace has been insignificant". After observing nearly a decade's performance of SF in Bangladesh, the Task Force (1987, p.1) notes that SF has "not produced satisfactory result in terms of engaging and sustaining the active participation of rural people". Against the common rhetoric and grandiose promises which surround SF plans, Hussain (1987, p.22) in a realistic assessment on the prospects of SF in Bangladesh comments:

> Although, the relevance of SF as a programme for ecological and economic restoration of rural areas is beyond question, one should not be too optimistic about

> the outcome of the present programme. Socio-political and institutional factors may severely affect the success of the programme.

Thus, there have been some, though limited, reports which signalled that SF is not an unmixed blessing after all; and more in-depth explorations are needed to make a realistic appreciation of SF.

The Research Problem

The above background sets the stage for my study, which explores the political economy of forest use in Bangladesh, and the status of SF within the surrounding political-economic environment. The rationale of the study is as follows. First, SF is not merely an isolated, static or technical strategy for afforestation, but rather a part of the wider historical process of forest use. The history of forest use, which is marked by a progressive emphasis on commercialization of forestry and the consequent alienation of local communities, provides the context for a meaningful analysis of current SF programmes. Second, lately SF has been the single concept used to present the prospects for sustainable forestry development in Bangladesh. It has received substantial support from the country's political leadership and the international donor agencies. In contrast to the uncritical propaganda attached to SF, this study attempts to assess the actual achievements of SF in the field, and the prospect of it realising its ambitious objectives in the face of unfavourable land tenurial arrangements, markets, government and local government institutions and policies, social relations and other similar political-economic constraints. Third, without knowledge of the surrounding environment, any attempt to appreciate the performance of SF may lead to an incomplete or superficial understanding. In general terms, Wood (1994, p.31) argues that any rural development strategy (which would include SF) in Bangladesh requires detailed studies on the rural social and political-economic setting; and at present there is "a sheer lack of knowledge" in this respect. In the same vein, in an evaluation of Integrated Rural Development Projects in Papua New Guinea, Crittenden and Lea (1989, p.114, cited in Rew, 1996, p.29) claim that

> the effectiveness of programmes and projects is related to the appreciation of .. the socio-political context and [the] understanding of the physical and cultural environment in which programmes take place.

Although there have been many studies on Bangladesh SF in recent years, they have been, for the most part, narrowly statistical in nature or concerned with specifically technical, economic or silvicultural dimensions (for example, Bangladesh Bureau of Statistics (BBS), 1990; Hoq, n.d.; Dalmacio, 1989; Thana Afforestation and Nursery Development Project (TANDP), 1993; n.d.; Manandhar, 1987; Abedin et al., 1990; Rahman, n.d.; 1987; Rahman and Khan, 1994; M.M. Bhuiyan, 1994). Social science studies are limited in number and are, with some exceptions, weak in empirical contents, methodology and analyses. These can be found in the forms of back-to-desk tour reports by project officials, academics and consultants (Chowdhury, 1993; Manandhar, 1986; Ahmed, 1994); literature reviews and descriptive reports (Anon., 1991; 1991a; Ahmad, 1990; Rahman, 1992; Bhuiyan, 1982; 1984; Ahmed and Bhuiyan, 1993); purely theoretical overviews (Salehuddin and Rebugio, 1990; Salehuddin, n.d.; Ahmed, 1991; n.d.; M.K. Bhuiyan, 1994); manifesto-type notes and papers on the "success" of particular SF projects and the potentials of SF in general (GOB, 1992; Chittagong Forest Division (CFD), 1994; Ahmed and Azad, 1987; Alim, 1988; 1993; Bhuiyan, 1994; Ahsan, 1991; Ahmed, 1987); and 'snap-shot' case-studies (Khan and Khan, 1988; BRAC, 1989; Khandakar, 1994; Miah, 1989; Kader, 1989; Hoque, n.d.; Khan and Khan, 1993). A very few comparatively in-depth empirical investigations on the political-economic environment of particular SF projects are available (for example, Quddus et al., 1992; Khaleque, 1984). It is this relative scarcity, to which this book responds. Fourth, the unprecedented advocacy by the protagonists (the government and aid-agencies) in favour of SF may lead one to wonder about their interests behind the support for SF. Beyond the rhetoric of stated goals, this study intends to provide some insight into the implicit interests of the protagonists in pursuing SF (these interests are referred to in the book as the "subterranean" purposes).

The book develops a number of inter-related themes. The first of these themes is the progressive commercialisation of forestry in south Asia, and the consequent alienation of local communities from the use of local forests. The history of forest use shows a continuous emphasis on commercial considerations. This emphasis tended to monetize forestry use; devalue community control of forest; alienate local people; and curtail local rights. The cumulative consequences of this process of commercialisation led to widespread public alienation. There has been intense conflict between state and local people; and the state efforts in forest management met with public opposition. At this point, there was a realization among governments and international donor quarters that in order to sustain the public forestry programmes, local people would have to be involved in the management of such initiatives. These realizations set the stage for experimentation with SF in the forestry sector throughout south Asia.

The second theme of the book is concerned with the rise, goals and achievements of SF projects in south Asia in general, and in Bangladesh in particular. In the above context of alienation, SF was launched with the stated goal of making room for local people in forest management. But SF's achievement in empowering local communities or ensuring participation of them in the design and management of SF projects has been insignificant. The evidence in this book suggests that SF, with few exceptions, is yet another "technical" and bureaucratically-managed afforestation programme, where local people are "invited" to perform strictly within the limits set by the government.

The third theme deals with questions like : why SF projects, albeit the great national and international eulogy, have only marginally achieved its stated goals? What factors could explain the insignificant performance? In response, Chapters 3, 6 and 7 of the book will argue that, performance of SF is regulated by political-economic factors of the surrounding environment such as markets, land tenure arrangements, social relations, power structures, and public policies and bureaucracy. These factors are referred to in the book as the "frame" of SF; and the "frame" is, in general, not conducive to the achievement of SF's lofty goals. The project goals and inputs are filtered through the "frame", and the ultimate outcomes are often quite remote from the initial formulations.

The fourth theme centres around an enquiry: why do the protagonists still pursue such unpromising goals of SF, when the general experience has shown that SF projects largely fail to live up to their ambitious goals? Chapter 7 examines this question. It shows that, although the stated goals largely remain unrealised, SF seems to be furthering some strategic but tacit interests of it's mainly national government and international donor protagonists. Through SF projects, the government, for example, can attract substantial external funds, reclaim the encroached state-lands, and reassert control over local communities. The donors can also exercise control over the national government, test their ideas, and promote a missionary image.

For better understanding, the themes and the major arguments of the book are simplified and summed up in Table 1.4.

In line with the four broad themes, the book will proceed with a number of specific objectives. The first three objectives are concerned with the broader theoretical and regional experiences of forest resource use and SF in south Asia. First, the book will attempt to conceptualise the evolution of forest use in south Asia (from the experience of selected states) in an attempt to illuminate the process of commercialisation of forest, and the consequent alienation of local people. Second, it will trace the rise, development and performance of SF programmes from a literature review of some selected south Asian countries. Third, it intends to shed light on the surrounding "frame" wherein SF projects operate; and to identify the major political-economic variables which influence and regulate the functioning of SF.

The rest of the objectives concentrate on the empirical context of the specific case-study areas of the study (in Bangladesh). Here, at first, the research will trace the historical policy changes and current practices which have led to commercialise forestry, and caused alienation of local people. Second, it will explore the implications of the surrounding variables of the "frame" on the performance of SF in the case-study areas. Finally, the book aims to assess the performance of the studied SF projects in light of their stated goals and initial formulations. It will also, briefly, shed light on the strategic but tacit interests of the protagonist of SF.

Organisation of the Book

This introduction, is followed by 7 chapters. Chapter 2 reviews the evolving process of forest use in south Asia by conceptualizing three stages: "the pre-colonial and pre-commercial era", "the advent and proliferation of commercialisation", and "the post-independence era". The evolution concerns the progressive commercialisation of forests and an undermining of local rights, whereby local communities have been alienated from neighbouring forests. The major "interests", which are competing for access to forests, are traced. These interests, mainly, are: the local people (especially the poor and weaker sections of local communities who depend on forests for subsistence requirements); the industrial and commercial forces; and the state with its national and international allies. The chapter reveals the inability of local people to resist the commercial pressure on forests and the consequent deprivation of their rights. The concept of "alienation" is used to summarize the process of deprivation, along with other impacts of commercialization.

In the context of public alienation and the growing conflict between state officials and local communities, Chapter 3 examines the rise, growth and achievements of SF as a strategy for people's participation in forest management. It reviews the experiences of selected south Asian states. It also examines the surrounding "frame", wherein SF projects function; and identifies the political-economic variables which largely regulate the performance of SF. The variables analyzed are: land ownership and tenurial arrangements; the role of public forestry organisations and policies; market opportunities; the role of local organizations and cooperative structures; social relations and power structures; and the role of international donor agencies. The identification and analysis of these factors also set the stage for Chapter 6, which in context of the case-study projects in Bangladesh re-examines the implications of these factors in the field. This chapter concludes that these factors are not conducive to the process of empowerment of local people; and therefore, the role of SF as a participatory development strategy has remained marginal.

Table 1.4 A simplified presentation of the central research themes and arguments of the book

Theme 1 : *Commercialization and Alienation* (Discussed in Chapter 2 in the theoretical and south Asian contexts. The theme is then followed and pursued in Chapter 5 in relation to the field study-areas of the research)	
Commercialization of forest resource use and management	Alienation of local communities
Commercialization includes such processes as state appropriation of forests; collapse of local patterns forest use; intrusion of commercial forces into the local economy and forests; revenue, commerce, and profit maximization motives of the state forest administration.	The processes of commercialization led to widespread public alienation. The major manifestations of alienation included the progressive diminution of local rights and control over forests; the exploitative terms of exchange and employment between commercial forces and the locals; the dispossession and displacement of local communities by central commercial interventions; feeling of estrangement among local people; conflict and mutual distrust between state officials and local people.

Theme 2 : *Social Forestry (SF)* (Chapter 2 provides an overview of the goals and achievements of SF; and a regional review of some selected south Asian cases. Chapter 7 evaluates the achievements of SF in context of the case-studies)	Theme 3: *The "Frame" of SF* (The "frame" is identified in Chapter 2; and re-examined in Chapter 6 in context of the case-studies)	Theme 4 : *The "Subterranean" purposes of SF* (Discussed in Chapter 7)
In response to public alienation, SF emerged to ensure participation of local people in forestry; and to redress the sores of alienation. But, it is argued that SF's performance has been insignificant and marginal.	Why SF only marginally achieved its lofty goals? It is argued that the performance of SF is regulated by political-economic factors of the "frame" including land, market, public policies and institutions, social relations, and power structures. These factors are not conducive to materialising the goals of SF.	SF persists despite limited achievements, because the protagonists have some strategic but tacit interests, for example, attracting foreign fund; regaining encroached land and state authority; promoting a philanthropic image among people.

Chapter 4 sets out the research agenda for the study, and the reasons for selecting the particular case-studies in Bangladesh. The methodology of the field-study projects is also described, together with their backgrounds.

Chapters 5 through 7 present the empirical findings of the study. In line with the evolutionary stages of forest use (as identified in Chapter 2), Chapter 5 portrays the historical changes in the philosophy and practice of forest management in the specific study-areas. While Chapter 2 reviews the evolution of forest use at a more general level (in the context of south Asia), Chapter 5, focuses on the field-sites in Bangladesh. It traces the trends which over the years have caused commercialization of forestry, and the consequent alienation of local people. Then, in light of Chapter 3, Chapters 6 and 7 assess the functioning and performance of SF as a strategy for creating avenues for public involvement in forest management, and for redressing the "sores" of alienation. Chapter 6 shows the implications of the surrounding "frame" on the performance of SF; and reiterates that political-economic realities of the study-areas are not generally favourable to achieving SF's stated goals.

Chapter 7 examines the achievements of SF against its stated goals. It reaffirms that in an essentially unfavourable "frame", the achievements of SF clearly fall far short of the lofty initial formulations. It then raises and analyses the subsequent question: why the protagonists of SF still pursue such unpromising goals? The argument developed is that, SF is effectively serving certain strategic but tacit interests of the donors and national governments. For the government, for example, SF provides opportunities for attracting greater foreign funds, managing and reasserting control over local people, and reclaiming lost and encroached state-lands. Besides, SF's populist image helps the state officials to serve as official providers in the country side. At the same time, for donor agencies, SF offers avenues for fostering their ideas, innovation and technology, exercising control over the national government (the unequal partner) and influencing departmental behaviour, and conveying a philanthropic image at the country-side. It is argued that despite the marginal achievements of stated goals, SF will prosper as long as it fulfils these interests of its protagonists.

Chapter 8, summarises the central problems and findings of the study. It starts with a review of the processes of commercialisation of forest use; and the consequent alienation of local communities. Then it evaluates the performance of SF as a "compensatory dividend" in appeasing the scars of alienation; and concludes that SF has not been able to regulate the major factors which causes alienation. This chapter also offers some food for thought for prospective future action in making the observed SF projects more effective. On the research side, it focuses on the study of alienation; and underscores the usefulness of this rather unexplored perspective in addressing the issues of forest resource use in general, and SF in particular. A list of some probable research areas on Bangladesh SF is also included.

2 Forest Resource Use in South Asia: A Conceptual Overview

Introduction

This chapter reviews changes in the social and economic use of forest resources in south Asia. The political-economic and historical processes which have commercialized forest use, and caused a systematic alienation of local people from their neighboring forests are traced. The account provides a context for the development of social forestry (SF) programmes and for the Bangladesh case-studies. The chapter concentrates on forest use in Nepal and India because of the depth of experience and commentary in these countries. However, frequent references have been made to other Asian countries.

Three major stages in the evolution of forest use are identified: the pre-colonial and pre-commercial era; the advent and proliferation of commercialization; and the post-independence era. The evolution concerns the progressive commercialization of forests, and an undermining of the rights of local communities. There are a number of "interests" in conflict over use and access to forests. These interests, mainly, are: the local people (especially the poor and weaker sections of local communities who depend on forests for subsistence requirements); the industrial and commercial forces; and the state and its national and international allies. The use of forests is shaped by the complex interplay of these contending interests, and the conflict between the traditional use-pattern of local people and "the commercial-centralized goals of distant accountability and whimsical market forces" (Burch, 1990, p.xv). The chapter reveals the inability of local people to resist the commercial pressure on forests, and the consequent deprivation of their use rights. The concept of "alienation" is used to summarize the process of deprivation, along with the other impacts of commercialization.

The alienation of local people from forests has led to growing tension between state officials and local communities. The inability of the state to manage forests in isolation of people has stimulated the development of participatory approaches to forestry which are often known as social forestry

(SF). The character and performance of the SF initiatives in south Asia will be examined in Chapter 3.

In the following pages I review the accounts of the general pattern of evolution of forest use in south Asia by examining the regional literature. The overall conclusion in these literature is that local rights and access to forests have become increasingly restricted as commercialization of local forests and economy has proceeded.

Approaching the Study of Forest Resource Use

In the face of intense competition for forest products the study of forest use has increasingly used the perspectives of "Political Economy" (see for example Nadkarni et al., 1989; Fernandes et al., 1988; Chambers et al., 1989; Blaikie, 1988; Anderson and Huber, 1988). Theoretically, the use of a particular forest should be about reconciling the specific ends with scarce means, as suggested by the neo-classical economics. Nadkarni et al. (1989, pp.19-21) provide an account of the applicability of the neo-classical views of economics to the present-day situation of resource use. They conclude that, with the diverse interests competing to establish their own rights on forests, the problem and process of such reconciliation become "no longer purely economic and hardly amenable to neat solutions of neo-classical economics" (Nadkarni et al., 1989, p.19). Then they (Nadkarni et al., 1989, p.20) distinguish the "more complex" approach of political economy from that of neo-classical economics and comment that

> a problem in the economy of resource use transforms itself into a problem in the political economy of resource use, when the ordering of ends in terms of their importance is done not by a given decision maker but through a struggle between vested interest groups.

Blaikie (1988, p.141) combines the concerns of ecology with political economy to "encompass the constantly shifting dialectic between society and land-based resources and also within classes and groups within the society itself". He terms his approach a "Regional Political Ecology". It links farmer and the context of his/her decisions over forest land-use to the complex social relations of production and exchange among similar decision makers, and to the state and the world economy.

The issue of competition as core is articulated in some studies from one perspective of the "Competition for Scarce Resources". The main assumption made here is that people compete by a variety of methods for scarce resources in the Third World and the winners in these competitive relationships improve their living and position in the society at the cost of their fellow beings. Primarily elaborating on the theoretical framework of Barth (1981), Jansen's (1986) study on rural Bangladesh exemplifies the application of this perspective. Though not done exclusively on forest resources, his study nonetheless reveals the complex "relationship between the acts and events and the aggregate pattern of behaviour and social systems" (Jansen, 1986, p.9) in the pursuit of scarce resources (such as land and water). White's (1992) socio-anthropological study of gender and class in rural Bangladesh, provides an insight into the competitive process of access to resources by "looking at connections across different aspects of village life [and by examining] how closely the economic and social relations are intertwined" (White, 1992, p.4). The application of this approach to problems of forest resource use may be of benefit in revealing the diverse interests and the political-economic variables that determine forest exploitation.

Another approach, called the "Socio-Ecological Perspective in Environmental Management", drawing primarily on the basic sociological tools, attempts to examine forest use by a sociological examination of the "differential stakes of various classes" and "interests" in the resource concerned (see for example Gupta, 1986, p.305). This approach enables the researcher "to deduce...the resource use options" and the operation of "economic enterprises that different classes of farmers" apply (Gupta, 1986, p.306).

Any analysis of resource-use over time (or history), almost certainly makes use of the "Historical-Analytical" approach. This facilitates sequential analysis and comparison between the situations at different stages of history. Nadkarni et al. (1989), for instance, have conceptualized particular historical stages of forest use in India to analyze the dominant trends in forest use. Table 2.2 summarizes this analysis.

Drawing on some of the selected advantages of the above theoretical approaches, I will address the study of forest resource use by combining the following considerations: first, historical analysis of the dominant political-economic trends in forest use and management. One of the major assumptions behind this consideration, as Frank (1992, p.109) argues, is that "the economic, political, social, and cultural institutions and relations we now observe there [i.e. "the underdeveloped world"] are the products of the historical development of the capitalist system". Hence "the approach is dictated by the belief that explanation is a statement of process and its historical determinations" (Magubane, 1979, p.7). "Every sociology worthy of the name," professed Mills (1959, p.146), "is historical sociology". The second consideration is the political-economic analysis of relations among the heterogeneous and unequal interests in forest use. The third is the examination of major political-economic variables of a specific resource-use setting. The focus on such factors is important because the "social relations and forces of resource use" are conditioned by the surrounding "political-economic realities" (Nadkarni et al., 1989, pp.25-27; also see Blaikie, 1988, p.141; Brokensha and Castro, 1984, pp.23-26).

In view of the above considerations, the approach applied in this text is essentially that of a "political economy". At this point, however, further clarifications on the concept of political economy are necessary, because the term is often riddled with diverse theoretical connotations. Initially, it was used synonymously with the study of economics and regarded as a branch of statecraft (Gilpin, 1966, p.160; Roberts, 1971, p.160). The concept received a much fuller explanation in the works of Smith, Ricardo, Marx and Engels. For example, two major works which specifically deal with the concept of political economy are by Marx (1951) and Engels (1951). By political economy, these works have focussed on the relational network among different classes involved in a particular resource use or production setting; and also on the forms and impacts of capitalist exploitation. Engels (1951, p.339) writes:

> Political economy begins with commodities...it is a commodity solely because a relation between two persons or communities attaches to [it]....Economics deals not with things but with relations between persons, and in the last resort, between classes; these relations are, however, always attached to things and appear as things.

In context of forest use, such a "thing" may take the form of a specific forest resource base, whose use is determined by an interplay of the relations among the interests ("classes").

The works of Marx and Engels, coupled with other major works of the "radical political economic school" (for example, those of Robert Owen, Lenin, J.S. Mill, the Fabians), provided a much broader analysis of the themes of political economy (see Burkitt, 1984 for a critical account of the radical school of political economy). Consequently, in its recent use, the concept transcends the purely economic approach; and focusses on such issues as "the linkage between political and economic factors in public policy" (Roberts, 1971, p.160), "the politics and economics of class relations" in a resource use (Jahangir, 1989, p.146; also see Nikitin, 1983, pp.13-20), the varied "noneconomic influences such as social structures, political systems, and cultural values as well as such factors as technological change and the distribution of income and wealth" (Wilber and Jameson, 1992, p.xvi). It has also a special insight into the "historical phases of (capitalist-commercial) exploitation and present interlocking structural problems" in the underdeveloped countries (Rita O'Brien, 1979, p.13; also see Frank, 1992; Baran, 1992).

The concept of political economy in this text is influenced by the above works and their theoretical considerations. In a nut shell, as Brokensha and Castro (1984, p.23) have argued, "political economy...considers how resource distribution, political policies, and market forces influence the allocation and use of resources". It is concerned with the analysis of political-economic trends in the evolution of forest use and their implications; the analysis of conflicting interests and patterns of competition for forests; the forms and forces of commercial exploitation; the changes in the social and productive structures and relations of forest use; and the examination of major political-economic factors which shape the access to, and the benefits from, forests. In varied contexts, many studies have used political economy as an approach, broadly, in a similar sense. For example, Jahangir (1989) has adopted this approach in the analysis of class relations and contending interests in the operation of local government institutions in Bangladesh; Nadkarni et al. (1989) and Fernandes et al. (1988) in the context of forest resource use in India; Gerry (1979) in assessing the impact of capitalist production on the petty producers in Dakar, Donal-O'Brien (1979) in showing the struggle between the ruling class and peasantry in context of a monocrop economy in Senegal; Klein (1979) in the analysis of social and structural changes as an impact of the "colonial intrusion" and the resultant "incorporation of a local economy into the world market"; Magubane (1979) in examining the linkages between colonialism and the development of race-class inequality in South Africa; and Baran (1992) and Frank (1992) in their respective treatment of the historical implications of the capitalist interventions for underdevelopment in the Third World.

The analysis of forest use in south Asia is a mammoth task by any standard. For convenience of subsequent discussion, the following section identifies various interests in forest use, and conceptualizes the historical evolution of the process of forest use.

The Evolution of, and the Interests in, Forest Use: A Conceptualization

There is a sizable literature on the principal "interests" which compete for access to, and use of, forests (see Fernandes et al., 1988; Nadkarni et al., 1989; Guha, 1989; Shepherd, 1992; Fernandes and Kulkarni, 1983; Chambers et al., 1989; Gadgil, 1989). Though there are considerable regional and structural variations among these "interests", a general pattern can be identified, as summarized in Table 2.1.

Table 2.1 The "interests" in forest use

Interests	Characteristics and nature
The Locals	a.forest dwellers; b.look upon forest as the sources of land, fuel wood and other daily needs; c.their use of forest takes place out side the mainstream of commercial market; d.the poorest get the marginal share of benefits; e.the advantage a class derives from forest is determined by the magnitude of assets it controls.
The Industrial Commercial interests motives	a.look upon forest as a source of raw and material; b.predominantly profit and commercial of action.
The State	a."Regulator" among diverse interests; b."Protection" (conservation); c."Productivity" (extraction and meeting public demands).

Source: Based on Nadkarni et al., 1989, pp.158-162 and Fernandes et al., 1988, pp.340-344.

The nature of these "interests", obviously, changes with the process of historical (political and economic) evolution. There seems to be three dominant and easily identifiable "interests" in south Asian forests. First, the livelihood demands and subsistence needs of local people (including the forest tribes), which can broadly be termed as the "local" interest; second, the commercial use and exploitation of forests, primarily motivated by the principles of money-economy and commercial profitability (the "commercial" interest); and third, the interest of the "state" as a mediator between the dominant local and external interests, by applying its constitutional powers. However, the performance of the state, as will emerge, is influenced by dominant commercial interests which generally "wins in the competition" (Chambers et al., 1989, pp.61-75) and the state tends to tilt towards the "winners".

The evolution of forest use in south Asia, can broadly be characterized by some historical stages (with distinct manifestations of each stage). Again, there are regional variations and these stages are often overlapping. For the benefit of discussion and theoretical conceptualization, I have tried (in Table 2.2) to identify a general sequence of this historical evolution, following Nadkarni et al.'s (1989, pp.23-28 and pp.160-164) discussion.

Table 2.2 The stages of forest use and management: a conceptualization

Stage	Exhibits
Pre-commercial and pre-colonial era	Local patterns of resource-use; local consumption; largely free from direct dependence on external "market"; less penetration of outside commercial forces into local areas; no major commercial exploitation; subsistence-oriented livelihood; no major threat of resource degradation.
Advent and proliferation commercializat-ion: Colonial intrusion	Hard wood extraction and revenue maximization for colonial purposes; massive profit-oriented commercial felling of forest; rapid degradation of resources; impairment of local/traditional patterns of forest use and customary rules; progressive diminution of local rights; state monopolization of forests; intrusion of commercial interests in local areas; appropriation of resource by commercial forces and consequent alienation of local people.

Consolidation of competition and commercializat-ion: The post-independence era	Similar trends to those in the previous stage (Commercial era): further intensification of commercial exploitation and centralized control over forests; "Growth-led planned economic intervention" which has caused mushroom-growth of state-patronized forest-based industries, displacement of local communities, increased forest clearance for technological and agricultural expansion and infra-structure building; intensification of competition for increasingly scarce resources; systematic exclusion of locals from forest use; increased hostility between local people and the state; belated realization among the national governments (supported by international agencies) about alienation of the locals and the difficulty in managing the forests singlehandedly.

These stages can be explained in terms of the degrees of integration of a "local" forest-use system into the "wider" economy, market and culture. This also refers to the gradual process of commercialization and bureaucratization (through state appropriation) of the indigenous patterns of forest management. The following sections will elaborate these major evolutionary stages of forest use.

Forest Use in the Pre-colonial and Pre-commercial Era

Documented historical evidence on pre-colonial forest use in south Asia is relatively scarce. The available literature vary in terms of quantity and quality of information. This variation, combined with the wide regional diversity of south Asia, make generalizations very difficult. However, a substantial literature, as will be shown in the following discussion, argued that the forest-based locals in different parts of south Asia had a tradition of growing trees and managing their neighbouring forests without the stimulus of outside interventions. Elwin (1963, p.51, cited in Anderson and Huber, 1988, p.36), among many others, has gone as far as to argue that "to a vast number of local people the forest is their well-loved home, their livelihood...their very existence". The underlying argument of this school of thought is that there was a close link (what Fernandes et al., 1988, p.150 called a "symbiotic relationship") between local communities and their neighbouring forest; people had sufficient access to forests to fulfil their subsistence requirements; and they practiced a localized, informal, and culturally determined pattern of resource use which ensured a balance between local demand and sustainability of production (see Guha, 1989; Anderson and Huber, 1988; Fernandes et al., 1988; Nadkarni et al., 1989; Fernandes and Kulkarni, 1983; Gadgil, 1985; 1987; 1989). Many regional case-studies have attempted to present evidence in favour of this argument. From Sri Lanka, for example, Wickramasinghe (1992, p.4) reports that "traditional forestry has been practiced over generations [and] it is a part of the survival system for the locals". In Nepal, Tuladhar (1991, p.103) notes, the traditional agroforestry systems practiced by Hill and Terai farmers "have effectively met local consumption demands for generations". In Bangladesh, as Dalmacio (1989), and Abedin and Quddus (1991), among many others, argue that the indigenous forest management, mainly in the forms of shifting cultivation (*Jhuming* and *Taungya)* practiced by hill communities, and the mixed production of perennial trees and annual vegetables in the homesteads, have been effectively practiced for centuries. Gujral (1991) maintains that some of the earliest examples of organized forest plantations, especially of local homestead agroforestry systems, have been found in Myanmar. Examples of indigenous forest management may be drawn from literature on other regions of Asia as well. In northern Thailand, as Ganjanapan (n.d.) reports, some of the community-managed "watershed forest" (*Pa ton nam*) and "sacred forest" have survived the process of "alarming deforestation" and provided livelihoods to local communities. Similarly, many researchers, including Foley and Barnard (1984), Kartasubrata (1991), and Gujral (1988) maintain that the "home gardens" (*Pekarangan*) and "mixed tree-garden" (*Kebuntalun)* in Java, and the "multi-storied Agroforestry gardens" in Sumatra are notable manifestations of indigenous forest use and its efficacy in meeting local demands. Therefore, it seems that local communities of Asia have a considerable heritage of traditional forestry practices.

The above studies have applauded the informal and localized pattern of forest use primarily on three counts: their effectiveness in meeting local demands, their adaptability to local contexts, and their sustenance over a reasonably long period. Some analysts, such as Gray (n.d., p.5 and back-cover), went as far as to conclude that local forest-based people

> have demonstrated that they are the best conservers of their environment which they use and manage according to their own cultural premises. In addition indigenous people consider themselves as custodians of their territories which have been passed down by their ancestors and have to be conserved for the generations to come....In spite of so many threats to their survival, [they] have managed to remain on territories which have retained a sustainable production potential for

many hundreds...years.

Though such observations may contain some elements of romanticism, from the regional studies it appears very likely that local patterns of forest use in the pre-commercial time had efficacy in meeting local demands, in its adaptation to particular contextual settings, and in ensuring some degree of local people's involvement and share in forestry.

It is difficult to ascertain the exact scope and scale of such localized indigenous forest use systems. Gadgil (1989) maintained that there were primarily two systems of resource use in the pre-colonial Indian subcontinent: the tribal system and the agrarian system. According to him, in the tribal system relatively homogenous tribal groups controlled a certain land area as a common property, and there were culturally determined pattern of resource use. The second system was that of agrarian village communities. Here too, there were community-controlled supply of forest products and religiously protected safety forests attached to each village. Both systems, Gadgil (1989) argued, had some notable characteristics in common: first, the resource flows were largely restricted within local territories; second, the community regulations ensured fulfillment of local demands and sustenance of the resource base; and third, the locals had a stake in conservation and propagation of the resource. Similar arguments were also presented by many others, including Fernandes et al. (1988), Guha (1989), Fernandes and Kulkarni 1983, Haimendorf (1982). Though there is a debate on the extent and nature of such common property (such as forests) in pre-colonial time, major studies suggested that a significant land-area, especially forest and watersheds, was most likely under community control in many regions of south Asia. Sharma (1965, Chapter 4), in his extensive study on the early medieval Indian feudalism, presented historical evidences from ancient texts and scriptures that in early medieval india (500-1200 AD), especially the pasture grounds, forests and water reservoirs could not be "partitioned"; and communities enjoyed "communal rights". Arnold and Stewart (1989, p.i) argued that even until the (early) 19th century up to two thirds of land in India was under community control. Singh (1986, pp.2-10) noted that prior to colonial forestry legislation of the 19th century, at least 80% of India's natural resources were common property. Poffenberger (1990) also offered evidence of traditional forest management as common property in some parts of southeast Asia. The above studies suggested that local communities might not have owned forest lands, but had a considerable involvement in their management, including access to forests for meeting their subsistence demands. To identify the nature and characteristic features (recognizing the regional variations) of these indigenous forestry practices, two (geographically and contextually) different cases are examined.

The Traditional Conservation Systems of the Himalayan Peasants in India

The information on this case is primarily drawn from Guha (1989, pp.28-34). His analysis is widely referenced, supported by an extensive fieldwork, and echoes similar findings of the major studies in the area (such as Haimendorf, 1955; 1982).

The "traditional conservation systems" of the Himalayan forests (specifically of Tehri Garhwal and Kumaun divisions), according to Guha (1989, pp.32-33), were notable manifestations of the "successful local practices" of a closely knit network of forestry, agriculture and animal husbandry. Under this management system, village and cultivation sites were reported to be judiciously chosen to make best use of the neighbouring forests, grazing lands and riparian fields. In such a situation, Guha argued, "all crops could be raised to perfection" (Guha, 1989, p.28).

It is said that forest-use was planned to ensure optimal utilization of forests in a variety of ways. In permanent hamlets, the oak forest supplied fodder and fertilizer for agriculture and livestock. The broad-leaved trees provided for fuel and agricultural implements. The forest of lower hills especially served pastoral purposes. In addition, forests were the primary source of medicinal herbs, and of food in times of dearth.

Guha, drawing on a large number of studies and on his own fieldwork, presented evidences of a socio-religiously determined pattern of resource use in the region. He showed, for example, that to shun over-exploitation and to regulate use, village communities had drawn "a protective ring" (Guha 1989, p.29) of social and cultural norms (as embodied in the religion, folklore and tradition) around forests. These traditions contributed to the conservation and propagation of forests. There were "magnificent" (Guha, 1989, p.30) plantations of sacred trees (especially of the *Deodar* tree) in all parts of the mountains. Some fields were left uncultivated which were dedicated to the "fairies of forests", and some forests were allowed to be used "only during culturally specified time" (Guha, 1989, pp.29-30).

Besides the religious and cultural regulations on resource use, Guha (1989, p.31) provided evidence of some "informal institutional practices" of forest management. For example, the amount of grass cut by each family was strictly regulated, and there were community rules prohibiting lopping of leaves in hot weather. "With the planting of timber trees a fairly common phenomenon, the jungles preserved within the boundaries were zealously guarded by villagers nearby" (Guha, 1989, p.31).

The pre-British governments did not seem to have a particular interest in forests (except for collection of routine taxes or cess and some other restricted usages). This enabled the hill peasant communities to enjoy, according to Guha, "untrammelled use" of, and "near-total control" over forests (Guha, 1989, pp.31-32. A detailed account of the pre-colonial forest use in Bengal will be provided in Chapter 5).

Traditional Forest Management in Nepal

Tuladhar (1991), Blaikie (1988) and Johnson et al. (1982), among many others, maintained that farmers in many regions of Nepal have traditionally built a productive-network of forestry, agriculture and livestock resources which ensured fulfillment of local demands and sustainable use of the resources.

Historically, in some parts of the country, it is reported that forests and pastures were collectively managed by particular ethnic or clan population (Sharma, 1991, p.102 and 107; Adhikari, 1990, p.261). For example, Sharma's (1991, pp.107-108) observation on the *Khakra* system of the Sherpa people manifested some features of the traditional system:

> [The system] used a set of management practices and rules to protect and share fuel wood, timber, fodder and grasses. The sherpas, who are illiterate, know exactly how many cattle can be supported on a particular khakra or grazing area. Many areas of healthy forest were managed by religious or community trusts.

Generally, most forests in Nepal had been traditionally managed by the *Talukdari* (1769-1951) and *Kipat* (during the *Rana* period, 1847-1951) systems (Barlett and Malla, 1992; Khan and Hasan, 1995). These traditional systems, it should be pointed out that, were not always "equitable" or "egalitarian" especially in the distribution of benefits; but they were said to be efficient in meeting the subsistence demands of local communities. For example, the *Talukdari* system was "conservative" (Khan and Hasan, 1995, p.53) and oriented towards meeting the interests of hereditary Rana regime (Gronow and Shrestha, 1991, p.2). But, as far as local rights and access were concerned, there seemed to be little interference by local (for example, the Talukdars) or central rulers (for example, the Ranas) into the survival strategies and subsistence needs of the locals (see King et al., 1990; Adhikari, 1990; Khan and Hasan, 1995; Khan, 1996a). Gronow and Shrestha (1991, p.2), therefore, concluded that these traditional systems were "unfair [but] effective". Many studies, including Bartlett (1992), Adhikari (1990),

Bartlett and Malla (1992), Gilmour (1989), Khan (1996a), and Fisher (1989), have argued that the involvement of local communities in forest use and management was not a new or recent phenomenon in Nepal; and that there existed indigenous forest management systems in different parts of Nepal.

Adhikari (1990, p.262) maintained that traditional rulers of Nepal in fact encouraged community management of forests, and even made provisions for handing over state-owned forests to local communities. The striking feature of this transformation was said to be the detailed community-guidelines for management of forest, called the *Lalmohar*. It contained regulations for collection of firewood, timber extraction, protection of birds and animals, and related other issues of forest use. These were, as Adhikari (1990) argued, environmentally sound and economically sustainable. For example, there were prohibitions on clear felling or burning of forest (if people needed firewood, they could only use the dry and fallen trees). There were also effective check and balance on timber extraction and utilization. The different patches of forests were managed by different ethnic groups, who obtained their rights either through the *Sanad* (the agreement for handing-over the forests through the *Lalmohar*) or through the *Bijan* (the registration of land) (Adhikari, 1990, pp.261-264).

There seemed to be a reciprocity of responsibility between the ruler (*Talukdar*) and the ruled (village communities) under the traditional system. This is evident from the *Lalmohar* which writes: "if the *talukdar* needs timber, he should ask the people and if the people need timber, then they should ask the *talukdar*" (Line 37 of *Lalmohar*, cited in Adhikari, 1990, p.262).

Some analysts, including Johnson et al. (1982) and Blaikie (1988), reported that farmers possessed a considerable knowledge of resource use, and in the face of environmental crises (such as droughts and cyclones) they could judiciously act by evaluating the available options. They were aware of methods of maintenance and repair of equipments; and thereby, could minimize damages.

Though there are regional and cultural variations, the above cases broadly echoed the general pattern of forest use in pre-commercial era. The core points of forest-use strategy as observed from these cases are as follows: first, the multifarious use of forests which was often a part of the general livelihood strategies of local people. Second, management and protection of forests were linked to socio-cultural and religious traditions. Third, the informal practices of forest use and conservation by local communities seemed effective (under ideal conditions) primarily in three ways: in meeting local demands and involving the communities, in their adaptation to particular contexts, and in their sustenance over a reasonably long period. Fourth, local communities seemed to enjoy a reasonable degree of access to forests. This, as many studies suggested (Gray, n.d.; Guha, 1989; Haimendorf, 1955; 1982), had instilled a sense of "assurance" of their rights, which in turn might have motivated them to act "collectively" in the use and conservation of forests (put simply, in a particular resource-use scenario, "assurance" refers to the institutional assurance that the members of a group would cooperate, and that the right of an individual member is safeguarded; see Gupta, 1986, pp.317-319; Locke, 1995, pp.10-15; Runge, 1986, pp.628-629). Fifth, indigenous farmers, whom Gray (n.d., p.22) referred to as "resource managers", had shown some scientific knowledge in their use of forests which was manifested in the selection of cultivation sites with careful attention to geographical and geological factors; in the cultural restraints on over-exploitation; in the optimum utilization of forest products in a variety of ways; and in building links between forestry and other resource systems like agriculture or livestock (see Chambers et al., 1989a for an extensive collection of experiences in farmers' indigenous knowledge, innovations and practical capabilities. Mellink et al., 1991 and Scoones and Thompson, 1994 have also focused on indigenous knowledge).

Overview

As the preceding review suggests, there is a substantial body of arguments in favour of the existence of a culturally determined and localized pattern of forest use in the pre-commercial era. The literature, in sum, emphasized three major features of the traditional system: it responded to local demands; it fostered a reasonable degree of access to, and rights on, forests by the locals; and it was largely adaptive to particular contexts (Gadgil, 1989; Sharma, 1978; Guha, 1989; Fernandes et al., 1988; Gray, n.d.; Haimendorf, 1982). It seemed likely that the locals, especially in the tribal regions, had considerable degree of control and right on forests until, as will emerge in the subsequent sections, local territories witnessed organized commercial interventions through the colonial rule. Guha (1989, p.185) argued, "prior to the reservation of forests [which was started by the colonial governments]...village communities [had] control over the means of production and over the resources needed to reproduce itself".

However, in the relative scarcity of documented historical and empirical evidence, the issue of pre-commercial indigenous community management of forests remains a controversial area of study. There is a "skeptical school" which disputes the idea of community management of forest in the pre-commercial era. Nadkarni et al. (1989, p.34), for example, from their research on Uttara Kannada (India) argued that "the popular impression that forests were treated as a community resource in the old days, equitably shared by all, did not appear to be true". Watson (1989, p.67) noted that "actual functioning of traditional systems may not be so egalitarian or idyllic as is often portrayed". Locke (1995, p.40) suggested that the "reduced rates of degradation" in the pre-colonial Uttara Kannada, was more due to the "low demand" of resource than to the efficiency of community management. Indeed, any serious observer will concede to these views to the extent that the pre-commercial era neither represented an "idyllic stage" (Nadkarni et al., 1989, p.35) nor was it egalitarian in distribution of benefits. Nadkarni et al. (1989) and Stebbing (1923) reported incidence of some deforestation in some of parts of Indian subcontinent in the pre-commercial period. As the Nepalese case suggested the distribution of benefit was not necessarily egalitarian or equitable. Gadgil (1989, p.5) reported that the pre-colonial agrarian society of India "was a highly inegalitarian system with the lower caste groups suffering exploitation". However, in line with the preceding review, Gadgil (1989) and Nadkarni et al. (1989) did suggest that as far as local rights, access, and the degree of commercial intrusion were concerned, the pre-commercial era seemed to offer some advantages to local communities. It, as noted before, seemed to ensure a reasonable degree of access to forests for the locals, and the fulfillment of their subsistence demands. The systematic commercial exploitation of forests, the widespread intrusion of market forces, and the consequent alienation of local people -- the trends which were to become the characteristic features of the colonial era -- were largely absent, or at least, did not become dominant until this period. Nadkarni et al. (1989, pp.32-33) agreed that "the locals had free access to forest land" in the pre-commercial era and that "the demand from the larger economy on the forest resources was marginal". Similarly, Locke (1995, pp.41-42), in her study on Uttara Kannada, presented evidence of the adverse consequence of colonial intrusion on forests with the result of, what she termed as the "colonial alienation", which included such trends as the "increased strictness" on the local people's access and the "progressive extension" of state reservation. Moreover, her argument that the "levels of demand [were] too low to cause [resource] depletion" (Locke, 1995, p.40) -- also seems debatable. Reporting on coastal region of Uttara Kannada (the district on which Locke remarked), Nadkarni et al. (1989, p.34) noted that "the district had much less forest and much more population pressure even in those days". Researchers like Gadgil and Iyer (1989) argued by presenting historical evidence that sustainable traditional structures of forest use existed in the area prior to the British colonial rule. Some analysts have attempted to show that culturally determined patterns of collective forest use persist even today in some localities in India and Nepal (see for example Gadgil, 1985; Gadgil and Iyer, 1989; Fisher, 1989; Gilmour, 1989). From the above discussion, it seemed very likely that there was no significant alienation of the locals from forests under the informal and localized pattern of forest management. Nadkarni et al. (1989, p.32, emphasis added) noted:

> There was *no alienation of the locals from the forest in spite of the state ownership of forest land during the pre-capitalist stage .. The locals had free access to forest land for grazing their cattle, fuel wood, wood required for agricultural implements and house construction*....[The traditional practices like] shifting cultivation was not considered illegal and improper.

Nevertheless, many questions are asked about the nature of traditional community management structures and their functioning. More commonly, these skeptical views centre around the following queries: how could the locals enjoy autonomy when the village communities were integrated into centralized political structures of power and authority? How the community forests could possibly avoid the most likely tendency of being over-exploited? Can a resource use system exist beyond the ideas and operation of a "market"? How could local people possibly enjoy freedom when forests were not under their ownership and the social structures were inherently inegalitarian? These inquiries demand further elaboration.

Some analysts depicted local communities as "little kingdoms": an undifferentiated, closed, isolated and self-sufficient entity. Maine's (1905, p.262) comment on Indian village is typical of these "idyllic" presentations:

> An Indian village is an organized society, and besides providing for management of the common fund, it seldom fails to provide, by a complete staff of functionaries, for internal government.

In the same vein, Hayami (1980) also identified village as the basic unit of rural life in Asia which could mobilize collective action to supply public goods essential for the security and survival of its members (also see Haimendorf, 1982; Marx, 1951a; Gadgil, 1989). These ideas have come under heavy attack from a number of major studies. Ludden (1985), Fox (1971) and Stein (1980), for example, argued that village communities in the medieval Indian subcontinent were certainly not an isolated entity, and were linked to wider and comprehensive units of political organizations through various reciprocal relations and exchange mechanisms. A full-scale analysis of traditional social formations is beyond the scope of this text. Suffice here to present a brief picture of the situation, and of the related rudimentary arguments with a particular focus on forest and pastoral lands. Stein (1980, p.264), in his probing study on the nature and formation of traditional state in medieval south India, noted that the state (called, *Chola*) was a

> pyramidally segmented type of state, so-called because the smallest unit of political organisation -- for example, a section of a peasant village -- was linked to ever more comprehensive units of political organisation of an ascending order (e.g., village, locality, supra locality, and kingdom) for various purposes.

Fox (1971, p.165) similarly argued that in the pre-industrial India village communities were actively connected with central political organizations including the state; and "at times, if the state was strong it imposed administrative organization directly on village communities". However, these studies, like many others, also suggested that though village communities were enmeshed in wider political-economic structures, they maintained their own identity and distinctive characteristics. In other words, perhaps the locals were not assimilated into a unified central authority, surrendering their peculiarities. Stein (1980), explaining the nature of medieval south Indian agrarian order and the state-locality relations, developed the conception of "segmentary state". He (Stein, 1980, p.275, emphasis added) explains:

> The conception...pertains to the ways in which *relatively self-sufficient, enduring and often quite ancient localized societies* can be linked together to form a state. Such a state is not an amalgamation or absorption of localized units into an organic greater unit such as implied in the unitary state, but is an arrangement in which *the local units -- segments -- retain their essential being* as segmental parts of a whole. One reason why each of the *segmental units remains autonomous* is that each is pyramidal, that is, each consist of balanced and opposed *internal groupings zealously cling to their independent identities, privileges and internal governance.*

Stein's view was supported by Ludden (1985) and Fox (1971) in their work on the pre-industrial south and north India respectively. Fox (1971) noted that though a village cultivator might be linked to the state through many levels of local, regional and central power holders, still he could participate in political decision making and governance processes. He (Fox, 1971, p.55, emphasis added) argued:

> Just as distribution process gave many different groups access to the cultivator's grain heap, so too the *political institution allowed many different levels and personages a share in decision-making and political management.*

Gadgil (1989, pp.3-4) noted that the agrarian village communities of pre-colonial India had their respective territories and a structure of internal administration. Local villages seemed to be ritually integrated into a ceremonial authority -- the king. Fox (1971), Ludden (1985), Sharma (1965) and Stein (1980), among many others, maintained that the kingship was a sacral institution; "it is dependent upon ritual, not administrative, incorporation" (Stein, 1980, p.275). The king collected tax from local communities through a chain of political mechanisms; but as Ludden (1985, p.37) argued, in the context of medieval south Indian local communities, "most payments to kings and chiefs would have had more the character of tribute than of taxation". He remarked that the medieval states were marked by "localism" where the "kings were apparently not directly involved in either extraction or investment outside their own core areas" (Ludden, 1985, p.30). Similarly, Sharma (1965, p.144) believed that in the Indian subcontinent "the king [might] have acquired rights over the soil as the representative of the community, but he was never conscious of it in early medieval times". Therefore, it seems that the institution of kingship was not a formidable interference on local communities as far as their rights and access to resources (especially forest and pastoral resources) were concerned. Being religiously and culturally determined, the institution of kingship or chieftainship enjoyed acceptance and reverence of local communities (see Stein, 1980 and Ludden, 1985. Fox, 1971, however, argued that on some occasions the kings in collaboration with the central states exercised coercion in realizing taxes). Some evidences of organized collective effort were also noted in the literature. Sharma (1965), as mentioned earlier, argued that forests and pastoral lands were mostly under collective use in the medieval Indian sub-continent. Gadgil (1985; 1989) and Haimendorf (1982) also maintained that there were community-controlled forest which ensured fulfillment of local demands, and its use was regulated by socio-cultural traditions. Ludden (1985, pp.34-37) noted that collective management of irrigation work in paddy cultivation existed in south Indian "settled-agriculture" villages. Another feature of traditional structures of resource management, as literature suggests, is their endurance. Ludden (1985, p.201) went as far as to remark that "the village community that typifie[d] agrarian social order [of medieval south India was] among the world's most enduring social forms". The above features of traditional communities seemed particularly relevant to the tribal and hill peasant populations of south Asia. Most of the above characteristics -- including the reasonable degree of autonomy and access in resource management, durability of local institutions, religiously determined pattern of resource use, reciprocity and exchange in localized mode of resource production and distribution -- seemed to have been more clearly observed in the tribal localities (see Jones, 1978; Saha, 1986; 1988; Guha, 1989; Haimendorf, 1955; 1982; Anderson and Huber, 1988; Fernandes et al., 1988; Fernandes and Kulkarni, 1983). Researchers like Haimendorf (1982, p.80) and Sharma (1978, p.64) presented evidence that tribal communities in India, enjoyed a kind of "de-facto ownership of forest", and they had a localized indigenous organization for resource management. In more general terms, Saha (1986, pp.287-288) noted that tribal communities in Indian sub-continent had a "high degree of social autonomy" and "communal organisation for production" and "control on principal productive assets, i.e., land and forests". But as the process of central intervention to "integrate" the tribals into central economy and society proceeded, "tribal society lost its economic autonomy and eventually its territorial base [and the tribal became] a marginalised alien in his own homeland" (Saha, 1986, pp.287-288; also see Saha, 1988). As regards forest resources, Jones (1978, p.52) noted that the Indian tribals had "their largely autonomous, self sufficient methods of production" (which were eventually disrupted by the state during the colonial and post-colonial periods). On the people of Baster (India), Anderson and Huber (1988, p.34) remarked:

> [The] socioeconomic organization was autonomous. The resource available to tribal society, the land and forest in which the tribals had customary...rights, were left to be managed without interference....In turn, as a self-supporting system (albeit, more or less subsistential and located in an extremely difficult terrain), [the system] could exist free of significant influence or interference from larger, more powerful forces beyond its borders.

From the above discussion, broadly, two points may be made: first, it appears most likely that village communities, although linked to wider polity and society, enjoyed a reasonable degree of discretion in managing local affairs and in their access to forests. The degree of central interference and coercion on local communities also seems reasonably limited. Second, there were localized and culturally bound structures of resource use and distribution, with distinctive customary powers and privileges.

Gadgil (1985; 1989), Ganjanapan (n.d.), Fernandes and Kulkarni (1983), among many others, argued that forests under traditional management reflected a special type of common resource, which, by nature, was different from the conventional explanation of common property resources, entailing the oft-quoted "tragedy of the commons" (Hardin, 1968). The "tragedy" has been observed by researchers, including Poffenberger (1990), Jodha (1990), Shepherd (1990), Gadgil (1989) and Arnold (1993), in the current unrestricted use of open access resources in many parts of the Third World, where "future is totally discounted and everybody is motivated to extract as much as they want for the present without restraint" (Gadgil, 1989, p.9. For details on the theory and the practical implications of the "tragedy of the commons" see Hardin, 1968, pp.1243-1248; Wade, 1988, pp 201-205; Gadgil, 1989, pp.8-10, Locke, 1995, pp.10-12; Runge, 1986, pp.31-59; Gupta, 1986, pp.305-321). It is said that forests under community regulations were not an "open access resource of unrestricted use". As the case-studies presented earlier suggested, there were localized institutional structures and socio-cultural norms for regulating forest use, which, as Gadgil (1989), Guha (1989) and Haimendorf (1982) argued by presenting regional evidence, saved the resource from being over-exploited. They maintained that the informal and institutionalized patterns of control which regulated forest use had acceptability and credibility among local people, and ensured that, according to Gadgil (1989, p.3), people "had a real stake in the security of the resource base of their territory and evolved a number of cultural traditions to ensure its sustenance". Ganjanapan (n.d., p.2) further explained:

> In contrast to the open-access common property resources..., the community forests as common property imply an organisational control and management which is accepted locally...[and] demonstrate clearly that man and forest can be living together in a meaningful way and serves the purpose of conserving and development of both without offending the other.

Gray (n.d.) and Haimendorf (1982) reported that indigenous people had their own ideas of barter and trading. Hence, as they suggest, the phrase "non-market or non-commercial framework" does not imply that the locals had no idea about markets, rather it refers to the exchange of goods within the local communities, and between the communities and local markets in the form of trading posts. Some analysts (for example, Hodges, 1988, pp.4-6; Gray, n.d., pp.33-37; Gadgil, 1989, pp.3-4) have discussed about the "attributes" of a market (primarily as "exchange activities") in terms of its integration into or independence of the "broader industrial market" and "capital economy". These attributes are: first, the exchange of goods between local communities; second, the presence of local markets and trading posts; third, the exchange links between local community and outside "market economy" (the national, regional and international economies). Resource management in the pre-commercial era is said to be of "non commercialized character" (Nadkarni et al., 1989, p.160), primarily because, according to Gray (n.d., p.35), it had not been yet "plugged with" the outside "market economy" by direct "cash-flows", as characterized in the third attribute of the above list. The main argument presented was that the "flows of materials were restricted within the territory" (Gadgil, 1989, p.3); and intrusion of the commercial forces was not yet widespread (Gadgil, 1987). Ludden (1985, p.37) also argued that transactions concerning village communities in the medieval south India "did not involve resource transfers outside localities".

Many analysts, including Guha (1989) and Gadgil (1987; 1989), argued that the inegalitarian nature of south Asian village communities was not a major hinderance to the efficiency of traditional forest management. Apparently, there existed a "caste diversification of resource use"; and as Gadgil (1987, p.9) maintained,

> in spite of the highly inegalitarian nature of cast society...the different caste groups living in a village tended to retain common interests in good use of resources of their locality due to a variety of arrangement for resource sharing.

Referring to other major studies concerned, Gadgil (1989, p.5) argued that "all these [caste] groups were linked together in a web of reciprocity and each group shared in the well being of the resources of whole village territory". It is said that despite the heterogeneity and diversity of their social formations and economic activities, the forest-based population have performed reasonably well in the management of their surrounding forests because of some conducive factors including the simple demands of subsistence-oriented life style; the religiously determined functions assigned to each caste and clan groups; adherence to religious and cultural restraints and obligations; low degree of outside intervention (see for details Guha, 1989, pp.9-34; Anderson and Huber, 1988, pp.25-31).

As the above review has suggested, despite regional peculiarities, the pre-commercial era can broadly be summarized as a resource-use paradigm characterized by the followings: first, a reasonable degree of access to, and involvement of, local communities in forest use and management. The localized institutions of resource management, though linked to wider political structures, enjoyed some degree of autonomy. The extent of intrusion and interference (especially as regards, commercial exploitation) of the central authorities into local territories seemed to be considerably less than the later period (as will be elaborated in the following pages). Economic activities were carried out within a territorial framework, and were broadly independent of the industrial market economy. Second, the production and consumption patterns were subsistence-oriented, and seemed effective in meeting the modest local demands of a simple life style. Third, as many studies argued, the patterns of resource use were characterized by such attributes as the "diversification [of use] to minimize the severity of competition for scarce resource" (Gadgil, 1989, p.5); site-adaptability, responsiveness to specific contextual demands and local needs, "damage-reduction in the face of crises" (Blaikie, 1988, p.144) and the sustainability of the resource-base (Gadgil, 1989; Haimendorf, 1982).

Given the regional and historical differences, it is difficult to identify the precise time and duration of this paradigm; but for most regions of south Asia, it broadly reflected the pre-colonization stage.

The Advent and Proliferation of Commercialization: The Colonial Intrusion and Local People

Since the late 18th century, most parts of south Asian forests witnessed an expanding process of commercialization and state appropriation. The transformation occurred primarily under the umbrella of the political process of colonization which continued till the mid-twentieth century. In following Magubane (1979, p.3), I use the concept of colonization to refer to the relationship between people and resources of "a subjugated society with its alien rulers"; and also to the political-economic "structures of domination created within the colonized society by imperialist relationships". In the process, the pre-colonial system of forest use was "superseded by another system of forest management, propelled by powerful economic and political factors, which rested on a radically different set of priorities" (Guha, 1989, p.34). This turning point had the following immediate implications.

The Change in Community Regulation and Management of Forests

It is argued by many studies that most regions of south Asia, especially the Indian sub-continent, saw the breakdown of community use and management of forests in the very early years of British rule (Singh, 1986; Poffenberger, 1990a, pp.1-6; Arnold, 1993; Gadgil, 1989, pp.9-14; Guha, 1989, pp.35-61; Nadkarni et al., 1989, pp.37-49; Haimendorf, 1982, pp.79-96; Shiva et al., 1983, pp.56-59). Local customary rights, religious and cultural norms of resource use, and traditional skills of local people were ignored by the new imperial rulers. According to Guha (1989, p.55), the "underlying belief" which guided the new state authorities was that "collective use sanctioned by custom" represented a major "threat to commercial forestry". Accordingly, in subsequent years, there seemed to have occurred a qualitative change in forestry, mainly in the forms of commercialization of forests and of the loss of customary rights of local communities. The immediate consequences were at least three-fold. First, local people were alienated from land and forests, which hitherto provided their survival needs (Fernandes, 1983, pp.13-15; Prabhu, 1983, p.136). As Haimendorf (1982) and Fernandes and Kulkarni (1983) argued, the tribal population of south Asia was particularly hit by this process of alienation. Randhawa (1976, p.76, cited in Joshi, 1983, p.27) also noted that, as a result of the colonial policies "the tribals [were] reduced from virtual owners to tenants on sufferance". Similarly Haimendorf (1982, p.80) reported that "the traditional de facto ownership of tribal communities was now replaced by de jure ownership of the state, which ultimately led to the exploitation of forest resources with total disregard for the needs of the [local] economy" (also see Sharma, 1978, pp.63-64). Second, many researchers (for example Prabhu, 1983, p.138; Guha, 1989, p.58; Fernandes, 1983, p.13) argued that there was also an erosion in people's sense of dependence on and reciprocity with forests. The locals lost control, interest and stake in the protection of forests which now ceded to an alien authority (Fernandes, 1983, pp.13-15; Guha, 1989, p.38; Prabhu, 1983, p.136). Third, as Shiva et al. (1983, p.59) argued, "the breakdown of community -- with the associated collapse in concepts of joint ownership and responsibility -- trigger[ed] off the degradation of common resources [including forests]". It is reported that, the forests in which local communities had customary access, became "open access resources", and succumbed to the "tragedy of the commons" characterized by commoditization, monopolistic ownership, over-exploitation, unrestricted and unsustained use, and non-accountability of the users to any regulatory institution (Shiva et al., 1983, pp.58-59; 1987, pp.241-242; Gadgil, 1989, p.9; also see Jodha, 1990; Poffenberger, 1990). In fact, Gray (n.d., pp.22-24) noted similar trends in alienation of local people from natural resources (like land and forests), and the breakdown of traditional management systems in many parts of the world as a consequence of colonization.

The Transformation of the Philosophy of Forest Management

The early days of colonial rule, Gadgil (1989, p.9) argued, were propelled, more than anything else, by imperial interests of "rampant profiteering". Tucker (1988, p.91) noted that in India, for example, "the first era of massive forest destruction occurred after it was absorbed into the British Empire....Whatever the time for each area, from roughly 1770 to 1860, it was a period of new commercial possibilities [in forests]" (The philosophy and activities of the Indian colonial government in the forestry sector will be discussed in Chapter 5 with special reference to the case-study areas).

The forest sector philosophy of the colonial governments of south Asia, especially in the Indian subcontinent, centred around revenue and profit maximization, state monopolization of authority, and organized commercial exploitation. This philosophy was argued to be a part of the broader political-economic mission of the colonial government in India. Marx (1951a, p.320) described the broader mission in the following words:

> England has to fulfill a double mission in India: one destructive, the other regenerating -- the annihilation of old Asiatic society, and the laying of the material foundations of Western society in Asia....[Accordingly] they destroyed it by breaking up the native communities, by uprooting the native industry, and by levelling all that was elevated in the native society. (also see Edwardes, 1967, pp.47-70)

Forestry under the colonial government mirrored Marx's observation. There was, according to Fernandes (1983, p.13), a "radical shift" in the philosophy and practice of forestry with similar consequences as Marx observed above. First, as discussed before, the colonial authorities in many cases intervened more closely into traditional patterns of resource use, not only induced by commercial motives, but also by a "civilizing mission" (Anderson and Huber, 1988, p.35). For example, as Anderson and Huber (1988, pp.32-35) argued, in Baster (India), the British colonial policy had two ends: one was to pursue "imperial interests of commerce" and the other was to "civilize", whom the state considered, the "semi-barbarious tribes". In fact, as Magubane (1979, p.4) maintained, it was a common tendency of the colonial authorities in different parts of the world to "assert that the colonized are biologically and culturally inferior". The second impact of the policy shift, which appeared as a sequel to the commercialization process, was the emphasis on market economy and commercial functions. As Gray (n.d., pp.34-37) noted, a "market-frame" was "imposed" on local people, who had virtually "no control" over it. This sparked a gradual process of dependence of local people on "external" forces, and according to Gray (n.d., p.37), "this dependency [was] at the root of the destruction of indigenous cultures and society".

The Initiation of "Scientific Forestry" and Professional Forestry Cadre

The vehicle designed to propel the new philosophies in forest management was called "scientific forestry", staffed by a bureaucracy of technically qualified foresters. The very establishment of the forest departments largely stemmed from commercial and revenue interests. For example, the following excerpt from a colonial document of the British Indian government reflects the purpose of the forest department (Government of India (GOI), 1919, pp.40-41, emphasis added):

> *The chief needs of the Forest Department are undoubtedly ...the exploitation of forests on more commercial lines*; and the extension of research and experimental work which should...be carried out...under commercial conditions....[T]he forest estate which are favourably situated for export, for industrial exploitation or for the very important agricultural demand have [to be] utilised fully...on the most economical lines....*Government should...play a more direct part than hitherto in the exploitation of its own forest estate, in order to obtain a larger share in the yield thereof and to ensure a fuller use of the many valuable species hitherto untouched....*

Accordingly, In India the first task of the forest department, following its creation in 1864, was to promote organized commercial exploitation of all the major accessible forests (Guha, 1989, pp.35-38; Gadgil, 1989; Nadkarni et al., 1989, pp.37-49. Chapter 5 analyzes the evolution of the activities of forest department in some details). In other regions of Asia as well, the creation of state forestry organizations primarily reflected the revenue orientation of the states. In Thailand, for example, Feeny (1988) reported that the establishment of Royal Forest Department (in 1896) followed the commercial exploitation of northern teak forests in the late 19th century. The administration of northern forests was taken over from the local Lao chiefs by the central government to enhance the government's revenue base (Feeny, 1988). There were a number of immediate implications of the so-called "scientific forestry" for the nature and patterns of forest use, which included the following.

Compartmentalizing the forests Forests were compartmentalized as a routine prescription of "scientific forestry". In India for example, following the dictates of the first Forest Policy of 1894, forests were divided into three broad categories: Reserved, Protected and Village forests (Anderson and Huber, 1988, p.37). The reserved forests virtually denied any right of local people, and were in fact an exclusive domain of the forest department (except for the grant of certain minor concessions to locals, such as, gathering of fruits). The protected forests were also managed by the government, with the recognition of selected rights to local people. The village forests were communal property of villagers. But in the absence of any regulatory institution, and with the gradual abolition of communal control (as discussed in the earlier section), "village forest tended to be successively depleted" and became "inadequate" even as grazing lands (Anderson and Huber, 1988, pp.36-37). Therefore, the impacts of compartmentalization seemed at least twofold. First, it formalized the process of eliminating local rights. Guha (1989, p.55) argued:

> The reservation of large tracts of forest meant an effective loss of control over their habitat for forest-based communities. Inevitably, an increased pressure was felt on the forests that did remain to villagers, in many cases hastening their destruction.

Second, it segregated the "formal" forestry sector (government forestry operations) from the "informal" forestry as traditionally practiced by local people. Many researchers in different parts of south Asia have observed this transformation (for the Sri Lankan experience see Wickramasinghe, 1992; for India see Guha, 1989; Anderson and Huber, 1988; Fernandes et al., 1988; Gadgil, 1989; Nadkarni et al., 1989; Fernandes and Kulkarni, 1983).

Negligence of the value of indigenous and traditional knowledge Many analysts argued by presenting historical evidences that the colonial governments were largely heedless to the traditional knowledge and survival strategies of local people in the forested regions of Asia (see Shiva et al., 1987; Guha, 1989; Fernandes et al., 1988; Gray, n.d.). Anderson and Huber (1988, p.18), among many others, maintained that, intent on the "insatiable appetite for more revenue" and the self-styled "civilizing mission", the colonial states ignored local silvicultural and ecological knowledge as "unscientific" information; and attempted to replace them with "scientific forestry". Similarly, Gadgil (1989, p.7) noted that, to meet the "insatiable demands" of imperial rule, the Indians were

> brainwashed into accepting that all that the old Indian system represented was bad, all the new system being imposed by the British represented was good....All that the Indians were doing earlier was condemned as unscientific, superstitious, and all that British trying to impose as scientific.

Silvicultural strategies to promote commercial interests With their emphasis on "rationalized timber production" (Nadkarni et al., 1989, p.51), the colonial forest departments pursued a detailed forestry management plan, called the "working plan". The objective of the working plans was "to produce the maximum revenue" and "to realize the best possible return" from the existing stock and future plantations (Nadkarni et al., 1989, p.51, quoting from a departmental working plan in India). In most cases, the plans advocated increased production of commercially valuable trees and exclusion of local people through the reservation of forests. This again, as Gadgil (1989), Guha (1989), Fernandes and Kulkarni (1983), and Anderson and Huber (1988) argued, had a number of implications for the traditional forest use systems. First, by giving way only to commercially valuable species, it contributed to the erosion of natural biological diversity. Second, it largely ignored the subsistence requirements of local people. Third, it often clashed with traditional customary patterns of forest use, because of the essentially different nature of commercial forestry. For example, as Guha (1989, pp.37-61) and Nadkarni et al. (1989, pp.50-55) argued, in a logged area under the "scientific forestry" practice, regeneration warranted complete prohibition of access of humans and cattle; and thereby, forbidding even the minimal rights of fuel and fodder collection for local people. The commercial character of colonial forest management becomes clear from the Uttara Kannada (India) case, where, as Gadgil (1989, p.14) reported, teak plantation was given priority over the natural evergreen forest; and "Bamboo, of such great utility to the local people, was considered a weed". Guha (1989, p.191) summed up the implications of the above silvicultural strategies:

> Subjected to commercial exploitation under [the so-called] "sustained-yield silviculture", the forest [had] itself been transformed into an industrial enterprise run on capitalist lines.

Progressive Diminution of Local Rights

Anderson and Huber (1988, p.44) argued that the gradual attenuation of customary rights could be observed from the very terms used in government documents of colonial India. He showed that in 1894 the customary rights were termed as "rights and privileges", in 1952 these became "rights and concessions", and in 1963 these were reduced to "mere concessions". This also showed that the process of restricting local rights continued well after the end of the colonial rule. Apparently, vast tracts of lands (forests, cultivated and uncultivated lands) were taken over by governments, by virtue of their "imperial rights", which, according to Guha, (1989, pp.37-39) was the "right of conquest and against which there was no appeal". In the process, as Guha (1989) and Haimendorf (1982) argued, the expropriated people, mainly the tribals, were pushed into remote territories and harder livelihood. In his review of southeast Asia, Poffenberger (1990, pp.7-8) also found that the expansion of government authority meant, for all practical terms, an effective erosion of local rights, and increased hardship for local communities. As a case study, the following table presents the phases of progressive diminution of local rights in Kumaun Forests (India):

Table 2.3 The progressive diminution of local rights: the case of Kumaun forests in India

1815-78: The government concentrated on the submontane *Sal* forests of the foothills and demarcated other forests for subsequent take-over.
1878-93: All these demarcated forests, along with other tracts in Almora and Gharwal Districts, were notified as "reserved" under the 1878 Act.
1893-94: All the waste lands, not forming part of the measured area of villages or the forests earlier reserved, were declared protected forests. The process engulfed (besides the tree-covered lands) snow clad peaks, ridges and cliffs, river beds, lakes, buildings, camping and pasture grounds, roads and shops. Eight types of trees were reserved, including *Deodar, Chir,* and *Sal*. Lopping of trees for fodder-fuel and the villagers' claim for timber were strictly regulated. Trade by the villagers in any form of forest product was prohibited.
1911-17: These cumulative incursions resulted in carving extensive reserves; and by 1917 there was a reserved forest of nearly 3,000 sq. miles. Detailed rules were made in restricting the rights of the villagers, specifying the number of cattle to be grazed, the amount of timber and fuel wood allotted to each right-holder. The villagers had to indent in advance to the government, even for their basic needs of timber and agricultural implements. The indent proposals were to be scrutinised and (if found satisfactory, then to be) supplied by the forest department. The traditional practice of burning the forest floor was banned within one mile of the reserves.

Source: Based on Guha, 1989, pp.43-45.

The process of over-riding traditional rights and use had other implications as well. First, some researchers maintained that it not only curtailed customary rights, but also ignored local rituals and beliefs. Gadgil (1989), for example, reported on some Indian cases, where the erstwhile "sacred" forests and ponds were coercively taken over by the state as "reserved forests" without regard to local beliefs and customs. Second, this also had a psychological impact on the people: "The things that the people used to get free from the forest [had] have to be purchased. The very idea [was] shocking to the people and they [were] left bewildered" (Government of Madhya Pradesh (GOMP), 1980, p.67, cited in Anderson and Huber,

1988, p.72). Moreover, as Haimendorf (1982, p.80) reported on the general plight of Indian tribes, "while they were forbidden to take even enough wood to build their huts or fashion their ploughs, they saw contractors from the lowlands felling hundred of trees and carting them off".

The Insertion of Local Communities into the Broader Market

Many studies maintained that besides creating state forestry organizations and marginalising local rights and tradition, the "opening up and civilizing mission" (Anderson and Huber, 1988, pp.34-35) of colonial governments, paved the way for a host of commercial opportunists to infiltrate into local forests and economy (see Anderson and Huber, 1988, Chapters 1 and 5; Haimendorf, 1982, Chapters 1-4; also see Fernandes and Kulkarni, 1983). These commercial operators assumed different forms in different context. In tribal regions of India, for example, they were the money-lenders, land-grabbers, liquor-sellers, timber traders and contractors, minor officials, and landlords (Jones, 1978; Saha, 1986; 1988; Haimendorf, 1982). In Chittagong Hill Tracts (CHT) of Bangladesh, they were the migrant businessmen (in close alliance with military forces who regulate governmental activities in CHT), local traders and contractors (CHT Commission, 1991). In Malaysia, the operators represented the local allies of international (mainly Japanese) timber traders (Gray, n.d.). In the process, as Anderson and Huber (1988, p.39) believed, a "dichotomous relationship of the exploiter and the exploited" was established between this new set of commercial interests and local people. Diverse means of oppression and exploitation were institutionalized including expropriation of local lands, and various exploitative terms and contract of trade and business (for details see Haimendorf, 1982; Guha, 1989; Anderson and Huber, 1988; Saha, 1986). Besides the above commercial forces, larger and more organized industrial institutions also penetrated into the economy and habitat of local people. According to Tucker (1988, p.106), the tea industries in northeast India, for example, provided a "classic case of a foreign-dominated plantation economy that controlled a colony's land-use pattern and was highly sensitive to market in the industrialized world". He noted that in Assam there were 764 working tea estates by 1900, which brought in 400,000 alien workers and made them work in sub-human conditions. These estates contributed to the ethnic struggle for land and other resources between indigenous tribes and immigrant workers. (In chapter 5, I will further discuss the nature and impact of these organized industrial interventions in the local context of the study-areas).

In the name of "economic development", "land settlement" and "establishment of law and order" -- the state extended its authority to even the farthest peripheral areas. In tribal regions of India, for example, the colonial government launched such (commercially motivated) "development schemes" as building communication network, land settlement, and establishment of government offices. The consequences were manifold. Haimendorf (1982, p.34), mirroring the views of other major researchers (Fernandes et al., 1988; Gadgil, 1989; 1983; Joshi, 1983; Anderson and Huber, 1988; Saha, 1986; 1988; Guha, 1989), argued:

> The extension of a centralized administration over areas ...which had previously been virtually unadministered...deprived the [local communities] of their autonomy...and exposed [them] to the pressure of more advanced population....Traders and moneylenders could now establish themselves under the protection of the British administration. Often they were followed by settlers who succeed in acquiring large tracts of aboriginals' land.

Anderson and Huber (1988, p.21) noted that under the land settlement schemes, many tribal people either lost their land to the "land hungry immigrants", or were "ruinously taxed". As Haimendorf (1982, pp.34-35) noted, the displaced locals often

> [being] unable to resist the gradual alienation of their ancestral land,...either gave way by withdrawing further into hills and tract of marginal land or, if no such refuge areas were left, had no other choice than to accept the economic status of tenants, sharecroppers, or agricultural labourers on the very land their forefathers had owned.

This process of incorporation of local communities into the central authority and markets, according to Anderson and Huber (1988, p.23), "mark[ed] the end or beginning of the end" for their territorial and cultural integrity and the relative freedom in the access to forests. As a general rule, it has been argued by many studies that, where people do not have the power to regulate their relations with central commercial interventions, they tend to get into unequal and exploitative dependence on the comparatively powerful and more advanced external forces (Gray, n.d., Preface; Haimendorf, 1982, Chapter 2; Guha, 1989, Chapter 6; Anderson and Huber, 1988, pp.20-23). Gray (n.d., p.37), from his extensive survey of indigenous people from all over the world, comments: "[this] dependency is the means of shattering the right of self determination of indigenous peoples to control their own lives and future".

As the preceding discussions suggested, the commercial era apparently marked a transition from an indigenous, localized, and broadly informal (though there were some formal structures of resource use in the pre-commercial era) system of forest use to a highly professional, centralized, commercialized, and formalized one. The nature and implications of the transformation are explained by Fernandes et al. (1988, p.324):

> To the formal system, the forests and other natural resources are only a raw material meant for profit. Its ownership is accessible to those who have access to a written document and to the expensive litigation-based legal system i.e. only to the powerful. To the tribal and other traditional communities, on the contrary, forests and other natural resources are assets of the community. They are meant to be used sparingly and to be preserved as a life support system. The cardinal rule in this system is mutual support and not competition as in the formal system of a capitalist economy that is being encouraged.

The culmination of these commercial interventions created mutual hostility between the two parties (commercial forces versus local people) and a state of separation of local communities from their surrounding forests. In many regions of south Asia there are histories of sporadic uprisings by communities, especially by the hill-peasantry, against external commercial interventions. For instance, Guha (1989) provides a detailed account of historical resistance by the Himalayan peasantry against "the march of commercial forestry". Similarly in Uttara Kannada (India), peasants had been protesting against the British commercial policies since 1800 (Nadkarni et al., 1989, pp.56-57).

The next phase of the evolution of forest use (which is continuing to-date) represents the consolidation and fortification of the centralized-formal system of forest management.

The Post-independence Era: The Consolidation of Competition and Commercialization

By the mid 20th century, most countries in south Asia saw the apparent cessation of colonial rule, and acquired political independence. However, the tentacles of commercial interests remained strongly grounded as before and became active under the new national states. Thus, this period was not essentially much different from the colonial period; and was characterized by similar trends in commercialization of forest and the consequent alienation of local communities. In fact, this period marked further intensification of the above trends, including the proliferation of commercialization of forest use;

further changes in agrarian and productive relations (causing more inequality in access to, and distribution of, resources); and more competition for forest resources. These characteristic trends will be discussed under two broad headings: "the march and proliferation of commercialization" and "the intensification of competition for forest use".

The Proliferation of Commercialization of Forest Use

Although by the beginning of the 20th century the "dominance of the larger economy" in forest use was "firmly established", there was "a major technical change in the use of forest resources when forest produce was processed for modern industries" (Nadkarni et al., 1989, p.64). Private and public industries were highly patronized by governments, in consonance with new development strategies of the independent states which were embedded in the commercial rationale of "industrialization" and "economic growth". Jenkins (1992, p.19) noted that Industrialization became a symbol of "decolonization for the newly independent countries of Asia", and it "was identified with increased economic activity and increased standard of living". The trends in commercialization of forest was, in fact, an integral part of these "growth-led" economic pursuits. As regards the general nature and consequences of these pursuits, ADAB (1992, p.2, emphasis added) noted that, they

> *promote reckless consumerism,...centralisation of power...and people's resource, power and autonomy are usurped....Thus alienated, people become passive and find no meaningful way to participate....*[They] set in motion a "destructive circularity" which might begin with construction of large dams,...monoculture tree plantation,...spawning of new industries etc.. These causes destruction of forests....The traditional popular knowledge and culture which value sustainable use of natural resources are down graded and devalued in the name of science and technology.

Accordingly, in the forest sector, raw materials for the newly established industries were provided and secured by the states at ridiculously low (subsidized) rates. This had (and still has) a number of implications. First, as Gadgil (1989, p.5) noted on the Indian experience, "pampered with subsidies, sheltered from all outside competition, [these industries] have grown to function in a thoroughly inefficient manner" (also see Jenkins, 1992, p.19; Repetto and Gillis, 1988, Chapter 9). Second, the industries were often subsidized at the cost of community resources, which were hitherto available for local use. Many local forests in India, for example, were said to be appropriated by the state for feeding the industries. Major studies reported that the poor and weaker sections of local population, especially women, fuel wood collectors, various "non-timber forest product" gathers, were particularly affected by this appropriation; and were pushed into further competition for the increasingly scarce forests (Gadgil, 1989, pp.15-16; Nadkarni et al., 1989, pp.65-82; Guha, 1989, pp.138-143). Third, under this new arrangement, the user and consumer of the resource had virtually no major stake in its regeneration. For example, Gadgil (1989, p.15) showed that, after the independence of India, the government supplied bamboo to the paper industries at less than a thousandth of its market price. But the industries had no responsibility for regeneration of bamboo. This, as some researchers argued, led to a dichotomy between "resource use" and "resource generation" (Gadgil, 1989, pp.15-18; Nadkarni et al., 1989, p.66). This also raises the issue of "social optimality" in forest use. Put simply, it refers to over-exploitation of forests beyond the "socially optimum" point by a privileged section without a corresponding contribution to its protection and regeneration (see for details Feeny, 1988). A large number of studies, reporting on regional experiences, showed that as resources of one region or one kind diminished, resources of another region had been made available to the industries (see Guha, 1989; Richards and Tucker, 1988; Gadgil et al., 1982; Gadgil, 1989; Fernandes and Kulkarni, 1983; Fernandes et al., 1988). Gadgil (1989, p.15) has termed this process as one of "sequential exhaustion".

Other major consequences of "modernization" and "industrialization" efforts by the newly independent states, include further displacement and disruption of local people and economy. Haimendorf (1982, pp.320-321), for example, noted that in the Indian states of Orissa, West Bengal, Bihar and Madhya Pradesh

> the establishment of steel works in the very centre of the tribals' homeland led to a large-scale displacement of tribal populations....They were dislodged from their traditional sources of livelihood and places of habitats.

Similar consequences have widely been observed in many other regions of south Asia (see for example Joshi, 1983; ADAB, 1992; Fernandes et al., 1988; Anderson and Huber, 1988). They have also been observed in other parts of the world. Haimendorf (1982, p.321) noted that in the tropical forests of Brazil, the Solomon Islands and Melanesia the welfare of forest-based locals was "sacrificed to the interests of local or multinational companies exploiting mineral or forest resources". Similarly, Repetto and Gillis (1988, p.28) showed that in Ghana, Ivory Cost and Indonesia in response to governments' industrialization incentives, industries flourished "at a heavy domestic resource cost...and faster deforestation" (also see Barraclough and Ghimire, 1990). Furthermore, as Trainer (1989, p.65) argued, the

> market forces predominantly developed wrong industries in the Third World....Mostly the development of industries provide[d] crops and consumer goods for the small rich local elites or for export to the rich countries.

The steel works and mineral exploitation in the tribal zones of middle India, for example, hardly produced any significant benefit to the tribals and were almost solely oriented to the central and foreign markets (Haimendorf, 1982). Other major interventions, which came as a sequel to industrialization, were the establishment of "planned development" schemes like hydroelectric and irrigation projects. One classic case in point is the Kaptai Hydroelectric Dam project in Bangladesh. Gain and Moral (1994, p.19) reported that in a series of "negative multiplier effects on environment and the forest people", this externally-funded project, built during 1959-63, "permanently submerged 40% of the total cultivable land in the Chittagong Hill Tracts", displaced 100,000 people, predominantly tribal, and pushed the evacuees to remote forests, accelerating the rate of deforestation. Haimendorf (1982, Chapter 12) also noted similar consequences of hydroelectric and irrigation projects in India.

Today, forest policies in south Asia show a high degree of continuity with their colonial heritage and have continued to maintain the centrality of commercial and industrial interests. Guha's (1989, p.143) comment on the Indian forest policies echoes the general state of affairs in other countries of this region:

> Several features testify to the marked degree of continuity between colonial and post-colonial forest policies. While the pressure of industrial and commercial classes may have replaced strategic imperial needs as the cornerstone of state forestry practices, in both periods "successful" implementation of policy has been at the expense of the hill peasantry and their life-support systems. The use of silvicultural and other strategies of manipulation and control, designed to limit and carefully regulate the access of the surrounding population to the forest, has been remarkably invariant....The forest department, assigned the role of revenue generating organ by...the colonial state..., has continued to be a veritable money spinner for the government.

In line with the colonial experience, south Asian forest policies continued to emphasize mono-culture plantation, subsidy to the industries, commercial exploitation, and revenue maximization (see Guha, 1989; Gain and Moral, 1994; Gadgil, 1989; Roy, 1987). These studies also noted that public policies have centralized all the attention, authority and resources to government forestry, superseding traditional rights and communal authorities. But, as Roy (1987, p.46) noted, the "irony" remained that in Bangladesh, for example,

> the prime supplier of the forest produce for the country was not publicly owned vast tract of forest land which attracted all the attention and development resources in the past, but it was the village forest raised and managed by the rural people without patronage from the government. (also see Douglas, 1982)

Another major public policy impact, which deserves special mention, is the effect of state-sponsored commercial logging. The liberal concessions and patronage, offered by public policies in the Third World, enticed local and international loggers and led to substantial deforestation (see different articles in Repetto and Gillis, 1988 and Gray, n.d. for an overview of the world situation). In fact, the exploitation of the rain and tropical forests through commercial logging (often joined by logging for international export) is a world-wide phenomenon. In the 1980s, for example, as Ngaiza (1991, p.58) reported, the Tanzanian region of east Usambaras was transformed into "a semi-desert" through the "haphazard logging" by a commercial alliance between a private company (the Sikh Saw Mills), an international agency (the Finnish International Development Agency) and the government. Throughout southeast Asia commercial logging is reported to be a major reason for deforestation (Gray, n.d., p.12). Colchester (1989) noted that in Malaysia the timber trade propelled by the Japanese logging companies has been destroying the Sarawak forest, despite local resistance. Gray (n.d., p.12) argued that Thailand today has "become a net importer of timber", having virtually destroyed its own forests by indiscriminate logging. Myanmar, which cherished the history of some of the earliest attempts in traditional forestry management, has said to have become the "current target for deforestation by Thai companies" (Gray, n.d., p.12). Barraclough and Ghimire (1990, pp.16-17) referred to estimates which suggested that as much as 4.4 million ha. of tropical forests might be logged each year to supply to European, American and Japanese markets.

After the cessation of random transfer of forest wealth from the colonial peripheries to the centre, the system of resource drain of the decolonized peripheries was formalized and continued under the new form of "global timber trade". As noted in the above paragraph, commercial logging was often prompted by the demands of international trade. Richard and Tucker (1988, pp.3-5) argued that the timber trade was a part of the process of building increasingly "unified global economy", which has been a dominant feature of the 20th century. They (Richards and Tucker, 1988, pp.4-5) argue that timber trade, in essence, manifests a process of

> industrial economies tapping the developing economies timber resources to meet their consumption demand and private investors [often in alliance with local commercial interests] cashing in on the high short-term profitability of timber exports from capital-starved countries...[and it has to be seen] within [the] context of interactions between globally dominant power of Western capital and regional markets controlled in part by local entrepreneurs responding to regional opportunities for profit.

The Intensification of the Competition for Forests

The post-independence era also witnessed intensification of the competition for forest-use. In this section, the nature and impacts of the competition will be elaborated.

The use of forests is determined by the competition among interested parties, each with their "own priorities or valuation of different uses or functions" (Nadkarni et al., 1989, p.160; also see Curtis, 1991, pp.179-182). There are some features of this competition. First, the parties have different interpretations of their interest in forest-use, and they act accordingly. For example, as Anderson and Huber (1988, p.6) noted,

> to the corporation, a "tree" was a resource to be mined for cash flow and replaced; to the forest department, it was a capital to be reserved and protected from tribals; and to bankers and planners, it was a substitute for other trees as sources of paper. To outsiders, the forest was an "environment". To the tribal people, it was their "place", their home.

Second, the competition is essentially an exploitative struggle; the interest of the weaker parties, which "are not strongly backed are sacrificed in the process" (Nadkarni et al., 1989, p.160). Third, the inequality of the process often results in over-exploitation of resource by the stronger parties. The beneficiaries of forest use are often not the ones who are responsible for its regeneration and protection (which raises the question of "social optimality" as previously discussed).

Keeping in mind the above, and drawing on the preceding discussion, I will now examine some selected political-economic variables which shape the process of competition and largely determine the share each party is able to achieve from the competition.

Land ownership and tenurial arrangements In south Asia, the ownership and tenurial status of land is of crucial importance in determining access to resources in general. In an earlier section, I examined the process of "gradual centralization of [governmental] proprietary rights to forest lands..., superseding traditional rights of local authorities and communities" (Repetto and Gillis, 1988, p.387). This transformation of tenurial rights on forests brought major consequences for local communities including the denial of local rights, the intensification of competition for limited communal resources including forests (causing rapid degradation thereof), and the influx of a host of commercial operators into local economy. In the post-independence period these trends continued, and further aggravated. In India for example, it is reported that with the "breakdown of community management of resources" and "unrestrained use" (Gadgil, 1989, pp.13-14), today community land resource has reduced to an average of perhaps 20 ha. per village (Chambers et al., 1989), which is again "typically heavily degraded and under open access usage" (Arnold, 1990, p.3).

Land tenure policies also had other major consequences. First, they seemed to have encouraged expansion of agricultural land at the expense of grasslands and forests. The process started under colonial governments, who, as Richards and Tucker (1988, p.5) argued, were "adept at turning their colonies' lands to the production of [agricultural] crops for metropolitan markets". It continued steadily from the 19th century through to the latter half of the 20th century (Richards and Tucker, 1988, p.5). Agricultural expansion (especially for commercial farming) and the consequent pressure on forests have been observed by many studies world-wide (see for example Tucker, 1988 for the Indian, Feeny, 1988 for the Thai, Castro, 1988 for the Kenyan, and Poffenberger, 1990 for the southeast Asian experiences). Second, as noted before, in many forested regions of the Third World, governments' liberal concessions encouraged the loggers and other commercial operators to exploit forests. Similarly, the previously discussed state land settlement and colonization schemes contributed to the exploitation of the locals by a variety of commercial operators, mainly in the forms of cheap, bonded, or contractual labour; spatial displacement of habitat; grossly unequal terms of trade and exchange; and exploitative money-lending arrangements (Saha, 1986; 1988; Haimendorf, 1982; Guha, 1989; CHT Commission, 1991; Repetto and Gillis, 1988; Barraclough and Ghimire, 1990).

In sum, the concept of ownership and management of forest land underwent a major change over the century. Sharma (1978, p.64) explains the nature and implications of the change:

> In this way, there had been a steady and certain change from de facto ownership of forests by the tribal community through economic exploitation by the state with de jure ownership....[N]on recognition of tribals' right but granting a few concessions finally [leads] to a stage where forest resources acquire the fiction of corporate ownership, their working governed by input-output considerations and tribal economic needs are irrelevant and irritant to the system.

The role of governments and public policies Forest bureaucracies (the principal regulatory government agency on forest management) and forest policies in south Asia, had their origin in the colonial system with a predominant commercial-industrial orientation. After independence, as Guha (1989, p.143) noted, the forest departments remained "faithful" to their colonial legacy, and continued to serve their congenital role of "veritable money-spinners" for the respective governments. The major limitations of state forestry organizations and policies have been well documented by a good number of studies (for example Joshi, 1983; Shiva et al., 1983; Gadgil, 1983; Lai and Khan, 1992; Shepherd, 1992; Banerjee, n.d.; Poffenberger, 1990; Gronow and Shrestha, 1991; Khan, 1994; 1996a; Khan and Hasan, 1995; Fernandes et al., 1988; Guha, 1989; Nadkarni et al., 1989). The most common of these shortcomings, as these studies suggest, include the elitist and custodian-like values and the para military service structures; the purely bureaucratic and revenue orientation to forest management; undermining of traditional rights, local knowledge and resource-use practices; widespread isolation from local people; corruption, connivance and collaboration with local and external commercial operators. Anderson and Huber (1988, p.49) summarized the general attitude of forest officials towards the forest-based people:

> From the perspective of forest officials, the tribal is an incorrigible destroyer of forest wealth, an irremediable "primitive" whose innate weakness can only be expected to be exploited.

A vast majority of literature maintained that governments patronized the process of infiltration of commercial forces into local economy; and in most cases, there had been an alliance of vested interests between governments and commercial operators (see for example Haimendorf, 1982; Fernandes et al., 1988; Nadkarni et al., 1989; Anderson and Huber, 1988; Gadgil, 1989; Guha, 1989; Bangladesh Rural Advancement Committee (BRAC), 1986; Gray, n.d.). Many of these studies also noted that, throughout south Asia, there have been reports of harassment and extortion of local people by government officials, especially of the forest and police departments. These extortions range from such "extreme" actions as murder and pogrom, through "moderate" actions of arrest or physical manhandling, to "soft" and very common practices of collecting varied illegal fees. Foley and Barnard (1984, p.147) commented that "arrogant use of power by minor bureaucrats is a universal phenomenon". Haimendorf (1982, pp.93-96) noted that "the tyranny" of "the minor officials" is a historical reality in "Indian village life". His (Haimendorf, 1982, p.95) following observation seems to present a general picture of the "every-day" harassment:

> [T]he continuous irritation of illegal fees collected by forest guard from the villagers...is an old practice....Anyone who refuses to pay the illegal fees is certain to be harassed by the forest guard, who can prevent those in his bad books from collecting even legally permitted forest produce and may charge the defaulter with forest offenses which had never been committed.

In the same vein, public policies (as discussed before) generally contributed to the conversion of forest lands to agricultural and other commercial uses, to the influx of commercial forces into local territories, and to the growth of state-sponsored industry and commerce at the cost of public revenue and local rights.

On rare occasions, states attempted, mostly in vain, to control the activities of the commercial operators. In context of India, Jones (1978, p.48) notes that "the laws exist [to regulate the activities of commercial operators], but no government that wants to remain in power can afford to implement them". Therefore, the states seems largely ineffective in performing their "ideal" role as the regulator of interests among various parties in forest use (as envisaged in Table 2.1). The states, as the previous discussion suggested, have generally facilitated the intrusion and operation of external commercial forces. Nadkarni et al. (1989, p.167) noted:

> [T]he state, the most responsible, powerful and farsighted of the...parties involved [in forest use and management], failed in its task....The Forest Department was swept off its feet against the flood of pressure from the larger economy....

The role of present local organizations With the change in indigenous patterns of resource management, some alternative efforts were made by the states to establish organizations at local level, mainly under the umbrella of "local governments". Rew (1986, p.397) argues:

> The relationship between "indigenous" and "introduced" structures change through phases of colonialism and commercial and administrative penetrations. These penetration...have created derived formal institutions -- the cooperatives and the local government councils are examples.

However, these organizations seemed quite different from the indigenous institutions. The performance and character of such state-sponsored local organizations have been extensively surveyed by a sizeable literature (see for example Uphoff, 1982; Ahmed, 1990; 1991; Huque, 1985; Roy, 1983; Sobhan, 1968; Hye, 1985; 1991; Rahman, 1988; Rashiduzzaman, 1968; Sarker, 1990; Westergaard, 1985; Rahman, 1994; Attwood and Baviskar, 1988). As these studies have suggested, such organizations have been designed to maximize central interests of the state and its local allies at the sub-national level. Instead of collaborating with other indigenous structures, the state-sponsored organizations often try to gain control over, and restrict the growth of indigenous institutions. Mostly, they reflect local patterns of power-structure and their leadership is dominated by local elites. In sum, they are argued to be a carrier of powerful political and commercial interests to local territories, mostly under the guise of "rural development" and "decentralization".

In the name of popular participation in forestry, similar attempts have been made by the states to establish so-called 'decentralized" and "participatory" local organizations. These initiatives included some of the early projects under the title of "social" or "community" forestry. Forest management through the *Panchayat* system in India and Nepal is an example of such effort. Again, the performance of these efforts have also been thoroughly explored (see Foley and Barnard, 1984; Gronow and Shrestha, 1991; Hobley, 1991; Task Force, 1987; Khan and Hasan, 1995; 1995a; Khan, 1994; 1995; 1996a; Arnold et al., 1988; 1990; Banko Janakari, 1993). Drawing on these studies, three basic points can be made on the performance of these local organizations as regards forest use. First, most commonly, these institutions are riddled with local factionalism and party-politics. Foley and Barnard (1984, p.120) note:

> In general, local institutions tend to represent and maintain local pattern of power and interest and the wealthier members of the village usually have dominant voice in decisions.

Second, proper representation of classes and factions of a given community rarely occurs in such local organizations. Third, the plethora of rules and bureaucratic formalities which guide the operation of these organizations are normally beyond the skill and understanding of local people. This not only reduces their interest and chances of participation in the organizations, but also make them dependent on the more powerful groups (for example, local "brokers" and "middlemen") who are conversant with the procedural labyrinth.

Market interventions and changing social dynamics Many researchers including Gray (n.d.), Guha (1989), Gadgil (1989), Fernandes et al. (1988), and Fernandes and Kulkarni (1983) argued that commercial interventions and the "bureaucratization" of forest use bespoke some major changes in the life, economy and society of local people. The changes, according to above analysts, included the erosion of community regulations and culturally determined pattern of resource use, and the increased dependence of local people on commercial forces. From his probing study of indigenous people of Amazon, Gray (n.d., pp.36-37) summarized the impacts of uncontrolled market-intervention on local communities as follows: first, "genocidal" which refer to the consequences of externally-induced rubber production of 1894-1914 in the upper Amazon. It wiped out 90% of local population through displacement, disease and murder. Second, the destruction of cultural possessions and heritage of local people ("ethnocidal"). Third, the "imbalance" in local subsistence economy through the introduction of cash economy.

There is also a strong argument that the intensification of competition for scare resources reduced the chances of cooperation among local people and contributed to conflict, distrust and individualism (Fernandes and Kulkarni, 1983; Guha, 1989; Shiva et al., 1987; Khan, 1994a). In some cases, as Guha (1989, Chapter 3) and Fernandes et al. (1988, Chapter 9) reported, the severe struggle for survival compelled local people to destroy forests, which they hitherto traditionally revered.

The "Winners" in the Competition

There is a trend in the existing literature to identify the "winners" in the competition for forest use. The argument, in a nutshell, is that, the (above discussed) commercial forces and political-economic variables, working together, culminate in a situation where the "rich and wealthier sections" manipulate and control the process of resource use, and "gains" from it (see Chambers et al., 1989; Fernandes et al., 1988; Fernandes and Kulkarni, 1983). The following comment by Fernandes et al. (1988, p.344) on the general nature and result of competition for forest resources, mirrors the conclusions of many other studies (for example Fernandes, 1983; Joshi, 1983; Kulkarni, 1983; Gadgil, 1983; 1989; Nadkarni et al., 1989; Guha, 1989; Khan, 1994; 1995; Arnold et al., 1988; 1990; Chowdhry, n.d.; Shiva et al., 1987):

> Today the needs of the people and of the environment seem to be in contradiction with those of the industry and revenue....The interests of the more powerful

groups, mainly industry and state...predominate and the people ...suffer.

Given the regional difference, it is difficult to generalize the "winners". In south Asian context, it is, however, possible to trace a broad picture of the "winners". Chambers et al. (1989, p.68), for example, from their extensive review of experiences, conclude:

> [N]ormal process[es] of appropriation and access to government lands and programmes have meant that losers have been women, the poor and forest dwellers; and gainers have been industry, contractors and forest officials.

They present a summary of benefit-distribution from various afforestation programmes in terms of "gainers and losers" (see Table 2.4). Though it is a rather simplistic presentation, it certainly provides some insights into the common trends and outcome of the competition.

Table 2.4 Main gainers and losers from afforestation programmes

Land and Programme	Gainer	Loser	Reasons
Production Forestry in (state) Forest land	Industry, local officials	Forest dwellers, artisans	1. Deforestation increased the distance to be travelled for collection of livelihood goods. 2. Mixed forest which provided livelihood goods were replaced by plantation corps. 3. Industries were supplied cheap raw material at the cost of artisans.

Social Forestry in Community land	Panchayat elite, urban consumer	Graziers, weaker sections	1.Grazing land were taken over by Govt. and were planted with non-browsable species. 2. Species like *casuarina* and *eucalyptus* were classified as fuel wood species, although these provided cash to the rural panchayats, and fuel wood/poles to the urban consumer, thus bypassing rural consumers. 3.Distribution arrangements were not properly worked out allowing the powerful to walk away with the "cake".
Farm Forestry in Private land	Rich farmers, absentee land-owners, contractors and petty officials	Women, landless labourers, small farmers	1. Tree farming on agricultural lands reduced employment, esp. for women. 2.Stringent felling and sale restrictions helped intermediaries and petty officials at the cost of small producers. 3.Serious efforts were rarely made to reach the resource-poor farmers.

Source: Chambers et al., 1989, p.69.

In the same vein, Chowdhry (n.d.) identifies the groups who have benefitted from various afforestation programmes in India. His analysis is summed up in Table 2.5.

Table 2.5 Principal beneficiaries of afforestation programmes

Afforestation Programmes	Who benefitted?	Reasons
Farm Forestry (plantation in private lands by the farmers themselves, with logistic support from the government)	Larger farmers and pulp, pole and paper industries	1. Farmers with larger landholding could avail themselves of advantage of government's logistic support. 2. *Eucalyptus* - the species promoted in the farm forestry, was oriented to the pulp, paper and pole industries. 3. Larger farmers were converting crop lands into eucalyptus plantation, causing reduced employment of the landless poor and sharecroppers.
Strip Plantations (canal bank, roadside and railway side plantations by the government with the help of the community)	The urban construction industry, paper and pulp industries, contractors and petty officials	1. *Eucalyptus* trees ultimately found their way to the industries, and were hardly of any substantial benefit to the local poor. 2. An alliance of vested interest was formed among the large farmers (as the producers), the industries (as the consumers) and the local state-officials (as the regulators)
Village Woodlots (plantation on village common lands by the local *Panchayats*, under the government's supervision)	*Panchayat* leaders and few rich farmers	1. *Panchayats* were not properly represented by the weaker sections of the community and were dominated by the rich and the powerful. 2. Weaker sections (especially the women and landless) largely remained unaware of the programmes, and had virtually no involvement.

Note: Based on Chowdhry, n.d., pp.61-66.

As the preceding review and the above tables reveal, the competition for forest resources is manipulated and dominated by the rich and powerful sections of local communities. These "winners" commonly include farmers with large land-holding, local government leaders, state officials, timber traders and contractors, forest industries and their local agents. Chowdhry (n.d., p.65) notes:

> The benefits [of afforestation schemes] have largely eluded the poor, the rural poor have not participated except as wage labour, and the women have been largely left out.

Though there are regional differences, the basic reasons for the weaker section of local population to lose in the competition seem to include the loss of control over the means of production (especially land); the erosion in community regulations of resource use; the commercial orientation of afforestation programmes; local people's dependence on commercial forces on unequal and exploitative terms of exchange; the general poverty, hardship and low standard of living of the locals; the absence of any efficient platform for articulating their interest and organizing collective actions; and the adverse role and unrepresentative character of government and local government offices, and public policies.

Conclusion: A Process of Alienation

The above overview of the evolution of forest use points to a process of commercialization and bureaucratization of forest management; and gradual deprivation of the livelihood needs and rights of local communities. The process of commercialization was spawned by the colonial rule and continued after independence of the nations in south Asia. Consequently, there has been a systematic alienation of local people from forests. Much of the literature points to the process of commercialization and alienation (see Guha, 1989; Fernandes et al., 1988; Anderson and Huber, 1988; Haimendorf, 1982; Joshi, 1983). Anderson and Huber (1988, p.39), for example, could see a process of "disenfranchisement" of local people from forests, set in motion by the colonial rule in India, which they "summed up [as] the dichotomy of the exploiter and exploited". The above literature convincingly argued that the colonial capitalism brought about radical changes in the social life, living and traditional organizations of local communities. The most obvious changes and mechanisms, as discussed so far, include the denial of customary rights and access; the disruption of traditional community regulation of forest use; the imposition of a commercially motivated state forest administration; the uncontrolled exposure to commercial forces, and the diverse forms of exploitation and unequal exchange between commercial forces and the locals; and the intensification of competition for forest use. These changes, which were primarily introduced by the colonial capitalist state, have continued and further intensified in the post-colonial era in the form of "planned economic growth" and "industrialization". Guha (1989, p.195) wrote:

> More recently, the policies of post-colonial regimes have followed [and] continued to exploit the predatory nature of colonial capitalism, which has now assumed grotesque and brutal forms unheard of in its original European home ....They have embarked upon an ambitious programme of planned economic development....In this strategy small and landless peasants, and women and tribal minorities, are expected to bear the burn of the forced march to industrialization....The post-colonial policies have continued to undermine the social, economic and ecological basis of peasant agriculture.

"Alienation" is used here as a generic term to refer to the various adverse impacts and implications of the above changes on local people. It is necessary at this point to further clarify the conceptual interpretation of alienation, because this is one of the key concepts of the book and the subsequent evaluation of SF programmes is premised on the processes of commercialization and alienation.

Theoretically, alienation is "bedeviled with conceptual ambiguity and terminological abuse" (Geyer and Schweitzer, 1981, p.2; also see Rotenstreich, 1989, pp.3-16). At the initial stage of its theoretical development, alienation did not necessarily connote a negative or repressive process (Lubz, 1981; Rotenstreich, 1989). The English concept alienation goes back to the Latin *alienatio, abalienatio* and Greek *alloiosis*. These concepts designated the sale of a commodity, the cessation, transfer of rights and property (Ludz, 1981, p.22). At the philosophical and theological level, it denoted the transformation of man who had passed beyond himself -- to become submerged in God or in the One. Philosophers like Plotinus and Augustine conceived of alienation as a state where the human spirit or soul is elevated and reaches its goal of the union with divinity (Rotenstreich, 1989, pp.3-4). The classical concept of alienation underwent a radical change in the 19th century through the works of Hegel (see Rotenstreich, 1989, pp.6-9; Ludz, 1981). Hegel, described it as a mode of experience of self consciousness, without which the absolute mind cannot unfold. Whereas in classical expression of the concept, unity is achieved "beyond" oneself, to Hegel it is a transitional state leading to the achievement of unity "within" and not beyond the limits of "speculative knowledge" (Rotenstreich, 1989, pp.6-7; Ludz, 1981, pp.26-27). Up to this stage, the concept of alienation did not particularly entail negative meanings.

The present use of the term with essentially negative connotations has its roots basically in the works of Marx. Unlike the purely philosophic abstractions of the concept (entailing, for example, such explanations as "the elevation of human soul"), Marx proffered a down-to-earth explanation of alienation. He discussed alienation within the limits of inter-human relations and treats it as a distortion of the very "essence of man" (Rotenstreich, 1989, p.12). Marx had "a relational view of the society", and he believed that man "is to a remarkably high degree the product of his society and by changing his living condition we change him" (Ollman, 1971, pp.237-238). The concept is developed within the above understanding of the society (that is, the relations and the changes within the society and their impacts on individuals). It "is not solely an economic phenomenon" and is a major part of the discourse of Marxist political economy (Ollman, 1971, p.13 and 205). Marx's analysis, based on an economic-materialist explanation, views alienation as a negative set of economic processes which characterizes the capitalist economic order. Ollman (1971, p.131) astutely summarizes the Marxist explanation of the concept:

> The theory of alienation is the intellectual construct in which Marx displays the devastating effect of capitalist production on human beings, on their physical and mental states and on the social processes of which they are a part.

Marx's concept of *Entfremdung*, entailing the above connotations, entered in English as "alienation" (Ludz, 1981, pp.28-29; Rotenstreich, 1989, pp.12-13).

Thus the current use of the term largely stems from the Marxian analysis of alienation of the productive worker under the conditions of the industrial capitalism. Marx regarded alienation as a characteristics of the market ("commodity exchange") and the capitalist exploitation ("alienation of the labour-power without return of an equivalent") (Torrance, 1981, p.75). Further elaborating on the Marxist interpretation, some recent scholars have attempted to re-define the concept with more clarity. Schacht (1981, p.40), for example, defines alienation as a process which "involves the loss of autonomy and preclusion of self-determination in one's productive, working life". Here alienation is perceived in terms of the performance of the labour, and the production of goods and services under circumstances in which "control over what one does has passed into the hands of someone else, or under the sway of some external power" (Schacht, 1981, p.40).

Broadly, this book applies the term in the Marxian interpretation of "alienated labour" in the development of capitalist relations of production. For practical analysis, the book concentrates on Schacht's (1981) elaboration of the concept with emphasis on such characteristic manifestations as "control", "rights", "autonomy" and "self-determination". The examination of these manifestations may be of use in the understanding of the changes described, the issue of political economy of forest use, and the interplay of diverse interests therein. At this point, one faces the next question: whether the Marxist construct of alienation can be applied to a rather different context of forest resource use. Guha (1989, p.56) offers a cogent answer:

> While the application of Marx's theory, formulated in an entirely different context, must be done with some degree of caution, I would argue that in both instances we are dealing with the growth of a social system (industrial capitalism/colonialism) that replaces another (craft production/subsistence agriculture) whose social relations did not produce conditions of alienation and estrangement.

The previously discussed radical changes created by commercialization of forest altered the traditional structures and patterns of forest use. Marx (1951, p.316) himself, commenting on the "the British Rule in India", noted that the "free-trade" and the "brutal state inference" (which were also the two common manifestations of the process of commercialization of forest) resulted in the "disappearance of small stereotype forms of social organisms of the village communities". From the historical review in this chapter, one can see a sharp tension between an indigenous and more informal system of forest use based on reciprocity, with an invading formalized and commercialized structure of governance. As Guha (1989, pp.189-190) argued, the traditional pattern of forest use which, "through a mix of religion, folklore and tradition had drawn a protective ring around the forest", was subdued by the "imposition of an alien system of forestry management". These manifestations of the process of commercialization of forest, such as the alteration of traditional use patterns, the imposition of commercialized structures, and the resultant alienation of the locals, have been observed in many forested regions around the world (see Repetto and Gillis, 1988; Haimendorf, 1982; Gray, n.d.). Haimendorf (1982, p.321), for example, notes:

> Destruction or alienation of tribal land as a result of industrialization is not a

> process peculiar to India, and it is well known that in areas such as the Solomon Islands or Melanesia,or the tropical forests of Brazil the welfare of the primitive tribes was sacrificed to the interests of [commercial] companies exploiting forest resources.

Thus, the basic manifestations of commercialization of forest are broadly similar to those covered and analyzed by the Marxist concept of alienation (with reference to a different context -- "industrial capitalism"). Broadly speaking, the process of commercialization of forest, in essence, has the similar manifestations and consequences to those of industrial capitalism; and, as I have noted before, the process can also be explained as a part of the industrialization and capitalist economic growth. Hence the concept of alienation, with emphasis on the above characteristic manifestations, can be used in the context of forest use. In fact, as Ollman (1971, pp.244-245) suggested, the concept has a universal appeal; because the basic relations and the nature between the exploiter and the exploited, and the character of commercial-capitalist exploitation, as analyzed by Marx, "have not undergone many significant changes from his time to our own":

> Evidence of alienated relations can be found throughout capitalist society. In every area of life, man remains an object of manipulation by others and by his own product ....The resulting relations between people in the same class and between the classes are still dominated by competition after jobs, goods and money....People play no greater role in their political life than they did in Marx's time....So long as people, their favors and products can be measured in terms of money, so long will the theory of alienation be useful in explaining why this is so.

Torrance (1981) showed that, in later texts of Marx, his classic ideas of capitalist exploitation and the consequent alienation of labour were applied to different contexts and instances (especially in the context of pre-capitalist social formations). Marx (cited in Torrance, 1981, pp.74-75) also argued that the "exploitative dependence of a conquered population" generally combined the conditions of "social hostility with appropriation by violence". Besides, there can be hardly any doubt that the concept is a powerful tool of analysis and can accommodate many facets and consequences of the capitalist-industrialist process. Ollman (1971, p.233) noted that "Marx's relational view of reality,...and his theory of alienation are extremely useful aids for understanding nature, man and society".

Accordingly, the concept is used in this book as a platform for synthesizing the diverse impacts and manifestations of the process of commercialization and bureaucratization of forest use and management. In the context of forest use, alienation may be used to denote a state of mental (psychological) and physical (material) estrangement of the local people, resulting from the political-economic process of commercial exploitation and appropriation. Here it includes the following manifestations and affects: the gradual diminution of local rights and control over forests; the diverse forms and impacts of exploitation of the locals by commercial forces (for example traders, contractors, government officials, money-lenders and other local and mainly external parties); the distortion in cultural and religious norms and patterns of resource use; the dispossession and displacement of local people by commercial interventions; the intensification of conflict and competition among unequal contenders for forest use; the impairment of community protection and management of forests; and the feeling of deprivation and estrangement among local people.

Now I will briefly recapitulate the characteristic processes which have been identified in the previous discussion of evolution of forest use and will attempt to relate them to the above manifestations of alienation. The aim is to show that the impacts of commercialization of forest on local people can be conceptualized and explained by the concept of alienation.

Changes in Rights, Control and Autonomy

As the preceding review of literature suggested, in the pre-commercial time, local communities apparently enjoyed a reasonable degree of de facto (if not legal or de jure) access and control over forests. The process of commercialization, mainly in the forms of state-sponsored land reservation-colonization schemes and patronage for commercial forces, meant, according to Guha (1989, p.55), "an effective loss of control over their habitat for the forest based communities". The immediate consequences of commercialization were at least threefold. First, through the processes of reservation and privatization, to use Marx's (1975, p.235) words, "the customary rights of the poor" were turned into "a monopoly of the rich". On the importance of such customary rights of local people on the forests, Marx (1975, p.230) maintains:

> We demand for people a customary right, and indeed one which is not of local character but is customary right of the poor in all countries...a customary right by its very nature can only be a right of this lowest, propertyless and elemental mass.

Not only such rights were denied, but also local people became, according to Shepherd (1992a, p.19), "the notional thieves" in the eyes of governments, causing "a gross and injurious intrusion into the life of the citizen" (Marx, 1975, p.243). Second, many researchers including Jodha (1990), Gadgil

(1989; 1985), Fernandes et al. (1988), Shiva et al. (1983; 1987) argued that with state-appropriation and privatization, only few inadequate patches of "open access" land were available for meeting local demands, resulting in their rapid degradation. Marx (1975, p.235) also noted similar processes of "monopolization of common property" in relation to forest use. Third, the review has also suggested that the influx of commercial forces resulted in further intensification of competition for forest use; and local people, usually the weaker party, systematically lost out. The socio-religious customs which guided traditional forestry gradually waned with the severity of competition and the predominance of commercial interests. At a theoretical level, Marx (1975, p.265) notes that with the commercial intrusion and the predominance of commercially motivated laws and interests, a situation of "abject materialism" is created in forests, which is, he considers, a "sin against the holy spirit of the people and humanity". One can see such a state of "abject materialism" in the commercial values, laws and policies which have regulated forest use since the commercialization of south Asian forests. Today, under the commercially motivated system of forest management, Sharma (1978, p.60) argued, forests have become a

> part of a new specialized sector as a big steel plant...[which] produces certain raw materials...necessary for producing other finished good for satisfying certain specialized needs. The forests are a part of long assembly line.

Amidst such predominantly commercial outlook, it seems natural that local rights and access pale into the insignificance. Local communities are, Fernandes et al. (1988, pp.210-211) argue, "no longer an essential part of forests" and the local economy has become "at best irrelevant to this system and at its worst, inimical, that needs to be controlled" (also see Haimendorf, 1982, p.80; Anderson and Huber, 1988, p.44).

"Individualism"

Many analysts believed that with the process of gradual commercialization, the community feelings and socio-religious customs were taken over by an exclusively materialistic pursuit of "individualism" (see Fernandes et al., 1988, pp.224-228; Guha, 1989, pp.55-58). Individualism was encouraged by governments, because "individual use of the forest represented a lesser threat to commercial forestry than collective use sanctioned by custom" (Guha, 1989, p.55; also see Haimendorf, 1982). One of the direct outcomes of individualism was the change in the attitudes of the locals. Many analysts including Fernandes et al. (1988) and Gadgil (1989) argued that, being lost in the competition for forest use, local people in some instances, took resort to "destructive dependence" and started to maximize exploitation of whatever scant resource they had access to.

"Exploitation", "Conflict" and "Dispossession"

According to the Marxist explanation, "exploitation", "conflict" and "dispossession" are some of the ways in which "capitalist extortion" is "institutionalized", causing widespread alienation among "the exploited" (see Torrance, 1981; Rotenstreich, 1989). The earlier review in this chapter alludes to similar processes in south Asian forests. The commercial intrusion seemed to have institutionalized capitalist exploitation in the following ways. First, as discussed in the previous paragraph, local people were dispossessed of their socio-religious customs and norms which influenced their relation with forests. And "dissolving social bond of possession between a person/group [local people] and an object [forest]" (Torrance, 1981, p.74) -- is argued to be one major means of capitalist exploitation (and the consequent alienation of the victims). Second, the locals were exploited by a network of unequal relations and contractual arrangements. Third, the commercial exploitation caused "estrangement", "conflict" and "mutual distrust" among local communities, between local people and outsiders, and, of course, between local people and the state. The situation conforms to the predication of Marxian analysis, which maintains that, "capitalist exploitation...generate[s] hostility, and remain[s] a scene of conflict" (Torrance, 1981, p.75). Indeed, many forested regions of Asia seemed to be riddled with conflicts, caused by an interplay of diverse interests in the competition for forests, The "winners" in such competitions largely represent the commercial ends of a larger economy. The very title of Guha's (1989) book, "The Unquiet Woods", succinctly illuminates the inequalities, conflicts and unrest which surround south Asian forests.

This chapter has argued that the gradual commercialization of forest has caused a systematic alienation of local people from forest use and management. The present mode of forest management with its commercial-industrial orientation marginalises local needs and consolidates the power and interests of commercial forces. The state has generally collaborated with such "winning" commercial forces. Consequently, forests have become a scene of conflict and distrust between commercial interests and local communities.

It is in the above context of public alienation that there was an apparent realization of the need for a policy shift in south Asian forestry. Since the mid 1970s, there has been a consensus among the policy-planners of national governments and international aid-agencies that, "until forest communities...are incorporated into national forestry plans and use systems, conflicts and mismanagement will likely to continue" (Poffenberger, 1990, p.17) and that "alienation of locals from management and regeneration of forest has to end" (Nadkarni et al., 1989, p.177). These realizations have contributed to the introduction of participatory afforestation programmes, popularly known as Social Forestry (SF). Apparently, SF was introduced to remedy the process of alienation and make room for wider participation of local people in forestry. Against these apparent goals, the next chapter examines SF's performance in selected regions of south Asia.

3 Social Forestry Programmes in South Asia

Introduction

Chapter 2 reviewed the gradual transformation of forest use towards commercialization and the consequent alienation of local communities. In the 1970s, the far-reaching consequences of the processes of commercialization and alienation became matters of concern to national governments and international aid agencies. The "custodial and production forest management systems" were found to be "failing"; and there were "mounting conflict" between state officials and local people (Poffenberger, 1990, p.xvii and 3; also see Barraclough and Ghimire, 1990; Shiva et al., 1987; Fernandes and Kulkarni, 1983; Task Force, 1987; Fernandes et al., 1988; Poffenberger, 1990a). It was repeatedly argued that the "responsible agencies are failing to sustainably manage the extensive tracts of forest lands that they claim" (Poffenberger, 1990, p.278), and that "it would be impossible [for government efforts] to succeed without people's active participation" (Task Force, 1987, p.8). In Bangladesh, for example, the rise of SF can be attributed to "the admission that the government has insufficient resources to tackle the resource depletion problem; the rural population must assume a prominent role in forestry initiatives" (Task Force, 1987, p.2). Similarly, in India (SPWD, 1992, p.1) SF is argued to be the

> outcome of the realisation of a home truth that without the...active participation of communities living in and around forests, no programme to regenerate degraded forests would succeed. It was the alienation of these people...that brought the country's once verdant forests to their present sorry pass, in the first instance, and it was again peoples' virtual exclusion from all programmes for forest regeneration that saw the programmes perishing on the launching pad.

These realizations set the stage for the emergence of SF in south Asia. Accordingly, during the mid 1970s most south Asian countries witnessed a major shift in forest policy in the form of participatory approaches to forest management, more popularly coined as SF, which aimed "to redress past village exclusion and encourage rural development" (Ribot, 1995, p.1588).

SF was launched with great vigour and lofty philosophies. Its stated goals included increased participation of local people in the planning, implementation and benefit-distribution of forestry activities; fulfillment of local demands; promotion of self-reliance and social equity among local communities; increased income, productivity and standard of living for the participating farmers; and the growth of local cooperative institutions.

In light of these goals, this chapter examines SF's performance in selected regions of south Asia from a review of literature. It shows that SF's achievements lag far behind its stated goals. From the review of regional experiences, the chapter argues that SF's performance is shaped by the surrounding social and political-economic environment (which is referred to as the "frame" of SF). The major variables of the "frame", identified in this chapter, are: land tenurial arrangement, market, public policies, government and local government institutions, international donors, social relations and power structures. These factors will also serve as a basis for evaluating the performance of the empirical case-studies of my research (three SF projects in Bangladesh; see Chapter 6).

The Nature and Goals of Social Forestry in South Asia

It is difficult to identify precisely what types of forestry activities constitute SF. There are regional differences in the nature and magnitude of the tasks performed under the umbrella of SF. Moreover, all states in south Asia do not specifically use the term SF. For example, participatory approach to forest management is known as "Social Forestry" in India, "Community Forestry" in Nepal, "Community and/or Social Forestry" in Bangladesh. However, SF programmes have much in common in their underlying philosophy and theoretical consideration.

The principal emphasis of SF is on wider participation of local communities in the protection and regeneration of forest resources. As noted in the introductory chapter, SF primarily refers to those public initiatives which enable local people to participate in the planning, implementation and management of local forests for their own benefits. In most cases (for example in Nepal, Bangladesh, Bihar, Tamil Nadu, West Bengal) the term SF is used in the above sense. Some common objectives of SF programmes in these countries are: to empower local communities by involving them in the planning, implementation and benefit-distribution of forestry activities in order to cater for local needs (especially the subsistence needs of fuel wood, fodder, timber and poles); to promote self-reliance and social equity among local people; to augment their income, productivity, socio-economic status, and living standard; to check environmental degradation, conserve soil and water resources; and to foster the development of local cooperative institutions (Oakley et al., 1991; Task Force, 1987; West Bengal Forest Department (WBFD), 1989; Arnold et al., 1988; 1990; Alim, 1988; Khan, 1996a). The following quote from a governmental SF plan-paper (Government of Karnataka (GOK), 1980, p.9, cited in Shiva et al., 1983, p.54) is a typical representation of the stated goals:

> The purpose of social forestry is the creation of forests for the benefit of the community through active involvement and the participation of the community. In the process, the rural environment will improve, rural unemployment substantially cease....The overall concept of social forestry aims at making the villages self-sufficient and self reliant in regard to their forest material needs.

On plan papers, SF shows a clear bias towards the socially and economically vulnerable sections of local communities (for example women, tribal groups and landless farmers). In theory, it commits to bring about social equity, and equitable distribution of benefits and decision-making powers in forestry schemes (Salehuddin and Rebugio, 1990; Srivastava and Pant, 1979; Rebugio, 1985).

Therefore, SF is by no means a mere silvicultural or technical afforestation strategy. The emphasis on the word "social" suggests that SF programmes are anchored on, at least, two broad values. First, "people's participation" in forestry activities for their empowerment and welfare. Second, the promotion of "social equity" through equitable sharing of decisions in the projects and benefits arising thereof.

However, the emphasis on public involvement leads to another major controversy -- the concept of "participation", which needs to be defined; because, first, participation has been argued to be the "lifeblood" of SF (Bandhu and Garg, 1986; also see Ahmed, 1991) and second, it will be one of the principal criteria for evaluating SF's achievement in this book. The notion of participation defies any universal definition and entails widespread ramifications (for details on the concept, see Mathur, 1985, pp.128-135; Oakley et al., 1991, pp.1-24; Khan, 1993, pp.106-120; Oakley and Marsden, 1984, pp.17-36; Cohen and Uphoff, 1980, pp.213-235; Goulet, 1989, pp.165-168; Midgley et al., 1986, Chapter 1; Bhaduri and Rahman, 1982, pp.1-14; Picciotto, 1992, pp.2-8; Gow and VanSant, 1981, pp.1-9). One major theoretical perspective on participation is to see it as a "cycle for rural development activity" consisting of "participation in decision making, implementation, benefits and evaluation" (Uphoff, 1989, p.1; Cohen and

Uphoff, 1980, pp.219-222; Uphoff et al., 1979, p.219). This framework for analyzing participation has been fruitfully used by some major studies on rural development in Bangladesh (for example Zaman, 1984; Hye, 1991). Considering the simplicity and practicability of this approach, some recent studies on Bangladesh SF have also (broadly) used this framework (see Task Force, 1987, p.1; Magno and Ali, n.d., p.9; FMP, 1992, p.2; Khan and Khan, 1993; Khan and Begum, 1997). Accordingly, participation is described in this study as the direct involvement of farmers in a particular SF "project cycle": in planning, decision making, implementation, monitoring and benefit sharing.

A wide range of activities are carried out under the umbrella of SF. Though there are regional variations, the salient features and components of SF programmes in south Asia are as follows (Task Force, 1987; WBFD, 1989; Arnold et al., 1988; 1990; Alim, 1988; Arnold and Stewart, 1991; Chowdhry, n.d.): Encouraging farmers to plant trees in their own farms by distributing free or subsidized seedlings and offering technical supports by the government (commonly known as "Farm Forestry"); "Strip Plantations" along roadside, canal banks and other marginal public and community lands; "Reforestation of Degraded State Lands"' by involving local communities; "Community Plantation" on public or community lands with joint management and benefit-sharing arrangement between the government and local communities.

Social Forestry at Work: A Review of Selected Cases

This section examines SF's performance in light of its selected stated goals. Notwithstanding the wide-spread advocacy for SF in national and international quarters, field experiences suggest that its achievement of the goals of participation, equity, and empowerment has been insignificant. Reviewing India's experience, for example, Fernandes et al. (1988, p.278) maintain that

> social forestry as it is practiced today, does not create in the people any vested interest in the preservation of...forests, and does not result in the people coming together and generating sufficient confidence in the long-term regeneration of their community.

In Bangladesh likewise, it is reported that, despite its priority status in government policies, any significant degree of participation or empowerment of local people in SF is yet to be achieved (Task Force, 1987; Khan, 1994;

Asaduzzman, 1989, Khan and Begum, 1997). In Nepal, a number of studies conducted on different regional SF projects noted that rarely have people participated in any major planning decisions for forest management and were passive about the government's policy changes (see for example Gronow and Shrestha, 1991; Koirala, 1985; Khan, 1996a; Sharma, 1991).

What are the real nature and functions of SF projects as practised in the field? What is the extent of people's participation in such initiatives? Though there are regional variations, Vergara (1986) noted two broad operational approaches to SF in south Asia: first, the traditional-bureaucratic top-down mode of decision-making and implementation; and second, where the management and operative tasks are in control of farmers. He maintains that the former approach, by far, predominates SF programmes. In the same vein, Sen and Das (1987) have noted two alternative sets of tasks (and management-responsibilities) in SF (see Table 3.1).

Table 3.1 Two common sets of tasks in SF

Operative tasks	Management tasks
nursery-raising, land protection, watering, weeding, fertilization, protection, exploitation.	selection of land, planning and deciding what to raise, organizing planting operation, managing the plantation, distributing the products after exploitation, marketing the surplus.

Source: Based on Sen and Das, 1987.

Public SF programmes, as the regional experiences suggest, mostly show a clear bias to the traditional extension services which delimit people's role to a minimal (mostly in performing the "operative tasks"). Falconer's (1987) review of forestry extension also noted the essentially top-down approach to SF projects, and the emphasis on physical and quantitative targets.

Keeping the above general findings in mind, I will now review some regional cases from south Asia. A note on the selection of the cases seems imperative here. Since my ultimate focus will be on Bangladesh, I have attempted to investigate the status of SF in the neighbouring regions (Nepal, West Bengal, Bihar and Tamil Nadu). Historically, Bangladesh was a part of the Indian subcontinent, and shares many similarities in social, ecological and historical characteristics with India and Nepal. Besides, India has a particularly rich experience in SF. As Arnold (1990, p.1) notes:

> The social forestry programmes in India form one of the largest and most innovative experiments in participatory forestry anywhere; and also one of the largest interventions designed to improve productivity and use of communal land.

The state of West Bengal has the closest similarity to Bangladesh in terms of population, language, history of forestry, and ecology. Recently, West Bengal SF programmes have received special significance; and there are plans of expanding West Bengal's experience in other Indian states, especially in Orissa, Madhya Pradesh and Bihar (Poffenberger, 1990a, p.18).

The site and population characteristics of the selected Swedish International Development Authority (SIDA) funded project in Bihar (as described in Arnold et al., 1990, pp.11-12), seem to be very similar to those of the empirical cases of my study (three SF case-studies in Bangladesh; see Chapter 4). In more general terms, Blair (1985, p.1241) also noted close similarities between Bangladesh and Bihar:

> The state of Bihar, which was until 1911 like Bangladesh a part of the Bengal Presidency in British India, is still very much like Bangladesh in many respects. The climate, the agricultural cropping regime, and even the rural class structure is very similar to that of Bangladesh.

Tamil Nadu was one of the first Indian states to experiment with SF and its SF programmes have attracted much attention in recent years.

The Nepalese case offers the most recent experience of bureaucratization (through nationalization) of forests and the consequences thereof, including the processes of commercialization and alienation. Poffenberger (1990a, pp.31-32) notes:

> In the evolution of Asia's forest management systems, Nepal represents an unique case. Nepal was one of the last nations within the region to place forest lands under central government control....This occurred one hundred years after state authority was imposed in most south...Asian countries.

Furthermore, all the selected cases are widely reported to be "success stories" by the respective governments (for reference to government documents on West Bengal see Bhuiyan, 1982; Locke, 1992; Palit, 1991; for Bihar see Arnold et al, 1990; for Tamil Nadu see Foley and Barnard, 1984; Arnold et al., 1988; Task Force, 1987; for Nepal see Banko Janakari, 1993).

My purpose in reviewing the cases is to trace the general experience of SF programmes, and also to examine in reasonable details the practical nature of SF operation, and the level and significance of its achievements. For brevity of analysis and for facilitating a systematic discussion, I will follow a simple framework of analysis, within which each case will be described under three broad headings. The first section, namely the "achievement of physical goals", will focus on the quantitative and physical targets of SF. Second, the "achievement of participatory goals" examines the achievement of the more qualitative goals of SF, such as participation and social equity. The third heading, "the regulators of performance", will trace the major political-economic variables which shape SF's performance. The essence and applied distinction of the terms "physical goals" and "participatory goals" are borrowed from Poffenberger (1990b, p.9).

West Bengal

West Bengal, an eastern Indian state with an area of 88,752 sq. km., is characterized by the highest population density (<68m and 766 sq. km.) in India (Palit, 1991, p.342). The forest area is unevenly distributed and is officially estimated about 13.4% (11,870 sq.km) (Banerjee, n.d., p.4); but "actually no more than 5% is under productive forests" (Shah, 1988, p.2). The per-capita forest area is 0.016 ha., which is one of the lowest in the country (Palit, 1991, p.342).

SF in West Bengal emerged in the early 1980s out of growing crises in the forestry sector. Basically, "the crises [were] two-fold", explains Banerjee (1989, p.1):

> [first] the crisis of eroding resources and its larger social-ecological implications. The second [was] the crisis of confidence, the ever-widening gap between the professional forester and people.

SF was launched to arrest the rate of resource depletion by reorienting the alienated people and revitalizing their support towards state initiatives in forest regeneration and protection. SF was supposed to strive towards empowering local communities by their increased participation in the decisions and operations of forestry schemes, and by augmenting the supply of forest products in rural areas (Palit, 1991; 1992; WBFD, 1989). SF have four major components: "strip plantations", "village woodlots", "farm forestry" and "reforestation of degraded forests" (Bhuiyan, 1984). SF activities in West Bengal are carried out mainly in the forms of "Joint Forest Management System" (hereafter, JFM), and the community-based "Forest

Protection Committees" (hereafter, FPCs) (for details on these institutional arrangements see Palit, 1991; Locke, 1992). To date, there are 2,092 FPCs in the state, which are protecting 307,991 ha. of regenerating forest, and have a membership of a total of 215,708 households (Locke, 1992 p.3).

Achievement of physical goals The physical targets of plantation and production (especially in the *Sal* forests under the JFM) were, more or less, achieved. Plantations raised through community participation showed "a very positive return on investment" (Poffenberger, 1990a, p.11). It was estimated that participating villagers, receiving a 25% share, would earn Rupees 500 per ha. per year (Poffenberger, 1990a, p.11). Although limited, there was also an increase in the availability and collection of the non-timber forest products in the JFM areas (Palit, 1992; Malhotra et al., n.d.; Poffenberger, 1990a).

WBFD (1986, cited in Shah, 1988, pp.2-3) noted that the various components of SF (strip plantation, village woodlots, farm forestry and reforestation of degraded forests) reasonably achieved their respective physical targets. However, while the components of farm forestry performed particularly well, strip plantations and village woodlots fared comparatively poorly. The comparative success of farm forestry (62,024 ha. was planted by 1986, against a target of 52,000 ha.) was primarily attributable to two factors: first, the unambiguous land ownership status and the necessary support from the government; second, West Bengal's impressive record in distributing the "surplus vested-land" for farm forestry purposes (Shah, 1988). The performance of strip plantation and village woodlot was not impressive. The survival rates ranged between 40%-60%; and local people were generally indifferent, and in some cases, openly hostile to the plantations (Shah, 1988).

Achievement of participatory goals Notwithstanding the repeated official emphasis on participation, the extent of popular involvement in SF was insignificant. With their long orientation as "resource cop", foresters found their new role as "community developers" at odds with their values, training and perceptions (Locke, 1992; Poffenberger, 1990a; Banerjee, n.d., 1989). The following comment of a professional forester (Banerjee, 1989, pp.2-3) illustrates the real nature and level of participation in West Bengal SF:

> We approached the implementation [of SF programmes] mainly through poor traditional knowledge of silviculture, and with imperfect and mundane knowledge of the needs of forest communities. To many of us the concept of people's participation, was obscure; it meant merely to give some employment opportunities

> to the people; to give some low-priced forest product....By achieving this it was presumed that we had fully...ensured popular participation...;as a result, we often approached the issue with a "I do, you participate" attitude. In this circumstance the forest Department will prepare a plan, implement -- may be offering some unskilled or semi-skilled jobs to the people and they are then expected to participate in the program. Such lip service can not achieve lasting result.

Participation was mostly limited to wage-labour contribution, part-time employments, and membership in some forms of "committees". Palit (1989) found that, women (especially, firewood collectors) were suspicious and shy of field staff, and their role in meetings and discussions was passive. The Indian Institute of Bio-Social Research and Development (IBRAD, n.d.) also concluded that social constrains thwarted women's involvement and they did not participate in decision-making in the presence of men. In cases where JFM initiatives were taken up with local political *panchayats*, a host of political problems also surfaced, including elite domination of decisions and factionalism modelled on the local pattern of rural power structure. Participation of different social groups rarely occurred (Locke, 1992; Poffenberger, 1990a).

Besides the above, a number of other issues can be raised concerning the goals of welfare, participation and equity. First, in some cases it was found that the firewood-collector families were "displaced by JFM" and became "migrant labourers in the nearby town" (Locke, 1992, p.16). Second, in the early years of protection when forests were closed to grazing and cutting, "the already high burden of women's work increased and there [was] no guarantee that they [would] benefit from the returns to this additional work" (Locke, 1992, p.16). Third, as Poffenberger (1990a, p.16) anticipated, during the final harvesting of *Sal* trees, the benefits would likely to be "captured by village elites, corrupt officials, and middlemen". Fourth, equitable inter-village distribution of benefits did not occur. At some instances, the forest department attempted to enhance equity by evenly distributing the income from SF products among the FPCs. But, prior to the introduction SF projects, some families were more dependent on forests than others. Hence, "equal division of revenues [was] not justified [because] the past forest dependency level and family needs varied" (Poffenberger, 1990a, p.16). Fifth, it was observed that whenever a forest was closed for regeneration during the initial years, the weaker groups, such as the landless, low caste, and tribal families (who were more dependent on forests for their subsistence needs than the well-off households) suffered a sharp decrease in income and resources (Poffenberger, 1990a, p.15; Palit, 1992, p.449).

The regulators of performance There are many factors which condition SF's performance. From the above discussion, I will attempt to identify some of the crucial factors.

Land proved to be a major consideration for motivation and increased participation of local communities in SF programmes. In the cases of village woodlots and strip plantations, limited planting land and tenurial insecurity were primarily responsible for the poor performance (Bhuiyan, 1984; Banerjee, n.d.). Tenurial security seemed very effective in motivating farmers towards SF. For example, the 99-year land *patta* (leasehold) programme instilled a sense of confidence and feeling of ownership among the *patta* holders, which, in turn, contributed to increased production in SF (Shah, 1988). The relative "success" of farm forestry was linked to the efficient distribution of "surplus" public land and the unambiguous status of land *patta* (Shah, 1988; Banerjee, n.d.; Sharma, 1993).

There seems to be a consensus among the analysts that SF calls for a thorough reorientation of the role and attitude of concerned public institutions, especially the forest department (Shepherd, 1992; Lai and Khan, 1992; Singh and Khare, 1993; Poffenberger, 1990; 1990a; Khan, 1996b; Khan and Hasan, 1995). Poffenberger (1990a, p.47) argues that it requires a shift on the part of state foresters "from being resource policeman and commercial managers, to playing an organizational and technical support role for rural community". Government documents also clearly recognized the need for a "bureaucratic reform" for SF (see for example WBFD, 1989). Notwithstanding the emphasis, however, little change was noticed in the values and practices of forest officials. The major limitations of foresters in West Bengal, like in other Indian states, include inadequate training and skills for performing their new advisory and supportive roles as "community developers" and "extension agents"; continued allegiance to revenue collection and policing; widespread connivance, corruption, collaboration with rural elites and other commercial forces (see for example Fernandes et al., 1988; Nadkarni et al., 1989; Fernandes and Kulkarni, 1983; Task Force, 1987; Banerjee, n.d.; Khan, 1994).

Besides the above, other institutional limitations were noted. First, inadequate manpower, logistics and faulty staffing pattern proved to be general problems. Banerjee (n.d.), for example, showed that in one major Forest Division in West Bengal, which was engaged in large-scale SF activities, the total manpower consists of only 1 Divisional Forest Officer, 4 Forest Rangers, 12 Deputy Rangers and 20 Extension Workers. Their working-area spread over as many as 5 districts with a population of more than 20 million. Second, the planning pattern within the government, especially in the forest department, remained essentially top-down, with little room for accommodating local and contextual demands. Banerjee (n.d.) has identified a number of major problems of planning which hindered SF's progress. These included the lack or inadequacy of decentralized ("micro") planning, especially at the Range or Beat levels; inadequate data-base on local demands and resources; and inadequate resource allocation. Third, the operational climate within the government in general, and the forest department in particular, remained strictly hierarchical. What Chambers (1983, p.211) observed in the general context of Third World public bureaucracies, seems to mirror the West Bengal situation (WBFD, 1989; Locke, 1992):

> In meetings subordinates are upbraided, cajoled and given orders....Real problems of implementation or impact are repressed; appearances of achievement applauded. Senior officers do not learn from their subordinates and subordinates do not learn from their rural clients.

Fourth, foresters had inadequate "extension" and "communication" skills (WBFD, 1989), although these skills are said to be the "cutting edge of performance" for SF (Magno, 1986). Fifth, recently some attempts at "micro-planning" have been introduced in West Bengal under the JFM. But Locke (1992, p.20) observed that "most FPCs have not engaged in sophisticated planning exercises [and] the support activities appear to occur on an ad hoc basis".

Ideally, SF calls for a cooperative institution formed by farmers to articulate their interest and ideas. The FPCs were envisaged to serve these purposes with the support of local political institutions (the *panchayat*). However, a number of major problems were observed which hindered the role of FPCs (Chandra and Poffenberger, 1989). In many cases people remained unaware of the functions and advantages of the FPCs, and they tended to be skeptical about their share after the harvest. The processing and marketing systems of SF products were disorganized and farmers had little control over them (Palit, 1992; Banerjee, n.d.). Locke (1992, p.20) noted that though the FPCs performed reasonably well in forest "protection", they largely failed to develop an efficient "management" system for sustaining the resources. The "institutional" limitations of the forest department and the "dynamics of inequality at the village level" were primarily responsible for limited performance of the FPCs (Locke, 1992, pp.17-18). As regards the institutional constraints on the operation of the FPCs, Locke (1992, pp.17-18) observed that :

> [c]urrently the FPC is hostage to the BO [Beat Officer] and at the mercy of the DFO [Divisional Forest Officer]. The BO is the only person who can call FPC meetings...and the DFO can dissolve the FPC.

Social relations and power structure greatly influenced the performance of SF (Locke, 1992; Banerjee, n.d.; Palit, 1991; 1992). Palit (1991, p.346) noted that the FPCs worked under the heavy influence of local political institutions (especially the *panchayat*); and manifested "varying political shades and colours" which often delimited the neutral character and "welfare purposes" of SF. Therefore, Palit (1991, pp.345-346) concluded that, although support from political organizations were required, SF organizations, especially the FPCs, should be "apolitical".

The prospect and role of market seemed to be crucial for SF. In conformity with the general experience of south Asia, the (forces and dynamics of) markets remained beyond the control of SF farmers in West Bengal. The major market-related problems included the monopolistic control by local elites and external merchants; exploitation by intermediaries; unfair prices for producers (farmers) and widespread price fluctuations; unfavourable and inadequate policy and logistic support from the government; difficulties in transportation and storage; and a variety of complex bureaucratic rules and procedures (Locke, 1992; Poffenberger, 1990a; 1990b; Shah, 1988; Arnold and Stewart, 1991; Banerjee, n.d.; Chandra and Poffenberger, 1989; Task Force, 1987). Locke (1992, p.17), for example, noted that "few marketing networks have been developed" for SF; and farmers, especially the female non-timber forest collectors, were "frequently exploited by middlemen". Palit (1992a) also observed "signs of glut in the market" for SF and non-timber forest products. Shah (1988) noted a case from West Bengal where two influential local elites were the sole buyers of SF products at lower prices, and made huge profits by selling them in external markets. They monopolized control over the procurement and selling of SF trees. Later, the government formed some "marketing committees" to streamline the marketing system. But these were loosely structured institutions, and proved ineffective in preventing the growth of monopolistic trends and unwanted external market intervention (Shah, 1988).

Although there have been some policy and legislative changes in favour of SF, the overall role of public policies seemed largely ineffective. In many cases, policies failed to realize that people's perception might often be quite different from that of government planners and consequently, remained naive to local needs and views. Banerjee (n.d.), for example, noted that SF planners emphasized production of fuel wood, fodder, small timber for meeting local subsistence needs. But after a decade, it became clear that rural people perceived tree planting more as an "income-generating activity" than as a "means of meeting subsistence demands" (also see Arnold and Stewart, 1991, p.18). Legislative and procedural formalities often led to public harassment by government offices. The "custodial" policies, as Palit (1992, p.449) noted, contributed to a situation where government officials, especially foresters, "virtually lost all credence with the communities". Similarly, the performance of the FPCs suffered due to the delay caused by the procedural labyrinth, especially regarding the formalization of decisions, and the approval of funds and sharing-agreements between the state and the FPCs (Banerjee, n.d.). Moreover, some regulations left no room for flexibility in responding to particular local needs and regional peculiarities (WBFD, 1989).

Bihar

SF in Bihar, as in other states in India, started in the mid 1970s, following the recommendations of the National Commission of Agriculture. Farmers were encouraged to grow trees on their land ("farm forestry"); and various other categories of public lands were used for "block plantations". Besides, the forest department itself planted trees on private and communal lands. By 1984-85 about 5,100 ha. were being planted annually under this scheme (Arnold et al., 1990, p.15). However, an expanded programme of SF was taken up under an Indo-Swedish Bilateral aid. Though the initial discussions began as early as 1979, this SIDA funded project became fully operational in 1985, covering the whole of the southern plateau. The project became the most significant SF initiative in the state, and attracted substantial international funds and attention. As mentioned before, the operational context of the project (especially in terms of site, population characteristics and nature of activities) seems similar to Bangladesh SF. In the following discussion, I will mainly concentrate on the performance of this particular project. The information is basically drawn from the report of a SIDA evaluation team (Arnold et al., 1990), which thoroughly examined the project's achievements in 1990.

The Bihar SF project (Amended Project Document (APD), 1989, cited in Arnold et al., 1990, p.15) stemmed from an expanded definition and philosophy of SF; and it stated that SF must

> [develop] a production system which primarily supports and satisfies rural development demand [and it] should aim at directing its activities in such a way that the weaker sections of the society are assured participation.

The production system was supposed to be sustainable. It planned to create a marketable surplus as well as meeting local needs. The project had particular bias to the "poorer and weaker sections of the community" (APD, 1989). In other words, the project goals embraced all the positive attributes of SF including people's participation, social equity, welfare, increased income and production, and fulfillment of local needs. The main components were: "rehabilitation of degraded forests" (hereafter, RDF); "farm forestry" -- plantation by farmers on marginal lands unfit for agriculture; "strip plantations" on public road, canal and railway sides, with usufructuary rights given to the landless families in the vicinity. The mechanisms for ensuring participation were threefold: the "Village Forest Committee" (VFC) -- the collaborative management institution of farmers, and the link between foresters and farmers; the "Joint Management Plan" (JMP) - an "action plan", enshrining the duties and rights of the government and farmers; and the "Contact Person" (CP) -- chosen from villagers to liaise between the forest department and the community (Arnold et al., 1990, pp.18-20; APD, 1989).

Achievement of physical goals By 1990, it became clear that the project had largely failed to live up to its lofty goals. At the end of the first (three-year) phase, the area covered under RDF had fallen short of the target by 35%, and the number of villages supported by the farm forestry component, by 70%. The number of beneficiaries participated was far below the target and the number of villages treated was even lower. However, the number of seedlings distributed was close to the target. The project also raised intended number of nurseries and created substantial employment in the form of wage labourers (Arnold et al., 1990, pp.2-3, 22 and 81).

Achievement of participatory goals The project lagged far behind in meeting its participatory goals. The social and economic impact, as the SIDA evaluation team (Arnold et al., 1990) noted, were marginal. The project put particular emphasis on the involvement and welfare of the "weaker sections of the society", especially women. But it was observed that, first, the project did not take any particular step to ensure that the "weaker sections" would benefit from the employment generated by nurseries and RDF plantations; second, women were benefitting and participating less than men in "all aspects of the project"; third, farm forestry was "not focused on the poor"; and fourth, foresters hardly recognized "the value of upgrading human capital through extension and publicity activities" (Arnold et al., 1990, pp.7-8).

However, as noted above, some employment opportunities were

created, especially in the form of wage labourers in nurseries and plantations. Limited benefits and assets were also created by farm forestry, although the medium farmers reaped most of the benefits. On the welfare and equity side, there was no significant impact of the project on the "weaker sections", especially in terms of improvement of their quality of life, economic stability and social equity (Arnold et al., 1990, pp.7-8 and 47-64).

Similar to most south Asian countries, the core goal of SF -- people's participation -- remained largely unrealized. The SIDA team noted that the institutional mechanisms for ensuring participation performed ineffectively. The VFCs were "not functioning as a collaborative management body"; the JMPs were compiled to fit bureaucratic purposes rather than practical requirements of local communities; the CPs were only partially fulfilling their roles; and the target groups were improperly identified (Arnold et al., 1990, pp.47-64). For example, on the performance of VFC, it was noted that (Arnold et al., 1990, p.40):

> [its] formation [was] simply a bureaucratic requirement [It] could in no instance be regarded as a reflection of villagers own needs and motivation....Membership of the committee generally centred around a few influential village elites. Weaker sections...certainly felt they had no power.

Similarly, the principal micro-planning exercise of the project -- the JMPs -- "did not exist for many years [and] it appeared that they were seen more as a document for social forestry wing staff [of the forest department] rather than for the community" (Arnold et al., 1990, p.42).

The regulators of performance The centralized and bureaucratic planning practices and policies seemed to delimit SF's achievements. The selection of species, for example, represented the government's choice rather than local demand. It was found that local people, especially women, expressed their concern over the change from multiple species with many uses, to *Acacia and Eucalyptus* which they saw as a "once-off" investment for cash and poles. Moreover, often there were "too ambitious" planning targets, which were far beyond the institutional capacity of the traditionally trained and oriented foresters. These targets also caused land suitable for agriculture to be put under tree plantation. This, in turn, led to a reduction in land available for other purposes (for example, grazing) (Arnold et al., 1990, pp.2-7).

The role of foresters hardly changed and they continued to follow the traditional policing mode of resource management. The project document (APD, 1989) recognized the need for training to reorient foresters towards their "social role of community developer". But it was observed that, field foresters, especially the Divisional Forest Officers responsible for SF, were not specialists in the areas which they had been given responsibility for, and had difficulties in coming to terms with their new roles through their traditional professionalism (Arnold et al., 1990, pp.65-71). Besides, the forest department suffered from shortage of manpower, logistics and funds to execute their responsibilities. The "extension" services were also largely ineffective and inadequate. There was no regular incentive and reward system within the department to motivate field staff. The mechanisms for building up communication between the forest department and local communities were ineffective. The group meetings, for example, had been very poorly attended by people. The male forest staff could hardly communicate with women's groups. The inadequacy of extension and communication skills became evident by the fact that a significant number of local people were simply unaware of the SF activities in their villages (Arnold et al., 1990, pp.1-7 and Chapter 7).

Market was another major factor in regulating the performance of SF. In the absence of adequate market information, farmers largely failed to make judicious decisions about species; and project staffs were also ineffective in their part to provide farmers with information and logistics. There was also a lack of standardized arrangements for monitoring market trends, especially demand and supply flows for SF products. Furthermore, the bureaucratic rules and formalities were not conducive. For example, the restrictions on harvesting, transit and sale of wood from private lands, particularly under farm forestry, led to the producers' (farmers') discouragement. Farmers had very little control over market and their "market dealings" were channeled through intermediaries (Arnold et al., 1990, pp.56-57).

The changing land use and tenure policies, which came about through SF intervention, had some implications. For example, first, in many cases agricultural lands were brought under SF plantations, causing displacement of the poor share-cropper families and making them more dependent on wage labour. Second, SF sometimes led to a reduction in the amount of land available for grazing purposes, which in turn increased women's burden of work, as they were required to meet the increased demand for fodder (Arnold et al., 1990, pp.61-62).

Social relations and structures also profoundly influenced SF's achievements. Arnold et al. (1990, p.20) noted that "the social goals of equity and redistribution" were particularly difficult to "pursue in a structured society through the limited intervention possible under Social Forestry". Their (Arnold et al., 1990, p.47) following observation on Bihar SF applies to most parts of rural south Asia with equal cogency:

> Implementation of a participatory project with a resource poor focus is difficult under institutional and social conditions of rural Bihar....Society has become more fragmented, stratified and less dependent on non-monetary reciprocity and sharing of resources. Cooperation between the different interest groups...cannot be automatically expected, as experience with the VFCs has shown.

Tamil Nadu

The southern Indian state of Tamil Nadu is one of the earliest states to launch SF, following the recommendations of the National Commission of Agriculture. But even before that, in the early 1960s, the state carried out "farm forestry" which aimed at establishing plantations mainly on rural wastelands, beyond the traditional "reserved forests". Tamil Nadu's special interest in SF can be attributed to some factors (Arnold et al., 1988; Foley and Barnard, 1984): first, the government's recognition that the stocking and distribution of the remaining public forests were no longer sufficient to meet the state's demands, and that forestry needed to be expanded beyond the bounds of public forests; second, the realization that government efforts could not succeed without popular involvement; third, the response to the growing national and international support for SF.

Thus, Tamil Nadu had a long tradition of SF. Between 1960 and 1981 about 133,000 ha. were planted under various SF programmes (Arnold et al., 1988). However, at the beginning of the 1980s, SF programmes entered into a more ambitious and expanded era, primarily through a substantial (financial and technical) assistance from SIDA. All the "farm" and "community forestry" activities in the state were integrated into a new "Social Forestry Project" in 1981 (Foley and Barnard, 1984, p.203). It became the largest and most significant SF initiative. In the plan papers the project explicitly focused on meeting local needs by wider community participation. It aimed at ensuring equitable distribution of benefits, and helping, in particular, the economically weaker sections of the community. The project components were similar to other states which included plantation on degraded community and public lands, strip plantation and farm forestry (Foley and Barnard, 1984; Sharma, 1993; Arnold, 1990; Arnold et al., 1988). The following discussion concentrates on the performance of the project and draws generously on the report of the SIDA evaluation team (Arnold et al., 1988) which made a thorough examination of the project.

Achievement of physical goals The SF project achieved its physical targets (for example, number of seedlings distributed, land-area planted etc.) reasonably well. There was good progress in achieving planting targets under the components of farm forestry, plantation on tank foreshore, and strip plantation. Up to the end of 1983, a total of 10m seedlings were distributed under farm forestry and 660 ha. of fodder plantation was established with the involvement of the *panchayats* (Foley and Barnard, 1984, p.203). Similarly, during the period between 1981-82 and 1985-86, an impressive 112,629 ha. of communal woodlot was established (Arnold, 1990, p.2). But often these achievements were later on offset by the problems of low survival rate, rapid depletion, inadequate protection and public indifference (Arnold et al., 1988; Arnold, 1990).

Achievement of participatory goals The socio-economic impacts of SF were insignificant. Arnold et al. (1988, p.47), evaluating the achievements of participatory goals, concluded that:

> There is very little positive impact on target groups (i.e. poor and marginal farmers, landless labourers, women and scheduled castes)....The target groups are on the whole left outside the project.

However, two notable positive impacts were observed. First, SF activities created significant rural employment. For example in 1985-86, SF schemes generated 6.8m man-days of employment (Arnold et al., 1988, p.47). Second, as noted before, some successful plantations were raised by farm forestry (from which relatively richer farmers could accrue benefits). On the whole, the programme had minimal affect on the poor and weaker sections in terms of any significant degree of rise in their living standard and participation in SF, or, of reduction in social inequality. SF's benefits were not equitably distributed. For example, only an insignificant proportion of the total seedlings distributed reached the small and marginal farmers. The survival and yield rates were very low, particularly among the marginal farmers. SF products were not made available to local community; rather they were transferred to external commercial markets. The "scheduled castes", especially the *Harijans*, with the exception of some employment opportunities, "in reality fall outside the area of its impact" (Arnold et al., 1988, pp.45-49).

The nature of participation was also remote from the stated goals. The "Village SF Committees" seldom functioned according to plan and the "target groups" hardly participated in such committees. In many instances, women were simply unaware of the SF programme in all its components and they seldom took any active role in the committees. The choice of species was not based on local needs. Participation of scheduled castes was, at the most, formal (Arnold et al., 1988; Arnold, 1990; Arnold and Stewart, 1991).

The regulators of performance As the above discussion suggests, SF's performance in achieving the stated goals, especially the social and participatory goals, was poor. A number of common institutional and social problems seemed to have shaped the performance.

The planning process within the government continued to be centralized and non-responsive to local conditions. The ambitious goals were hardly in harmony with field realities (Arnold et al., 1988; Arnold, 1990). For example, mirroring Bihar's experience, the goal of "benefitting the weaker sections of the rural population with more participation and equitable share" seemed unrealistic; because it required quite fundamental social change which was far beyond the scope of the programme (Arnold et al., 1990; 1988). Moreover, the planners had major misconceptions about the nature and needs of the community. For example, they envisaged that rural community was a fairly homogenous entity, there was acute fuel wood shortage in the area, and the poor would automatically benefit from SF plantations in meeting their fuel wood demands (Arnold et al., 1988, pp.19-27). But soon after its initiation, the project (Arnold et al., 1988, p.1) found that

> the rural areas of Tamil Nadu proved not to be experiencing widespread fuel wood shortages; the shortages that so exist are concentrated among the weaker sections who control little land and those with land are growing trees not to meet subsistence needs but to generate income.

The selection of species reflected the forest department's choice rather than the local views. The low survival rates, for example, in strip plantations was mainly due to poor species selection (Arnold et al., 1988). Often, as happened in Bihar, plantations were raised on crop lands, rather than on lands which were "unsuitable for agriculture" (Arnold et al., 1988; 1990).

The forest department's role was also typical of the general south Asian experience. A management study found that the organizational culture and operation of the department were custodian-like, rigid, strictly hierarchical and technically (as opposed to socially) oriented (Om Consultants, 1986, cited in Arnold et al., 1988). Bureaucratic formalities, unnecessary and lengthy reporting system, top-down mode of communication and information flow, absence of a reward and punishment scheme, shortage of manpower -- were some of the major limitations of the forest department which were inimical to SF (Om Consultants, 1986; Arnold et al., 1988; Foley and Barnard, 1984; Sharma, 1993). Besides, the extension and public communication services of the government, especially of the forest department, were inefficient. It was found that the SF workers faced difficulties in living up to the demands of their new role. They often perceived their task as the "guardians of plantations", rather than as "extension workers". The frequent transfer of staffs affected SF's performance. Skilled personnel were transferred by central authorities from SF projects to more "lucrative" territorial positions, without even consulting the project management (Arnold et al., 1988; Om Consultants, 1986).

Land tenure arrangements, as in other instances, have played a vital role. Tenurial security and clear ownership status, for example, contributed to the relative success of farm forestry. In contrast, community lands remained sources of conflicts, owing to the ambiguous ownership status (which could partially explain the poor performance of the "communal woodlots") (Foley and Barnard, 1984; Arnold et al., 1988). Sometimes communities were not interested in providing their land for SF plantations. In these cases, forestry had to compete with other important uses, notably, grazing and fuel wood collection (Arnold et al., 1988).

No institutional attempt was observed to provide systematic market information to farmers. However, there was a growth of commercial market-oriented farm forestry by the middle and large farmers (although the poorer farmers were virtually left out this opportunity). In general, market prospects for SF products remained uncertain and threatened by such factors as price fluctuations, manipulation by intermediaries, and lack of institutional security and support. Moreover, as Arnold et al. (1988, pp.24 and 44) observed, farmers in Tamil Nadu were gradually being "integrated into the market economy"; even poor farmers, with their subsistence orientation, were also "deeply drawn into the market nexus'", exposing themselves to varied manipulations by commercial forces.

The rural life in Tamil Nadu was no exception to a general south Asian village -- a scenario characterized by diverse interests competing for limited resources; conflicting groups; party politics; and heterogenous and hierarchical social relations and power structures (Arnold et al., 1988). During the course of programme implementation, it became gradually clear that reaching the weaker sections of the community was a formidable task, given the socio-economic realities. There were cases of conflicts, even of violent clashes, once plantations were raised on community lands. In some instances, plantations on community lands increased the shortage of grazing lands; and the poorer classes were worse hit by the scarcity. In other cases, some influential and rich individuals encroached on SF plantations, especially, on tank foreshore. Worse yet, there were "cases of selective eviction", as Arnold et al. (1988, p.50) noted, where "poor people have been unceremoniously evicted, but [the] rich and powerful...have not been touched at all".

The role of local institutions, especially the *panchayats*, was largely unsatisfactory in ensuring popular involvement and in creating a cooperative organization for farmers (Arnold et al., 1988). The project aimed at gradually shifting the SF responsibilities to the *panchayats*, with the forest department receding to a supportive advisory role. But serious doubts were raised about the efficacy of local institutions (Arnold et al., 1988, pp.25-26):

> [L]ocal leaders are often biased in terms of class, caste and factions, and they have little capability of managing the joint property resources. That is one reason why the benefits of development programmes to such a small extent tend to trickle down to the target groups.

Generally, the *panchayats* in Tamil Nadu, as in many other states, had an elite-dominated leadership, scant revenue sources, limited institutional capabilities, and an inherently unequal pattern of functioning (Arnold et al., 1988; Foley and Barnard, 1984; for some thought-provoking discussions on the performance of the Indian *panchayat* system in rural development see Reddy, 1982; Johl and Mudahar, 1982; Hadden, 1982). They reflected the voice of local elites and affluent farmers. Arnold and Stewart (1991) found that, instead of meeting local demands, most *panchayats* were interested in selling the SF products to external commercial markets. Moreover, in some instances they were given responsibilities far beyond their institutional capacities. For example, the conditions laid out for the transfer of management responsibilities to the *panchayats* and the harvesting of SF plantations were clearly beyond their means and comprehension (Arnold, 1990; Arnold et al., 1988).

In sum, SF in Tamil Nadu largely failed to live up to its lofty goals, especially in respect of participation, equity and welfare. It failed to take into account the complex political-economic frame of a typical south Asian village and to tailor the goals accordingly. The unequal character of social formations in Indian villages, elite dominance in power structures and local institutions, the commercial alliances of vested interests, institutional incapability of local institutions -- were some of the local variables which have shaped SF's performance in Tamil Nadu; but remained far beyond the its control. Moreover, the SF project relied on the already overstretched and orthodox institutions, such as public bureaucracy and local *panchayats*, to carry out the challenging and innovative tasks of SF.

Nepal

Nepal has a land area of 147,181 sq. km. and a population of about 19 million (Kayastha, 1991, p.12). 42.8% of the total land area is either forest or shrub land (Bartlett, 1992, p.96). The natural forest was reduced by about 570,000 ha. between 1964 and 1985 and reforestation (as compared to the deforestation-rate) was a meagre 47,000 ha. (Bhattarai, 1990, p.68).

In Nepal, SF emerged mainly as a response to the failure of previous state policies and to the inadequacy of the orthodox role of the forest department. Traditionally, as Chapter 2 has discussed, local communities enjoyed a reasonable degree of access to forests in Nepal. The state rulers until about the mid-twentieth century showed little interest in forest management; and forests, in most areas, were managed by indigenous regulatory systems. Prior to 1950, the state's attitude towards forestry was of "general indifference" (King et al., 1990, p.2).

A turning point in the history of forestry in Nepal was the introduction of the Private Forest Nationalization Act of 1957 and the Forest Act of 1961 which brought all forests under state control, and provided legislation for bureaucratic administration of forests (Gronow and Shrestha, 1991; Adhikari, 1990). Forests were thus brought under the control of the forest department, which was assigned to perform a policing and licensing role. These laws were seen by local people as a means of destroying the traditional forest management systems and of increased exploitation of local forests (Adhikari, 1990, p.260). Furthermore, the Forest Preservation Act of 1967 defined a wide array of forest offenses and prescribed more stringent penalties, aiding the forest department to further intensify its policing and law enforcing role. In the process, mutual hostility and distrust sprang up between state officials and local communities. By the mid 1970s, the failure of nationalization policies became evident which was attributed to two reasons: first, the institutional incapability of the forest department; and second, local people's perception that their traditional rights of access had been curtailed by the government (Bajracharya, 1983; Adhikari, 1990; Gronow and Shrestha, 1991). These policies, in fact, replaced the traditional forest management systems by a "bureaucratic rule of state officials" (Gronow and Shrestha, 1990, p.2). In the last two decades, realizing the failure of the mass-scale forest nationalization and the government's limitation in managing its newly acclaimed territories singlehandedly, the efforts in decentralizing forest management and SF were intensified. For example, the Decentralization Act of 1982 empowered the *panchayats* to form local "consumer committees" for local forest management (Gilmour et al., 1989).

Against this background, SF in Nepal emerged, first, as a response to the failure of previous public policies; and second, to reorient local people who looked on forest nationalization as a means of deprivation of their rights in a resource which they enjoyed for generations; and third, to improve the relationship between foresters and local communities. SF in Nepal "formally commenced in 1978" (Bartlett, 1992, p.95), when the government introduced a set of radical rules (the *Panchayat* Forest Rules of 1978 and *Panchayat* Protected Forest Rules of 1978) to create community managed forest called the *panchayat* forest (PF) and the *panchayat* protected forest (PPF). Under these rules, any forest area which had two-thirds of the land requiring replanting could be designated as PF. All the income from such forest went to the *panchayat*. Forest land that required protection and enrichment was called the PPF. Each *panchayat* could have 125 ha. of PF and 500 ha. of PPF (Foley and Barnard, 1984, pp.215-216; Task Force, 1987, pp.75-77; Adhikari, 1990, pp.260-261). Subsequent legislation created further provisions for decentralizing forest management and for specifying the nature of community control over these forests. Since the political and administrative reshuffling of the country in 1990 and the abolition of the *panchayat* system, forests have been directly handed over to the "user groups" instead of the *panchayats* (Bartlett and Malla, 1992). The Forest Act of 1993 was the latest forestry legislation, which attached primary importance to SF. It enshrined detailed provisions concerning the hand over of (parts of) national forests as community forests (to the user groups); the formation and rights of the user groups, and its relation with the forest department; and the management and utilization of community forests (Kanel, 1993, pp.2-5).

Thus, there has been a major policy shift in favour of SF in Nepal. SF has "received the highest priority in the forestry sector in Nepal" (U.R. Sharma, 1993, p.9). It has also received corresponding support from international donors. Shrestha (1993, p.6) has observed that "there are more...donors involved in the [SF programmes] than in any other programme". A number of major SF and watershed projects were launched with assistance from various international organizations, notably, the Nepal-Australia Forestry Project, the Hill Forest Development Project and the Community Forestry Development Project (for a summary of the major SF projects in Nepal see Kayastha, 1991. Banko Janakari, 1993 also documents some major regional SF case studies).

Achievement of physical goals In line with the general experience of other south Asian countries, the physical targets, especially those of plantation establishment and seedling distribution, were mostly achieved (Foley and

Barnard, 1984, pp.215-216; Gronow and Shrestha, 1991, pp.2-3; Task Force, 1987, p.77; Magno, 1986, pp.56-58). Gronow and Shrestha (1991, p.3) noted that in various districts "progress [was] made in establishing new plantations". Similarly, Foley and Barnard (1984, p.216) observed that

> the Community Forestry Development Project made substantial progress....Virtually the whole of the planned 400 nurseries were operational. Some 3,700 hectares of panchayat forests had been established, together with 300 hectares of protected panchayat forest. Seedling distribution for private planting was just over 1 million, more than was planned for the whole project.

While the general progress in reaching the physical targets was satisfactory, a number of major problems were also observed including wrong choice of species; poor survival rate of seedlings (which was about 65% in the Community Forestry Project); and inadequate nursery-maintenance (Task Force, 1987, p.77; Foley and Barnard, 1984, p.216).

As regards the formation of the user groups, delegation of management responsibilities and negotiation of operational plans with the user groups (which constituted the primary focus of SF), "the progress on a national basis", as Bartlett (1992, p.98) noted, "had been rather slow". About 880 user groups were established by the forest department in 63 districts; but only 400 operational plans have been negotiated with these groups (Bartlett, 1992, p.98). About 100,000 ha. of government forests were handed over as community forests (out of 3.5m ha. of land potentially available for SF) (Tamrakar and Nelson, 1991, cited in Bartlett, 1992, p.98). Kanel (1993, p.4) also noted that SF programmes "are neither effective in handing over nor efficient in input utilization". In Terai region, Shrestha and Budhathoki (1993, p.24) found that, during 1991-92 only 7 out of 16 districts managed to achieve the target of user group formation; the achievement on operational plan preparation was only 32%, and hand over of forests was below 30% of the original target.

Achievement of participatory goals As regards participation, SF's performance was insignificant. People remained largely unaware of the changes in forest policies and the benefits of SF. One study, for example, found that 80% of villagers were simply unaware of the changes in forest legislation and 92% were skeptical as to who would ultimately benefit from SF (Koirala, 1985). From a survey of Terai region, Shrestha and Budhathoki (1993, p.26) found that "local people were still doubtful about [SF] programme [and they apprehended that] "the government might take the forests back during the period of harvest". The "weaker sections" rarely participated in forest committees. Baral (1993, p.14) notes:

> The participation of real users in general assemblies [of forest user committees] is always very poor. Very few women participate in discussion. Proposals mainly emerge from dominant users.

Hobley (1991, p.112) noted that the user committees were "largely ineffective, unrepresentative and had virtually no control or responsibility". The percentage of "dominant users" ("the rich and political persons") in the forest committee was "always high in comparison to others" and the government "listen to the ideas of the dominant users only" (Baral, 1993, pp.13-14). Bartlett et al. (1992), from their survey of 12 forest user groups, concluded that only 4 could be considered effective.

Poffenberger (1990a, pp.34-36) raised a number of equity issues as regards the operation of forest user committees. First, as discussed before, the committees were managed by dominant "clan groups"; and they did not represent all classes of villagers. Second, the sharing of benefits was not "equitable" in the sense that the benefits were not distributed according to the varying needs of the households and their level of forest-dependence. Third, similar to the experience of many other regions of India (see Poffenberger, 1990a; Locke, 1995; WBFD, 1989; Arnold et al., 1988; 1990), it was found that when forests were closed for SF, it had profound affect on poor households. In most cases, the poor had to migrate to cities in search of alternative income and women faced increased burden of work and responsibility. In the above context, Gronow and Shrestha (1991, p.3) concluded that:

> Progress [of SF] across the various districts of Nepal so far been disappointing. Progress has been made in establishing new plantations, but rarely have people participated in the planning decisions...or management of forests.

The regulators of performance At the forestry organization level, there was a continued adherence to the top-down bureaucratic pattern of forest management (see Banko Janakari, 1993). The "catalytic, technical and advisory" (Kanel, 1993) roles which the forest department was supposed to play in the promotion of SF remained a distant reality. Gronow and Shrestha (1991, p.7) found that, "most staff continued their policing role: trying to maintain some control over the forests, at times apprehending villagers (often the poorer) and harassing them". State officials still remained "tree oriented and not people-oriented" (Agarwal, 1986, p.112). Shrestha and Budhathoki (1993, p.25) also noted that "authoritarian attitude of the field staff", "wide gap between the local people and forestry staff" and "inadequate extension activities" were some of the principal problems of SF in Nepal. Mirroring the general experience of south Asia, the government, especially the forest department, suffered from some major institutional problems which constrained SF's performance including the purely technical and bureaucratic attitude to forest management; the "revenue and policing" mode of service orientation; lack of participatory environment within the government; strictly hierarchical flow of command; absence of a motivation system; erratic transfer and promotion systems; and corruption and connivance (Baral, 1993; Gronow and Shrestha, 1991; Agarwal, 1986; Khan and Hasan, 1995; Nurse et al., 1993; Shrestha and Budhathoki, 1993).

Considering its major significance for SF, the participatory management plan for the community-managed forests, known as the "micro planning practice", deserves particular mention. A number of major problems of planning were observed which were inimical to the performance of SF. First, it was noted that local people were rarely consulted during SF project design and planning; even in cases where they were consulted, only the voices of the rich and dominant section were heard (Baral, 1993; Hobley, 1991). Second, field staff (who were immediately responsible for implementing SF programmes) were inadequately consulted during project planning (Baral, 1993; Nurse et al., 1993; Gronow and Shrestha, 1991). Third, in many cases, SF planners failed to appreciate field realities, and consequently, policies were often devised on the basis of erroneous conceptions. For example, as Hobley (1991) noted, the Nepal Australia Forestry Project assumed that: "local community" and "local forest users" were synonymous; the poorer sections of the forest users would participate in meetings; and the forest committees would be able to determine and execute their duties independently. But eventually it was found that local communities were complex and heterogenous; the weaker sections virtually had no voice in the committees; and such committees tended to be biased towards the powerful and reflected the local power structure (Hobley, 1991). Similar cases of erroneous assumption in project design were also reported by King et al. (1990). Fourth, the "operational plans" at the user-group level suffered from delayed approval by the forest department, inadequate technical and logistic support, and a plethora of complex bureaucratic formalities (Baral, 1993). Bista et al. (1986, cited in King et al., 1990, p.2) also noted that "the government approved format for management planning was too technical for the village people to understand" and people had little "sense of ownership and belonging to the planning process". The broader macro-policy framework was also not conducive to smooth operation of SF activities in the field. A survey on 20 district forest officers showed that, all the officers considered "unclear and inconsistent policies" to be the principal problem for Nepal's SF (Shrestha and Budhathoki, 1993, pp.24-25).

The social frame and the role of local institutions also impeded the implementation of SF's goals. The Nepalese society, similar to other south Asian countries, is characterized by rigid class and caste differentiation, highly unequal social relations, and skewed pattern of asset and resource distribution. The forest committees in particular, and the *panchayats* in general, were reported to be dominated by the upper castes (*Brahmins* and *Chetries*), and reflected the interests of the rich and powerful (Sharma, 1991; Baral, 1993; Poffenberger, 1990a). The performance of the user committees has been already discussed. Regarding the role of the *panchayat* in SF, Poffenberger (1990a, p.32) noted that

> [it] was often not the most appropriate social unit for carrying out management tasks [of SF]. Many village panchayats lacked both experience and operational capacity to manage resources, while the composition of organization leaders did not...reflect resource users.

Access to resources was, to a large extent, shaped by patronage relations between the poor (weak) and the rich (powerful) (Hobley, 1991; Wignaraja, 1984; Gronow and Shrestha, 1991). Wignaraja opined that catalytic support was needed to help the people to break "this vicious circle" (Wignaraja, 1984, p.8). Even within the government bureaucracy patronage links were very much active. Bista noted that "the present bureaucratic culture is based more on a client and patron relationship than on work performance" (1991, cited in Kanel, 1993, p.4).

In sum, as far as the goals of participation, empowerment and social equity were concerned, very little headway was made in the SF practices. The only notable positive impacts were the generation of employment opportunities in various plantations and nurseries, and the major change in forest policies at the macro-level. The radical policy shifts, however, often failed to appreciate field-realities, and consequently, the policies hardly had any major impact on empowering local communities to any significant extent. Notwithstanding the stated goals and national support, the Nepalese SF in reality functioned basically as a routine government-sponsored rural development programme, administered by a collaborative alliance between government and local government offices, which produced limited benefits to people. Baral (1993, p.67) has aptly summed up the true nature of SF:

> Quite often Department of Forest (DOF) staffs simply view the community forest

> plantation programme as another form of state forestry. The intended beneficiaries are rarely consulted in any decision making. The DOF staff independently produce seedlings, select planation sites, plant the sites and finally provide a watcher to take care of the plantation....[T]he intended beneficiaries rarely know why and for whom the plantations were created.

Overview

I have so far reviewed the experience of SF in some selected cases from south Asia which are popularly cited as good examples of SF practice. If SF is interpreted as a development strategy which seeks wider participation and empowerment of local people, the above review suggests that the level and significance of its achievement have been insignificant.

However, as I found in the course of the literature search, there have been some "pockets of success" (Gronow and Shrestha, 1991, p.3) in SF. These are of pilot, experimental and demonstrative nature, and can best be regarded as exceptions. One such example was the Taralabulu Rural Development Foundation in Karnataka (India). This experimental project, designed by a professional group under the guidance of Professor Gadgil (a renowned Indian scholar), investigated into the ways and means of local people's involvement (Nadkarni et al., 1989). Similarly, in Kabhre Palanchok and Dhankuta districts of Nepal "successful" experiments in SF were carried out (Gronow and Shrestha, 1991, p.4). Saint (n.d., pp.11-20) has also reviewed some experimental "successful efforts" of community management and SF in India.

However, "on the whole", as Asaduzzaman (1989, p.15) concludes, SF "barely passes the acid test of people's participation". Given the generally unfavourable socio-economic "frame" which surrounds SF projects, its stated goals and lofty philosophies seem largely unrealistic. The political-economic variables (which I roughly referred to as the "regulators") seem to condition SF's performance. The next section elaborates some of such crucial variables.

The Frame of Social Forestry

A true understanding of SF's performance calls for an analysis of the particular realities of the surrounding environment. The literal meanings of the word "frame", relevant to my purpose, are "structure that gives shape and support [to something]", "to shape" or "to form" or "to bring about" a phenomenon (Samsad English-Bengali Dictionary, 1990, p.432; Oxford

Advanced Learner's Dictionary of Current English, 1987, p.342). My use of the word refers to the surrounding social and political-economic "structures" of SF which "shape and support [or thwart]" its performance, and "bring about" the end-results (achievements). In other words, the term "frame" is intended here to connote the surrounding environment of a particular SF programme and the political-economic variables therein, which broadly shape SF's achievements. Of late, seminal studies, reporting on the world-wide experience, have unequivocally established that any resource use strategy (including SF) needs to be appreciated and analyzed in context of the surrounding realities (the frame); and the success of a particular strategy depends on its ability to match with these realities (see Cernea, 1985; 1990; Noronha and Spears, 1985; Barraclough and Ghimire, 1990; Brokensha and Castro, 1984; Ives and Pitt, 1988). The rationale behind the emphasis on understanding the surrounding setting, in the words of Brokensha and Castro (1984, p.1), is that

> [strategies for] Natural Resource management cannot logically be considered except in relationship to their specific social setting. They are a part of a complex system of relationships and institutions: the emphasis is on system, which forces us to regard a wide array of relevant factors.

However, the variables (or factors) within the frame (or the setting), which constrain or bolster SF's performance, are too many and too wide to be accommodated in this brief analysis. Drawing on the previous literature survey and case-studies, I will now present an overview of selected major variables which seem to be particularly significant for SF.

Land Ownership and Tenurial Arrangements

The crucial implications of land (and tree) tenurial arrangements for SF have been well documented by a sizeable literature (some extensive and widely-referenced studies include Fortmann and Riddell, 1985; Fortmann and Bruce, 1988; Cernea, 1981; Barraclough and Ghimire, 1990; Raintree, 1987). The term "land tenure", derived from the Latin *tenere* (to hold), "refers to the possession or holding of the many rights associated with each parcel of land" (Riddell, 1987, p.2; Bruce et al., 1985, p.viii). Land rights are seen as "bundles of rights" which have at least three dimensions (Bruce et al., 1985; Riddell, 1987, pp.1-4). The first dimension is "people" which is concerned with the network of human interaction and social relationships; because land rights are the "result" and "expression" of such interactions. The second dimension of "time" also influences tenurial arrangements. For example, if trees with long gestation time (as compared to short-rotation crops) are planted in a land, these can effectively extend tenure (possession) of that land in favour of the planter. Third, tenure rights are associated with "space", and with the interactions and competition among different uses and users over a particular space. Therefore, land tenure is a complex issue; and SF planners need to be alive to different dimensions and implications thereof.

There are regional differences in land arrangements, which constrain generalizations. However, the importance of land tenure in SF is at least three fold. First, generally speaking, land is the most instrumental variable in determining the competition and access to resources in rural south Asia (Jansen, 1986; Jannuzi and Peach, 1980; Rahman, 1994). Second, it is also "the most obvious outward manifestation of power relationship" which determines people's share of "wealth, opportunities and power" (Barraclough and Ghimire, 1990, p.18). Third, it serves the socio-psychological purposes of incentive for and assurance of participation and investment in afforestation schemes. Moreover, land and poverty have intimate correlation. Barraclough and Ghimire (1990, p.18) argue that "rural landlessness or near landlessness, is almost synonymous with rural poverty in most developing countries". And poverty in turn is one dominant force which shape the political-economic strategies people adopt for production, access to opportunities, and ultimately, for their very survival (Beck, 1992). Jannuzi and Peach's (1980, pp.79-80) observation on Bangladesh mirrors the reality of many regions of south Asia:

> Control of land determines the structure of power...who will benefit from rural development programs and who will not....[It is] the critical variable in determining what is produced for whom, and for whose benefit.

Thus, land is of great significance for any resource use strategy in south Asia. It largely determines people's access to resources, benefits and power. It also shapes the survival strategies people adopt in their day-to-day competition for the limited resources.

A clear understanding of land tenurial arrangements and other related factors of the frame is imperative for SF (Fortmann and Riddell, 1985; Barraclough and Ghimire, 1990; Brokensha and Castro, 1984; Cernea, 1988; 1989; Bruce, 1989). The legal or formal tenurial arrangements must not be confused with the existing field realities. Cernea (1988; 1989) presented a case of a SF project in Pakistan which planted trees on village "common" land on the (erroneous) assumption that this would spread the benefit throughout the community. In reality, to the utter failure of the project, it was found that the so called "commons" were under effective de facto control of some local rich and influential families; and they were the ones who had benefitted from the project. Therefore, as Cernea (1981; 1989) argues, beyond the "paper status" of land tenure, SF planners are required to have an insight into the actual field situation. Riddell (1987, p.2) also warns that "we need to look beyond the de jure land laws to the de facto law in action of day-to-day land tenure".

In general, it has been argued that, ownership, or at least, reasonably long tenurial right of land (which instills a sense of security among the participating farmers) is essential for successful SF (Fernandes et al., 1988; Gujral, 1988; Jayaraman, n.d.; Shepherd, 1990; Rahman, 1992; Ahmed and Azad, 1987; Kayastha, 1991). In Karnataka (India), for example, "foresters found ...that they could interest villagers in tree-planting on their own land far more easily than they could start a village plantation" (Shepherd, 1990, p.2). Village plantations generally involve community lands which often pose complex or ambiguous tenure and use status. In contrast, lands under private ownership or individual use have clear tenurial status, and therefore, are less likely to cause conflicts. This largely explains the comparative success of the private or farm forestry components of SF in India and Nepal (Kayastha, 1991; Arnold et al., 1988; 1990; Shah, 1988; Banerjee, n.d.; 1989). In Orissa (India), villagers demanded title to land as a condition for participating in SF, because otherwise they felt unsure of profiting from the plantation when the trees would be matured (Jayaraman, n.d., p.10). Shah (1990, p.15), evaluating the West Bengal (India) SF programmes, comments that "99 year land *patta* [leasehold] helped to instil...a confidence and a feeling of ownership that encouraged [farmers] to invest effort in improving the productivity of their land" and to participate in SF. SF projects in Himachal (India) also attached importance to the security of land-tenure as a motivating factor (Divisional Forest Officer (DFO), Himachal Forest School, personal communication). In the Betagi and Pomora SF projects in Bangladesh (two of the case-studies of my research) long tenurial security of land was identified to be the primary motivating factor for farmers' participation (Rahman, 1992; Ahmed and Azad, 1987; Quddus et al., 1991). Asaduzzaman (1989), in his overview of SF in Bangladesh, also noted that one major reason behind the limited participation of people in SF was the absence of clearly defined tenurial rights.

Therefore, there is a strong body of literature which suggests that long-term security of land (or permanent title) serves as a motivation for farmers, and thereby, may increase the chances of SF's success. However, the status and implications of land tenure are complex issues, and need to be appreciated in context of many other variables and considerations. Some of such points of consideration are as follows. First, bringing fallow or degraded public lands under tree cover through SF is a welcoming change. But, in the process, handing over public lands to private individuals may not be "socially justified" (Nadkarni et al., 1989, p.173). Because such privatized allotment implies a privileged provision to a particular section of the community at the cost of reducing the rights of other equally deserving candidates. As mentioned in earlier discussions, public or community lands are being distributed for SF in many regions of south Asia. Such land allocation schemes may lead to the individualization of a public or community resource, by curtailing similar rights of other deserving parties. Second, besides tenurial status, land-use strategies and options are also of great significance to SF, but are often not paid much attention to by planners. In Himachal (India), for example, lands allotted for SF were mostly planted with low quality species which were "neither economically viable nor environmentally sustainable" (DFO, personal communication). Similarly, government-sponsored massive *Eucalyptus* (and other mono-culture) plantations in many regions of south Asia, without regard to local views and particular surrounding frame, also provoked suspicion and bitterness among farmers (see for example Fernandes et al., 1988; Khan and Hasan, 1985; Agarwal and Narain, 1985; FAO, 1989; Saxena, 1992). As noted in the regional case-studies, agricultural lands were used for afforestation purposes in some parts of south Asia, which could also be debatable as a general strategy. Third, it has been found that providing land ownership helped in particular situations; for example (Shah, 1988, p.15), where

> tree growers derived no value from land before it was planted with trees and had therefore learnt to subsist through other means. Income from trees is thus an additional bonus to them.

Put differently, this may suggest that despite tenurial security, farmers may not be very interested in afforestation, where they have other comparatively lucrative options of land-use available. Besides, farmers sometimes tend to look at afforestation as a subsidiary source of income and not as the primary activity. In these situations, the importance of SF may recede to background; and land allotted for SF may be used for quite different purposes, and thereby, upset the very motto of SF schemes. Indeed, as I will eventually discuss (in Chapter 6), a similar situation has been observed in my case-study projects in Bangladesh. Fourth, intimately related to land tenure, is the issue of tree tenure (see Foley and Barnard, 1984, pp.65-66; Riddell, 1987; Bruce, 1989; Bruce et al., 1985; Fortmann and Riddell, 1985; Fortmann, 1987); because "trees can like minerals and water be an object of property rights in itself" (Bruce, 1989, p.3). Sometimes trees can be used to establish rights to land (Bruce et al., 1985, pp.ix-xi; Riddell, 1987, p.3; Fortmann, 1987, pp.17-26). Thus, an understanding of tree-tenurial arrangements is also imperative for SF. Fifth, as noted before, land tenure arrangements are the outcomes of existing social relations and structures. Land tenure, therefore, needs to be understood in context of the broader social frame, which, in south Asia, is generally characterized by deep-rooted dependency relations, exploitative productive and exchange mechanisms, and sharp social inequality. Given these generally unfavourable social realities, farmers may not automatically be able to enjoy the benefits from a SF scheme, even when the tenurial status is unambiguous or favourable.

In sum, land tenure in SF is a complex issue which needs to be interpreted in relation to the particular surrounding frame. Even when farmers are accorded land ownership, they may not able to utilize productively the land owing to other unfavourable variables. For example, patronage relations and elite domination may deter a farmer from enjoying control over the land, albeit under his(her) formal ownership. Ambiguity over tenurial status and failure to relate it to the specific frame are proven causes of poor performance of SF (Cernea, 1988; 1989; Riddell, 1987; Barraclough and Ghimire, 1990; Bruce, 1989).

The Role of Public Forestry Organizations and Policies

Ali (1991, p.3) argues that public policies in general, and forest policies in particular, are "of major concern" to SF in any country. In Chapter 2, the general role of public policies and governments in shaping the process of forest use was discussed. The discussion suggested that concerned policies and state bureaucracies essentially remained centralized, non-responsive to local demands, elitist, and oriented towards revenue maximization. They fostered the process of forest commercialization, reflected the interests of commercial forces, and were characterized by widespread public isolation. Keeping in mind these general lessons, this section examines some implications of forest policies and forestry organizations for SF's performance.

Following their colonial legacy, forest services and policies in south Asia have continued to be oriented towards the commercial considerations of revenue generation and maximized resource exploitation. SF's performance, as manifested in the regional case studies, was measured by the achievement of physical targets. The deeper purposes of SF, for instance, its performance in promoting participation and social equity, were often ignored by governmental planners.

Policies, generally speaking, remained biased towards the influential commercial interests in forestry. As Barraclough and Ghimire note (1990, p.17), policies tilt towards "the interests of influential forces upon which governments are most dependent for crucial support in order to survive". Similarly, from an extensive survey of the Indian situation, Fernandes et al. (1988, p.268) have concluded that SF policies give "preference to industrial needs over those of the people" (also see Fernandes and Kulkarni, 1983; Guha, 1989; Khan and Hasan, 1985).

Policies, in turn, were profoundly influenced by existing social frame. Operating within a social frame characterized by deep inequality of resource possession and distribution, political instability, elitist power-structure -- policies and state organizations rarely represented the voice, demands and interest of the poor and weaker section of local communities. Brokensha and Castro (1984, p.41) have concluded that "in many countries, policies concerning forestry management are...uneven". In some recent policies (for example, the Forest Bill of 1980 in India), apparently, there was an attempt to address the interests of various parties in forest use: industrial, revenue, environmental, and forest dwellers' interests. But such policies failed to understand that "these interests belonged to unequal groups...;when [they] began competing for the same resources, the interests of the powerful sections...prevailed and powerless lost out" (Fernandes et al., 1988, p.284).

As the regional case-studies revealed, forest policies and forest departments often failed to play their catalytic and supportive role in facilitating the growth of local cooperative institutions for SF. In many cases, SF projects were managed by state bureaucracies. In other cases, local institutions concerned with SF had to function under the profound influence of central government machineries. Generally, bureaucratic control, procedural formalities, and continued central interference constrained the process of delegation of power and responsibilities to local institutions (Arnold and Stewart, 1991, p.19; Arnold, 1990, pp.7-9). Moreover, public policies and bureaucracies often accelerated the fall of the traditional patterns and indigenous structures of resource use (Poffenberger, 1990; 1990a; Arnold and Stewart, 1991; Jodha, 1988; Arnold, 1990). A study by Jodha (1988, cited in Poffenberger, 1990a, p.4), on 82 common property holdings in seven Indian states, found that the indigenous management systems gradually ceased to function, since the forest nationalization and land reform policies were launched in 1951. Similarly, in Nepal, the nationalization policies of the late 1950s paralysed the traditional resource-use systems. These policies also led to the alienation of local people from forests, and the consolidation of mutual hostility between the government and local communities (Bajracharya, 1983; Adhikari, 1990; Gronow and Shrestha, 1991; Khan and Hasan, 1995; Khan, 1996a).

In many instances, policies clashed with the goals of SF. The policy-emphasis on monoculture production of commercial species in India, for example, often ignored the local demands of fuel-fodder and remained solely oriented to industrial markets (Fernandes et al., 1988, Chapter 10; Fernandes and Kulkarni, 1983). Furthermore, some highly debatable industrial tree-species were promoted under SF schemes which allegedly caused environmental degradation and were of little relevance to local needs (for example, for debates over *Eucalyptus* plantations in SF projects in different regions of south Asia see Foley and Barnard, 1984, pp.151-160; Shiva et al., 1983; 1987; Manandhar, 1987, pp.9-10; Saxena, 1992, pp.53-58). As observed from the case studies, rigid policies and bureaucratic formalities often hindered the process of "micro-planning" and marketing of SF products. They also created opportunities for harassment of the people by the local level state officials. The plethora of regulations and formalities also discouraged local institutions (for example, the *panchayats*) to take the responsibilities of SF. Besides, as Singh (1988, cited in Arnold and Stewart, 1991, p.19) argued, the "legal policy status...is such [that]...the people are not in a position to actually negotiate the terms of contract with the Forest Department and are sometimes in conflict with the interests of the local government (*panchayat*)"

Forest policies, in other instances, were found to be vague and ineffective. The Bangladesh experience, for example, showed that the current national forest policy largely ignored many crucial issues of SF, including the provisions of community participation, specific principles for private forestry development, involvement of voluntary and non-government institutions, homestead forestry management, and the rights of forest dwellers and tribal families (Ahmed and Bhuiyan, 1993). Besides, forest policies and forest organizations often failed to take into account particular field-realities, local demands and contextual peculiarities. As found in the preceding case studies, often policies proceeded on wrong assumptions about the working grounds of particular SF programmes and their possible impacts in the field.

Forest policies, principally, find expression in the field by the activities of the forest departments. The south Asian public bureaucracies in general, and the forest departments in particular, have their roots in the colonial rule with predominant commercial considerations (Chapter 2). Their growth, as Poffenberger (1990, p.19) suggested, could be seen within the broader "process of bureaucratization of societies throughout the region". The formation of forestry organizations helped governments to regulate forest use to their vested interests. The forest departments' role essentially remained as "revenue collectors" for central governments and "protectors of forests" from local people who in the eyes of the governments were the "notional thieves" (Shepherd, 1992a, p.19; also see Fernandes et al., 1988; Fernandes and Kulkarni, 1983; Shepherd, 1992. The genesis and role of the forest departments has been discussed in Chapter 2 and will be further elaborated in Chapters 5 and 6 in the empirical context of Bangladesh case-studies). In the process, hostility and distrust flamed up between local communities and governments, which have continued unabated to date. Bureaucracies consolidated their position and power "by increasing their numbers and authority, extending government activities into the [local] economy, and combining bureaucratic and private economic pursuits" (Poffenberger, 1990, p.19). In fact, the "bureaucracy has attempted to appropriate an ever increasing share of societal surplus" (Evers, 1987, p.680, cited in Poffenberger, 1990, p.19). In this historical context of the state appropriation of resources and the commercial orientation of the forest departments, there seems to exist a clear difference in the priorities and goals of governments and that of local people. The World Bank (1994, p.292) noted:

> The revenue and economic return objectives of the states conflict with...the social objectives of providing fuel, small timber and fodder needs to the local people.

These conflicting priorities and the historical antagonism between local communities and governments were at odds with the values and goals of SF (which aims at promoting the interests of local communities).

In their attitudinal and value orientation, public bureaucracies in south Asia have continued to remain "elitist", "extremely powerful", "non-representative" and "exclusive" (see La Palombara, 1963 for an overview of the nature and role of bureaucracy in the Third World political development. Ahmed, 1980; Khan, 1980 and Anisuzzaman, 1979 offer some insightful perspectives on bureaucratic behaviour in Bangladesh and Pakistan). Indeed, as Ahmed (1980) noted in his analysis of bureaucratic elites in Bangladesh and Pakistan, bureaucrats were elitist and were isolated from people in terms of their origin, power and life-style. Even the comparatively lower level officials (for example *Thana* or sub-district officials in Bangladesh) also tended to be elitist in their outlook. Although they "came from rural background, yet they lack endeavour and commitment to the upliftment of rural masses" (Rehman, 1983, p.65). Therefore, it was not unnatural that development interventions (including SF), administered by such an elitist bureaucracy, were often marked by bureaucratic domination, and isolation or insignificant involvement of local people. Bureaucratic values led to the consistent tendency of blaming people for all problems in the forestry sector and to the imposition of "stringent" regulatory measures and "military-style controls" on local people (SPWD, 1992, p.39). Even today, in the days of great appreciation for participatory forest management, some quarters within the government in India, for example, "are worried that empowering forest communities will erode [the forest] departmental authority over natural resources" (SPWD, 1992, p.39). These facts suggest that there has been little change in the elitist values of public officials and their general inclination to bureaucratic administration. Under such bureaucratic governance, Almond and Verba (1963, pp.161-162) argue, people become "passive subjects" and not "citizens" ("citizens" meaning the "active participant[s] in the political decision making and governance process"). In their classic work, Almond and Verba (1963, p.162) describe the implications of such governance and the characteristics of "subjects" in the following words:

> The man whose relation to his government is that of a subject [becomes] a passive beneficiary or victim of routine governmental actions....[T]his relationship would exhaust what is expected of him. What the government does affects him, but why or how the government decides to do what it does is outside his sphere of competence....[H]e should be loyal and respectful of authority.

In Chapter 2, the basic limitations of the forest departments in working with local people were identified. From that discussion and from the review of the cases in this chapter, there appeared to be a consensus among analysts that the institutional structures and value-orientation of the forest departments were inadequate and ineffective in handling the challenging tasks of SF. In particular, foresters largely lacked the communication and extension skills needed for performing their new roles of "community developer" and "motivator" and for responding to local demands. Besides, little change was observed in their "custodian policing" and "para-military discipline" (Locke, 1992, p.7).

In sum, the role of public policies in forestry sector in general, and SF in particular, has been largely unsatisfactory. In Bangladesh, for example, "there was an increasing consensus that forestry sector policy...was ineffective in maintaining the sector's contribution to the national economy" (Task Force, 1987, p.28). SF projects generally worked under profound bureaucratic influence. The role of forest policies and forest departments in supporting SF's lofty goals of participation and equity was inadequate and unfavourable. In more general terms, Poffenberger (1990a, pp.3-5) sums up the role of forest policies and forestry organizations in south Asia by the following words:

> Policies and regulations have strengthened the authority of government managers and commercial interests at the expense of small rural user groups, eroding local management practices and creating conflicts among these groups....The growth of state control over forest lands has led some economists to suggest that feudal landlords have been replaced by a new powerful class of professionals and bureaucrats. It is apparent that forest communities have lost the authority to manage forest resource....

Market Opportunities

Chapter 2 discussed the broad-based impact of uncontrolled market on local communities and concluded that where people did not have sufficient control over market forces, the consequences were devastating. For SF too, markets are of crucial implication. In order to enable farmers to fully benefit themselves from SF, their control over market is essential. "Control" by local people is a wide and far-reaching process which can assume many forms. Some immediate manifestations of such control may be the ability of local people in reducing the monopolistic tendency in market-transactions; tackling middlemen and intermediaries; ensuring regular outlets for selling their products (and the freedom of choosing such outlets); ensuring fair price and establishing a reasonable control over price fluctuations. In general, SF farmers in south Asia seem to be far from establishing such controls.

As observed from the case-studies and from a regional review of literature (Fernandes et al., 1988; Shah, 1988; Saxena, 1990; 1991; 1992; Malla, 1993; Messerschmidt and Hammett, 1993; Quddus et al., 1992; Rahman and Khan, 1994; Locke, 1992; Poffenberger, 1990b), the major market-related problems faced by SF farmers are the following: lack of access to market-outlets for selling SF products; wrong choice of species (which often had no local value or prospect for selling); unequal and exploitative terms and conditions of exchange and contracts; monopolistic domination by the powerful commercial forces (and their agents); complicated bureaucratic rules and procedural formalities relating to marketing; bureaucratic harassment and corruption; lack of systematic market information; and inadequate advisory and logistic support from governments and other agencies. Many instances in the above studies testified to these problems. For instance, the West Bengal case can be referred back to, where two influential elites monopolized control over the procurement and selling of SF trees and made huge profit by exploiting the farmers (Shah, 1988).

Similar cases of manipulation by intermediaries, unjust prices, and farmers' lack of access to markets were observed in many parts of south Asia (see Saxena, 1991; 1992; Malla, 1993; Rahman and Khan, 1994; Shah, 1988; Quddus et al., 1992). In Bihar (India), farmers could sell trees only through the forest department. The small farmers could not secure a just price for their products, because the forest department was mainly interested in large-scale transactions, and brought with it a host of procedural formalities which were beyond the farmers' comprehension and made room for bureaucratic harassment (Arnold et al., 1988; Shah, 1988). Likewise in tribal areas of southern Maharastra (India), due to the lack of proper market opportunities, farmers used to get only about Rupees 20 for a log of Teak stem, which had an open market value of at least Rupees 400-500 (Shah, 1988). Furthermore, markets for SF products were often oriented towards and dominated by the "external demands of industry and urban consumers" (Foley and Barnard, 1984, p.26).

Harris (1989) noted that markets generally perform two broad functions: "allocative" and "exploitative". Saxena (1991, p.165) explains the functions and their implications:

> To the extent that markets facilitate commodity production, and integrate producing regions with consuming regions, they help the farmers in choosing the most profitable cropping pattern. Farmers allocate their resources in commodity production on the basis of signals they get from markets. Markets may also play a retrogressive role by coercing producers to sell at a low price through monopoly, credit and withholding of information.

The preceding discussion suggested that the role of market in SF was "more exploitative than allocative" (Saxena, 1991, p.165). SF farmers, in general, were unable to establish any significant control over markets and were exposed to varied manipulations by market forces. Governments (and other catalytic agencies) also largely failed in providing the necessary logistic support, and in creating a conducive atmosphere wherein farmers could tap the benefits of market. Barraclough and Ghimire (1990, p.21) have summarized the impacts of market on local people:

> Most indigenous peoples depending on forests in developing countries are...incorporated in national and wider markets. Markets everywhere and at all levels are subject to monopolistic manipulations....Market forces are much better at increasing the wealth and power of the strong than of the weak....Poor peasants have little purchasing power, scant assets and little political influence [to control and use markets to their benefits].

The Role of Local Organizations and Cooperative Structures

Chapter 2 briefly discussed the limitations of existing local organizations in empowering the people and in creating a cooperative platform for articulating their interest. It concluded that these organizations were hardly an effective substitute for the traditional institutions and indigenous resource management structures. These organizations generally tended to represent the interests of local influential classes and were dominated by local elites in service delivery and resource distribution.

As the case studies revealed, the *panchayats* and the forest-user committees in India and Nepal, for example, were unrepresentative of the interests and voices of the poor and weaker section of local communities. These organizations were often sabotaged by local power-politics and were practically taken over by the elites. Reviewing the south Asian experience with the *panchayat* system, Uphoff (1982a, p.35) concluded that

> on balance, the system appears nowhere to have lived up to...expectations in terms of impact on development, promoting social equity or achieving broad popular participation.

Similarly, Fernandes et al. (1988, p.283) found that in Orissa (India) a committee in charge of decisions and operation of a tribal SF project ultimately "turned out to be a committee of powerful persons like money-lenders, most of them non-tribals". As Arnold et al. (1990, p.40) argued, the formation of forest committees, in many instances, was

> simply a bureaucratic requirement...and could in no instance be regarded as a reflection of villagers' own needs and motivation....Given the nature of the society ...this is to be expected. Membership of the committees generally centered around a few influential village elites. Weaker sections...certainly felt they had no power and often did not attend meetings that were held.

In general, these organizations were externally designed and enthused, and rarely reflected local demands and regional conditions. Consequently, as Hobley (1991, p.112) argued, they

> had not been set up out of an endogenous desire to manage forest collectively, but from an external intervention....[T]hese committees were ineffective, unrepresentative and had no control or responsibility.

Rural development interventions such as SF, channelled through these organizations, as Blair (1978, p.65) argues, have an "almost universal tendency to benefit the dominant elites" and to "serve a relatively few at the top of the economic structure". Under these initiatives, there is generally a growth of alliance among central elites, their local allies, and the bureaucracy in the management and distribution of benefits and resources. Consequently there is a "natural proclivity of elites to take over and run" local organizations; and governments, almost as a rule, tended to use these institutions as their sub-national "power-base" (Blair, 1982, pp.392-393). The bureaucratic domination of decisions in the so-called "participatory organizations" and the alliance of vested interest between bureaucracies and local elites have been proven beyond all doubts (see Uphoff, 1982; Ahmed, 1991; Rahman, 1994; Sarker, 1990; Attwood and Baviskar, 1988). The political regimes in south Asia seldom made genuine attempts towards empowerment and institution-building in favour of common people; primarily because such initiative would go against their local allies -- the local elites -- on whose "support, compliance and cooperation" political regimes survived (Blair, 1982, p.459; also see Blair, 1978; Rahman, 1994; Ahmed, 1991; Wood, 1994). "In sum", as Blair (1982, p.460) concludes, in the eyes of governments and their allied local organizations "the mass of subsistence farmers, landless laborers and rural non-agricultural workers simply do not count politically and can be ignored". Therefore, in this unfavourable context, it is no surprise that the role of local institutions in promoting the goals of SF remains limited and ineffective.

The above "sobering facts" (Blair, 1982, p.462) of south Asia also deterred the growth of any genuine (as opposed to rhetorical) cooperative structure through which farmers could articulate their own interests and demands. In Bangladesh, for example, Tepper (1976, p.29) has noted that "the relative absence of institutional groups and the difficulty in creating them is at the heart of [the country's] development problem". In general, as he argues, "the existing rural social and political organization is segmented, diffuse and in flux" (Tepper, 1976, p.32). "In fact" , as Blair (1978, p.74) concludes,

> the history of cooperatives in South Asia has been as melancholy as that of the other rural development institutions. Cooperatives also have almost invariably been taken over...by village elites in their own interest ...;[they] have served to channel funds into the hands of the rural rich.

The high blown expectations from cooperative actions in south Asia frequently produced "correspondingly bitter" results (Attwood and Baviskar, 1988a, p.1). Cooperation among local people has been rarely achieved. The very limited successful cases of cooperation observed in south Asia were mostly in the forms of "selected and negotiated alliances" (Attwood, 1988, p.77) and "not a system of communal solidarity" (Attwood and Baviskar, 1988a, p.4). From his empirical survey in Western India, Attwood (1988, p.84) concludes that people opt for cooperative alliance only when "such alliance is made necessary by inescapable technical factors in the production process". Similarly, Khan (1994a, p.107), from a extensive review of literature on the contextual condition for cooperative initiatives in south Asia, concludes that "since the general socio-administrative context of south Asia is not conducive to cooperative endeavours, cooperation is most likely to be selective".

In the above context of limited prospect and selective nature of cooperation in rural south Asia, some studies have attempted to identify the conducive variables (or conditional factors) which may facilitate cooperation (see Wade, 1988, pp.215-217; also see Khan, 1994, pp.101-107; Poffenberger, 1990a; Korten, 1987; Attwood and Baviskar, 1988; Uphoff, 1982). One of the first condition is the states' commitment and support in promoting the growth of cooperative structures. However, the south Asian states generally tended to follow a centralized approach to cooperative development -- which Hyden (1988) has called the "blue-print approach" -- often ignoring indigenous organizations, local opinions and regional conditions. In some instances, the states got into conflict with local cooperative organizations, fearing them to be prospective rival bases of power (see for example Bandyopadhyay and Eschen, 1988 for an Indian case in point). As has been commonplace, the imposition of centralized control and using local institutions as sub-national support-bases for ruling governments often resulted in widespread corruption, waste and subsequent distortion of local cooperative structures (Uphoff, 1982; Attwood and Baviskar, 1988; Rahman, 1994). The second condition -- the "control" structure -- is important for the success of cooperative endeavours. Such control mechanisms can regulate behaviours of the concerned groups and can serve as the dispenser of Olson's (1971) "selective inducements". The "control" functions may take diverse forms including a system of punishment for breach of common rules and the detection of the cases of "free-riding". Third, the nature of particular resource and the concerned users is also an important consideration for cooperation. In a situation of common-resource use, cooperation may be facilitated by a clear demarcation of boundaries of the resource itself and its users. The problems like "free-riding" can easily be detected if the resource is a clearly defined one and the users are aware of the utility of the resource (Wade, 1988). Fourth, cooperation is also conditioned by the nature of the production-process. Experience suggests that, in the cases of agricultural or technical productions, concerned producers may readily cooperative, if they find that cooperation is imperative to sustain the production process (in other words, where cooperation is a "technical necessity") (Attwood, 1988, p.84). For example, Appadurai's (1986) study on the sharing of irrigation-wells in an Indian village noted that the small farmers cooperated because they had no other choice in sustaining irrigation and they could not afford to construct their separate wells. Similarly, Wade (1988) showed that there were severe demand, scarcity and risk associated with irrigation and common grazing in south India; and cooperation occurred because the risk could only be reduced by acting collectively. But in the case of SF, cooperation may be limited and selective. For example, a social forestry plantation on a common land requires expenditure and restriction on use at present in return for the prospective benefits in future (may be 10 years hence) (Wade, 1988, pp.212-213). In this situation, where net collective benefit is not immediate, people may not be easily motivated to cooperate. Fifth, the leadership of a cooperative institution is required to have an acceptable mode of accountability to their members. Without a credible structure of accountability, "personal or factional antagonisms and suspicions might easily burst the organization asunder" (Wade, 1988, p.195). Accountability is based on such factors as proper representation of concerned interests and groups in the institution; "publicness" and exposure of the institution's performance; and an effective rule-enforcement system.

However, the regional literature on the performance of cooperative institutions for rural development in general (see Attwood and Baviskar, 1988; Wade, 1988; Korten, 1987; Uphoff, 1982) and SF in particular (Cernea 1988; 1989; 1985; Noronha and Spears, 1985) suggested that these conducive conditions rarely existed in rural south Asia. Consequently, the growth of endogenous (as opposed to externally imposed) cooperative organizations was retarded; and even when established, these organizations largely failed to play their desired role of empowering local communities.

Social Relations and Power Structures

The implications of social relations and power structures for the performance of a particular SF project are profound and manifold. The most common causalities between social relations and SF's performance are, at least, threefold. First, local people's access to SF project design, management and benefits is largely regulated by such relations. Second, these relations may constrain or bolster the growth of local level institutions (for example, a farmers' cooperative) which are, in turn, crucial for institutionalizing the process of popular involvement in SF. Third, these relations can serve as a communication-link between governments and the people, and among the different groups and interests in a particular resource-use context.

Generally speaking, in the agrarian and subsistence production systems of south Asian villages, social relations and power structures play vital roles in determining who get a chance for survival, in what terms, conditions or rights, and to what extent one is going to reap the benefit from any development programme. In Bangladesh, for example, Jannuzi and Peach (1980) note that the agrarian structure is defined by a complex set of relations and arrangements, which governs the relationship among people, land and power. In other words, these social relations largely influence the link between people and their entitlement to (scarce) resources, and thereby, shape their strategies of survival (Jansen, 1986; Beck, 1992). Wood (1994, pp.78-91), from his extensive field-work in Bangladesh, notes that one dominant phenomenon of the "political order of a peasant society" is the "patron-client relations" which largely influence the "factional pattern of competition" for resources . Furthermore, he (Wood, 1994, p.98) concludes that

> the socio-economic relationships within the village are not conducive to poor peasant and landless political organization within the village itself. Here, the process of class differentiation -- encouraged by the introduction of new agricultural technologies [SF can also be seen as an innovative technology] -- maintains the poor in an increasingly weak...condition.

The regional cases from south Asia offer broadly similar experience and lessons. Access to resources is, to a large extent, shaped by patronage relations between the poor (weak) and the rich (powerful). In the context of Nepalese SF, Hobley (1991, pp.116-117) observed that, those farmers who had "personal relations with the patron" and were "closely allied with local power structure" could benefit themselves from SF programmes. These social realities led Hobley (1991, p.116) to conclude that:

> In a country [Nepal] where personal relations with patrons determine access to resources and power, a committee composed of low castes and women and so, with no access to higher levels of bureaucracy, could not function...without [a certain degree] of social change.

Wignaraja (1984, p.8) also notes that "a truly participatory development process [such as SF] cannot be generated spontaneously given the existing power relations at all levels and the deep rooted dependency relationships".

Similarly, as observed in the Indian states of Bihar and Tamil Nadu, membership in the village forest committees (the primary local organization in charge of managing SF activities) centered around few influential elites (Arnold et al., 1988; 1990; also see Khan and Hasan, 1995; 1995a). They dominated decisions and activities of the committee through their respective clientele. "Weaker sections...certainly felt they had no power" (Arnold et al., 1990, p.40).

There are different manifestations of social relations and power structures with different degrees of influence on SF. One dominant feature of social relations, as mentioned above, is patronage. The economic and social transactions in the agrarian production and distribution systems of south Asia occur through a "highly personalized patronage network" (Zaman, 1984, p.18; also see Jansen, 1986; White, 1992). In one of the seminal works on patron-client relationship, Scott (1972, p.92) defines it as

> a special type of dyadic (two persons) ties involving largely instrumental friendship in which an individual of higher socio-economic status (patron) uses his own influence and resources to provide protection and/or benefits for a person of lower status (client) who for his part reciprocates by offering general support and assistance, including personal services to the patron.

Zaman (1984, pp.19-21) notes that in Bangladesh patrons provide leadership and serve as "brokers" between villagers and the "outside world" including mediating the distribution of governmental goods and services. In distributing such "public goods", patrons exert their influence and control over local governments and other local organizations, and "treat their followers preferentially" (Zaman, 1984, pp.20-21).

The patronage relationship is "essentially unequal in character, involving domination and submission" (Zaman, 1984, p.20). Scott (1972, p.93) also maintains that though "some reciprocity is involved" in the relationship, "the balance of benefits heavily favors the patron". Similarly, Clapham (1982, p.4) notes that "inequality" is the "key-element" in a patron-client relationship and it involves "the superiority of one over the other". It has been repeatedly argued that these relationships allow local elites to reinforce their power in the countryside and appropriate the principal benefits from rural development interventions (see Mashreque and Amin, 1993; Jahangir, 1982; Ahmed, 1989; Momen, 1993; Burch, 1988; Jansen, 1986; Blair, 1978; 1982). These relations also help to strengthen the vested alliance between central elites (especially the ruling governments) and their local partners (Blair, 1978; 1982; 1985; Jansen, 1986; Rahman, 1994). Recent studies on Bangladesh have re-emphasized the significance of understanding the profound implications of patronage relations for analyzing development programmes and their surrounding social setting. White (1992, p.37), for example, finds that "the key to understanding social and economic inequalities" is "to grasp these relations" through which they are "reproduced". Similarly, Jansen (1986) suggests that "competition for scarce resources" occurs largely within the patronage network.

Caste structures and fictive relations are the two other notable social structures which substantially influence rural development projects. These social structures are complementary and mutually-reinforcing. The caste-hierarchy, for example, provides a structured framework and channel, which facilitates the growth and smooth operation of patronage links. Although not as rigid in nature as the traditional Indian caste-system, in rural Bangladesh too, Tepper (1976, p.33) has observed "caste-like characteristics" among the majority (Muslim) population (what he terms as the "Muslim status-hierarchy"). In his empirical study on farm land afforestation in north-west India, Saxena (1990, p.6) for example, found that caste was "a greater barrier to tree growing than a lack of land". It was observed that the higher caste farmers had performed well both in tree planting and sale of products. There may be several caste-related factors which influence SF's performance. First, the higher caste groups have greater access to education and to the channels within the bureaucracy which enabled them to tap the sources of credit, markets and public services necessary for SF. Put differently, the higher caste groups can use their superior caste-status to negotiate and liaise with the bureaucracy in order to appropriate the lion's share of public goods and services. Second, the high caste affluent families can exploit their patronage network to diversify their income sources. They do not have to depend solely on agriculture or SF; and have access to varied non-farm businesses which enable them to wait till the trees are matured and commercially harvested from SF plantations (Saxena, 1990, pp.7-8). Third, the high caste groups, through the patronage network, can form their own respective support-base of clients. This in turn enables them to dominate local level institutions. Baral (1993) and Sharma (1991), among many others, have noted that local "participatory" institutions concerned with SF in Nepal represented mainly the dominant ethnic and caste groups, such as *Brahmin* and *Chetri*. As the case studies revealed, biased leadership in local organizations, and high class or caste domination in the decisions and benefits of development projects (including SF) have been widely observed in many regions of south Asia.

Besides caste or status-hierarchy, fictive relations also help the dominant rural elites to consolidate their power and to sabotage the benefits of rural development programmes. These informal relations enable them to gain greater loyalty and allegiance from clients, to command increased number of clients, and to enjoy higher social status (Mashreque and Amin, 1993; Sarkar, 1980).

SF's performance, especially its effectiveness in achieving the goals of benefiting the disadvantaged sections of local communities and promoting social equity, is also constrained by the widespread social inequality. As the case-studies revealed, SF projects often failed to ensure equitable share of benefits to the weaker population and to involve them equitably in project decisions and management. This failure is, to a significant extent, attributable to the unequal social frame of rural south Asia (the working-ground of SF) which is characterized by severe inequalities in income distribution and asset ownership; highly unequal access to resources; inequality in social structures and institutions; and severe competition among unequal contenders for scarce resources. These dimensions of social inequality and their impacts on development programmes have long been matters of concern for major analysts on Asia. Myrdal (1971, Chapters 7 and 10), for example, in his classic work on Asia has observed that "the vicious circle of inequality [which] is bolstered by the caste system, the ethnic discrimination, nepotism and the general set of social and religious taboos..., -- impede the rise of social, regional, and occupational mobility and perpetuate the persistence of segmentation in social and economic life" (Myrdal, 1971, p.89). In this context of pervasive inequality, what Myrdal (1971, pp.89 and 121-122) considers as "the major obstacle" to development and economic growth, "the actual development usually goes towards increased inequality"; and the development efforts in the Asian countries "are ruled by compromises and accommodations within and between the upper class and the various groups that constitute the bulk of the upper class". Even after 25 years, his observation neatly applies to many south Asian countries, where the current development programmes function amidst inherently unequal social relations and structures, and consequently, are being sabotaged by the rich and powerful. A large number of studies, concerned with the process and consequences of rural development efforts in general, and SF in particular, testify to this (see Barraclough and Ghimire, 1990; Chambers et al., 1989; Fernandes et al., 1988; Brokensha and Castro, 1984; Anderson and Huber, 1988; Bruce, 1989; Uphoff, 1982; Blair, 1978; 1982; 1985). Barraclough and Ghimire (1990, pp.20-21), for example, note that within the existing social relations of the Third World societies, "market forces" help to "increase the wealth and power of the strong". Given the widespread social inequality, as Chapter 2 and the preceding case-studies have suggested, the richer farmers have usually benefitted from SF in line with the general experience of other rural development efforts (for example, the "green revolution" programmes. It was widely noted that the rich farmers were the major beneficiaries of "green revolution" around the Third World. For an extensive and widely referenced analysis of "green revolution" see Hazell and Ramasamy, 1991). Brokensha and Castro (1984, pp.23-26), in their survey of natural resource management, argue that there is a growing process "institutionalizing inequality" of opportunity, whereby benefits of development "remain frozen at the upper and at times middle layers" of societies in many parts of the world. The regional case studies of south Asia in Uphoff (1982) re-affirmed that the skewed pattern of benefit distribution of the government sponsored development efforts had further strengthened the richer sections and widen the social inequalities.

The process of social inequality is also fortified by other dominant features of social structure, such as (the above) patronage and caste stratification. In rural south Asia, the combined impact of these mutually reinforcing structures is manifested by the mammoth inequality of asset ownership and benefit distribution, the elite dominance in the development programmes, and the consolidation of alliance among commercial and political interests. The International Labour Office (ILO, 1977, pp.19-20) showed that in many Asian countries, including Bangladesh, India, Indonesia, Malaysia, Pakistan, Sri Lanka, Philippines, the richest 20% of the households receive about half the income; whereas the poorest 40% receive between 12% and 18% of total income. The bottom 20% receives only between 3.8% to 7%. In this context of inequalities, "benefits of technological and other innovations" do not "trickle down", rather "serve the elites" (Brokensha and Castro, 1984, pp.24-25). Besides, Briscoe (1979, p.2) has noted skewed pattern of ownership of trees and related assets. In Bangladesh, he observed that 16% of the families owned 80% of trees, 55% of cropped land and 45% of cattle. This also meant that assets, which produce fuel-energy, were controlled by an elitist section of the society. Indeed, as Fernandes et al. (1988, pp.281-284) warned, the failure to take into consideration these inequalities could lead (and in some cases has led) SF initiatives towards the same fate as that of the "community development" programmes of the 1960s and 1970s, which often resulted in greater inequalities and strengthening of the already powerful sections.

Thus, the complicated web of societal relations and structures in south Asian villages, which cuts across social and economic groups, largely shapes the impact and performance of development programmes such as SF. The reality is that since these relations are largely unequal and exploitative, the benefits of development interventions tilt towards the upper echelons of communities who control (or are in alliance with those who control) the resources and power. The weaker sections are largely precluded from any major benefit-sharing arrangement, and have to remain happy with the meagre benefits which they struggle to manage through their respective patronage links and other connections.

The Role of International Donor Agencies

The south Asian SF projects are predominantly supported (both financially and technically) by international agencies. In fact, the very growth of the ideas of "participation" and "community involvement in forestry" can, to a large extent, be traced back to international concern and propaganda. Some analysts have gone as far as viewing the concepts of SF, empowerment and participation as "western impositions" on the Third World (see for example Fox and Fisher, 1990, pp.116-117). In this context, it is no surprise that SF activities in south Asia are under profound influence of the donor quarters.

In the last two decades, there has been a cogent exhortation for participation and community involvement in forestry by the international aid agencies. The underlying argument was that local communities must assume a central role in order to face the crises in the Third World forestry sector. The Brandt Commission (Brandt, 1980), for example, suggested that the focus of development should not to be on "machines" or "institutions", but on the "people". FAO (1978) also called for people's "wider involvement in forestry management"; and the World Bank has oriented its policies and shifted its emphasis towards "rural forestry" and "forestry management with greater involvement of rural people" (Spears, 1978). In the same vein, in 1986-87, FAO (cited in Oakley et al., 1991, p.41) further noted that the central issue of forestry development was how to activate the involvement of local people in forestry activities. The World Resources Institute's (WRI, 1987, p.27) comment summarizes the international concern for participation in forestry development:

> It is widely recognized that governments and development assistance agencies alone cannot solve the deforestation crisis. Success will depend...on...the active participation of the millions of small farmer and landless people who daily use forests and trees to meet their needs.

There are a number of major implications of the above international support and concern for participatory forestry development. One major implication is that the support provided opportunities for attracting foreign fund for SF by the national governments. Shiva et al. (1987, p.245) have argued that the "availability of foreign funds" has become the "starting point for social forestry". Similarly, in his review of forestry development in the Asia-Pacific, Rao (1989) commented that during the last decade SF has become "a multi-million dollar activity". In south Asia, between 1977 and 1986, some 60% of the World Bank lending for forestry were for SF and related projects (Bhattarai, 1990, p.66). In India alone, over the past 15 years the World Bank has invested US$ 500 million, mainly on SF projects (World Bank, 1994, p.294). Thus, the prospective external funds, as I will eventually explore in more details (in Chapters 6 and 7), contribute to the "special interest" of the south Asian governments in SF. Some recent analysts (for example, Sharma, 1991) have started to argue that the case for the rise of SF and its priority position in governments' agenda is "symptomatic of the problem of lobbying for foreign aid" (Sharma, 1991, p.121). In Nepal, for example, it was observed that the government "earns its foreign income not by exporting forest products but by lobbying for more aid in the name of its forest sector" (Sharma, 1991, p.121). For the poor, over-stretched and exhausted governments of south Asia, SF seems to offer a fertile ground for exploring foreign funds.

In the field offices, donors' pressure (channelled through the central line authorities) contributes to the tendency of chasing physical and quantitative targets of SF. As the case studies revealed, local foresters' concern for SF primarily centres around the achievement of physical targets which is the dominant criteria for evaluating SF's success. The insistence on physical targets has, in turn, a number of implications. First, being absorbed in the chase of quantitative physical goals and "staying on schedule", field officers often pay less or no attention to the more important goals of SF, such as participation and equity. Poffenberger (1990b, p.9) notes:

> Since forest programs are generally assessed in terms of the achievement of physical targets, participatory goals often take second priority. Involving communities in decision making and management planning takes time. To stay on schedule, it is easier for the forester to take the lead in making plans for the activity including what species to plant, where, and when....[Forest] departments have often held onto all decision-making responsibility and management.

In fact, Colchester (1994, p.90) has gone as far as to argue that notwithstanding their support for "participatory" development, "historically, the development agencies have fought shy of...bottom-up approach". Second, for the purpose of central monitoring of financial aid, a cumbersome load of paper work (concerning routinized bureaucratic procedures of financial control and record maintenance) is devised for SF projects. Field officers also do not seem to mind the formalities, partially because of their "risk aversion behaviour" (Blair, 1978, pp.73-74). In the general context of rural development projects, Blair (1978, p.73) noted that public officials are "vitally concerned to minimize risk to his own career by pleasing [their] superiors". For example, in the distribution of agricultural credit, as Blair (1978) observed in Bangladesh, officers lent the money to the local rich and wealthier persons who had the capacity to repay. Thereby, the "wise bureaucrat" followed "a responsible policy in making the loans, i.e. lent money only to those with a clear capacity to repay them"; and "minimize[d] risk to his own career" by following a safe policy and clear paper work (Blair, 1978, p.73). In the case of SF, since field officers' performance is evaluated primarily by efficient achievement of the physical targets, they endeavour to produce results which are visible and quantitative. As the case-studies revealed, considering the relative "success" of farm forestry as compared to other components of SF (for example, communal woodlots), foresters showed more interest in farm forestry and minimized their own professional risk by helping the programme which was achieving its physical targets. Third, preoccupation with physical targets often led central planners and field officers to neglect regional conditions and contextual realities. Sharma (1991, p.120) noted a Nepalese SF project which "was an example of textbook management with no base reality". Similarly an Indian experience (Madhya Pradesh State Forestry Development Corporation (MPFDC), 1978, p.3, cited in Anderson and Huber, 1988, p.22) showed that

> the government's insistence on the fulfillment of specific physical targets...without any knowledge of local conditions led to expenditure of funds on projects of little utility....[H]ouses were built only to be washed away in the rainy season, basketry centres started where there are no bamboos, and bee keeping established where there are no flowers.

Thus, as Singh and Khare (1993, p.282) note, "donor's role" in SF is often one of "he who pays the piper calls the tune", and therefore, "people's processes" and "ground realities" are "getting only cursory attention".

Conclusion

This chapter examined SF's performance in light of its selected stated goals. The review of the popular cases of south Asian SF suggests that the role of SF, especially as a participatory development practice, has been insignificant. Though there have been some progress in achieving the physical and quantitative targets, the crucial participatory goals (such as participation, empowerment, equity) have largely remained unrealized. Therefore, SF's performance in compensating and redressing the process of public alienation (by ensuring participation of local people in forestry activities and equitable distribution of benefits arising thereof) has also been insignificant.

The chapter has argued that the performance of SF is shaped by particular political-economic variables of the surrounding "frame" within which SF projects function. It has elaborated some of such crucial variables. The analyses of these variables suggest that the surrounding setting (or the working ground of SF) is not generally conducive to achieving SF's lofty goals.

Public SF programmes have largely been a bureaucratic version of community participation where people have virtually no significant bearing on policy decisions, and are invited to join and perform strictly within the limits set by centralized authorities. SF, in this sense, seems to be yet another technical land-use and afforestation strategy, where "people" are, at best, a new "input".

As regards the structure of the book, this chapter provides the conceptual basis for the subsequent analysis of the empirical case-studies in Chapters 6 and 7. On the basis of the contextual variables identified and analyzed here, Chapter 6 will examine the working-ground of the case-study projects in Bangladesh. Then, Chapter 7 will further elaborate the performance of the projects in light of the general experience acquired in this chapter. As a preparation for the empirical investigation and field work, the next chapter sets the research agenda and methodology of the study.

4 Selection of the Bangladesh Case Studies and the Research Methodology

Introduction

The past three chapters reviewed the conceptual-theoretical framework and rehearsed the core arguments of the study. They attempted to show, from comparative evidence, the transformation of forestry towards commercialization, and the consequent alienation of local communities. The stated goals and performance of SF (as a strategy for redressing alienation and involving local people in forest management) were also examined. The rest of the book will examine these trends and issues in selected localities of Bangladesh. This chapter, therefore, states the general research methodology for the field-work and the criteria used in selecting the case-study areas for detailed empirical investigation.

The methodological aspects of research set the stage for acting on theories and carrying out empirical investigation. Methodology is defined as a "system of explicit rules and procedures on which research is based and against which claims for knowledge are evaluated" (Nachmias and Nachmias, 1992, p.14). It serves some vital purposes of research. First, it facilitates communication between concerned parties (mainly, researcher and respondents) in a particular research. Second, it serves as a basis for logical reasoning. Third, it helps to verify empirical findings of other studies (and methods used in them), and to facilitate replication. Methodology is, therefore, at the core of any scientific research; simply because "what sets the scientific approach apart from other modes of acquiring knowledge [is] its methodology" (Nachmias and Nachmias, 1992, p.14).

There is no ideal or universal methodological prescription for research. The suitability of research techniques and approach is "essentially relative" (Isokun, 1985, p.81) and is influenced by the particular research topic, its demands and contextual setting; level of competence of the researcher; and structural and logistic limitations of the research. There is no method which is universally acceptable or absolutely flawless. One logical way to minimize the weaknesses of a particular method is to administer a combination of methods. This has been popularly known as "methodological pluralism" (see Ahmed, 1991, pp.166-167) -- combining various qualitative and quantitative methods to complement each other. In this study I have relied on "methodological pluralism" for two reasons. First, it provides more flexibility to researcher by offering choice of alternative techniques and useful combination of methods to obtain optimal results. Second, it reduces the adverse affect of any particular method.

This chapter introduces the case-study projects in Bangladesh, together with an analysis of the criteria for their selection. It also describes the methodological considerations and techniques concerning this study, including the ways of data collection and sources of information.

The Cases and the Criteria for Their Selection

To examine the major research themes of this book (for example, forest commercialization, public alienation, frame and performance of SF; see Chapter 1), I attempted to identify suitable empirical cases, which more or less mirrored the characteristic processes of forest use, as reviewed in the preceding chapters. More specifically, I was looking for the following characteristics in the probable field-cases. First, the cases should broadly reflect the major trends in the evolution of south Asian forestry (as conceptualized in Chapter 2). Second, they should also contain the typical "interests" in forest use, as identified in Chapter 2. Third, since I was primarily concerned with public initiatives in afforestation, the cases should fall within the scope of governmental policy and programmatic interventions. Fourth, the case-study sites should more or less represent the typical socio-economic and geo-physical characteristics of the country (especially in terms of population growth rate, size and concentration; literacy rate; land holding pattern; exposure to state services like health, education and social services; and climatic and geological factors). Besides, the management and administrative set-up of government offices (especially of those associated with forest management, namely, the forest, police, and land revenue departments) should be similar to the country's general pattern.

In light of the above criteria and considerations, I purposely selected three public SF projects from two geographical localities in Bangladesh for conducting the field-work (The Betagi and Pomora SF Projects at Rangunia *Thana* ("sub-district" -- previously known as *upazila*) in the District of Chittagong, and the Chandra Agroforestry Research and Demonstration

Project in Kaliakair *Thana* of Gazipur District). The duration of the fieldwork was little over 11 months -- between September 1993 and August 1994.

I had a number of interests and advantages in these particular cases. These projects enjoyed great deal of attention from national and international quarters. The publicity and popularity of the projects made me curious and inquisitive. The Betagi and Pomora projects, for example, are reported to be the "first systematic efforts" in SF in Bangladesh (Rahman, 1982, p.6) and are widely acclaimed "success stories" by a sizeable literature (see Ahmed and Azad, 1987; Alim, 1984; 1988; Rahman, 1987; 1988; 1991; Bhuiyan, 1982; Bently et al., 1987; Fortmann and Bruce, 1988). The following quotes manifest the extent of reverence the Betagi-Pomora projects enjoyed, especially from the donor quarters:

> So many reforestation have been made, so much money spent, finally for nothing, all being destroyed. It is time to change....It [the Betagi SF project] is the solution. (Dumont, 1984, cited in Ahmed and Azad, 1987, p.45)

> The project has born fruit and has become a grand success. This is now a model for multiplication in other areas. (Remarks of A.M. Malik, FAO consultant, dt.8.9.85; Source: Inspection Book, *Bhumiheen Samity*, Betagi)

> I am inspired! The use of denuded land in this manner can solve many problems faced by the people of the world....Wherever I go, I will take with me much of the wisdom of Betagi. (Remarks of the Agroforestry Expert, Winrock International, Washington, USA, dt.30.1.87; Source: Ibid)

> This effort is most inspiring! (Remarks of the Consultant, UNDP/FAO project: BGD/29/077, dt.27.1.86; Source: Ibid)

> I was particularly happy to see this excellent work with previously landless poor people and previously unproductive denuded land. I hope the lessons learned will permit rapid spread to other parts of the country. (Remarks of Charles Bailey, Ford Foundation, Dhaka, dt.17.1.86; Source: Ibid)

The third project -- Chandra -- is relatively new and have not yet received widespread publicity. However, the official records were full of impressive statistics of its "successful achievement of targets" (Office records of Chandra Beat Office). Besides, selection of Chandra helped me to make comparative analyses with Betagi and Pomora, which were different in terms of geographical location, year of initiation, and the extent of popularity. Consideration of these diversities and regional peculiarities, I hoped, would help me in making a realistic assessment of SF.

These projects constitute a major part of the mainstream public SF initiatives in the country.

From methodological perspective too, these cases offered some advantages. First, a number of surveys on the Betagi and Pomora projects were already available. These studies (for example Rahman, 1987; 1991; 1992; Task Force, 1987; Miah, 1989; Hoque, n.d.) offered limited "base-line" information for my study-purposes and facilitated longitudinal comparisons of selected issues (for example, rise in living standard and income, frequency of state support-services to farmers). Second, the number of participating farmers in these projects were reasonable enough to enable me to conduct the research even within the time and resource limitations. Third, since studies were already conducted in the Betagi-Pomora locality, farmers had some idea about the culture of research. I expected that this would help in administering the data-collection methods, such as interviews, questionnaires and group discussions. However, this did not turn out to be an unmixed blessing. On some occasions, I found in the course of field-work that farmers' familiarity with surveys prompted them to give biased or artificial responses. My vigilance and long-acquaintance with the sites helped me to identify and avoid recording of such artificial responses. In Chandra, little systematic investigations were done before and it offered a relatively unexplored ground for research.

Two of the cases (Betagi and Pomora) are located in Chittagong, where the major public forestry training and research institutions of Bangladesh are also situated. The Bangladesh Forest Research Institute (the country's only specialized forest research organization), Institute of Forestry, Bangladesh Forest Academy (the only training institution for officials of the Bangladesh Civil Service: Forests) are placed in Chittagong. Any research study on Bangladesh forestry almost inevitably needs to consult these institutions. Selecting cases from Chittagong enabled me to keep in regular touch with these institutions.

Besides, I had some personal advantages of working in these selected sites, especially in Chittagong. First, the sites were known to me for long and I knew the local dialects fairly well. Second, I had personal contacts with and access to the above forestry institutions and offices. Third, I had family-ties with the key-forestry officials concerned with the selected projects. This enabled me to gain access to the otherwise "invincible" bureaucracy. For my study, the support of government offices was vital. The extensive use of historical and departmental records in this book, and the long informal discussions with senior government officials -- are the two clear results of this support. Fourth, my acquaintance with the sites (and the people around) also enabled me to arrange some logistic supports (such as transportation and accommodation) on personal initiative. Given the major resource constraints of the study, this was a strong consideration.

Thus, combining the methodological and personal advantages, I felt that the field-work sites of Betagi, Pomora and Chandra could best serve the study-purposes.

Chandra

The locality, where the Chandra project (Agroforestry Research and Demonstration Project) is located, forms a significant part of the central *Sal* (*Shorea robusta*) forest tract of Madhupur Garh and lies in Kaliakair *Thana* of Gazipur District (see Figure 4.1). It is under the purview of Chandra Beat of Kaliakair Range in the Dhaka Forest Division. A brief description of the general ecology and geography of Chandra *Sal* forest is reproduced here from Anon. (1994, p.8):

> It lies at 24 [degrees] North-90 [degrees] East....The forest areas are high lands varies from 4 to 10m [metre] from the surrounding plain lands. The mean annual rainfall is 2500mm most of which falling between June to October. The mean annual temperature is about 77F [degrees Fahrenheit] and the mean monthly relative humidity is 76%...;soils are pale brown to yellow red in colour...pH is highly acidic.

The central *Sal* forest tract, which includes Madhupur Garh, lies in the greater Districts of Dhaka, Mymensingh and Tangail. It spreads between the Brahmaputra and Jamuna rivers, with a length of about 96 kilo metre (km) and a width of about 8km to 24km, running from north to south (Anon., 1994, p.1). The Chandra *Sal* forest, which constitutes a major section of the Garh, is about 12,138.45 acres of forest land (Anon., 1994, p.1). The area is the home for a sizeable tribal population, mainly consisting of *Garo* and *Kooch* (*Barman*). It is estimated that in Madhupur Garh there are 21,500 *Garo* and 2,000 *Kooch* inhabitants; and they mainly practice *Jhum* (shifting) cultivation (Task Force, 1987, p.49). *Barmans*, the successors to *Kooch* tribe, have been living in Chandra for generations (Ahmed, 1994, p.7).

The historical background of the area will be discussed in Chapter 5. Here, I will briefly touch upon the situation which led to the emergence of the concerned SF project in Chandra. The control of the area passed from the private *Zamindars* (land-lords) to the state forest department during the early 1950s. In the following years, the area witnessed progressive commercialization of forests and the resultant alienation of local people.

Consequently, there was, as the Task Force (1987, p.50) noted, an "evolution from a subsistence economy to a market-oriented one". The commercial forces gained predominance; and illegal logging became "the single most dominant factor" contributing to rapid deforestation in the locality (Khaleque, 1984). "A substantial portion of these forests [ended up] without any vegetative cover" (Agroforestry Research and Demonstration Project (ARDP), n.d.(a), p.3). The relationship between the state and local people became embittered; and the government's attempts in reforestation failed because of intense public hostility. The area, though officially under the purview of the forest department, was actually encroached by local people under patronage from a powerful group of local elites.

In context of the above, the government and donor quarters (ARDP, n.d.(a), p.4) came to the realization that:

> It is not feasible to evict the people or exclude them from any tree production programme. Any tree production activity on such land should involve the present occupants and a mechanism for maximizing production from these lands through a participatory approach needs to be evolved.

The realization led to this experimental SF project, involving 300 acres of land in five sites (Chandra, Kachighata, Kalidas, Sagardighi and Dokhola) in the Districts of Gazipur, Tangail and Mymensingh (under the jurisdiction of Dhaka and Tangail Forest Divisions) (ARDP, n.d). The project was fuelled by a Ford Foundation grant of Taka 78.39 lakhs or US$ 245,000 (ARDP, n.d.(a), p.2). It became operational in May 1992. Up to 1994, the project covered 240 acres of (mostly) encroached state forest land, involving 130 households (with a total population of 623) (Ahmed, 1994, p.5). In the Chandra site, the project spreads over 60 acres of land, involving 40 households, with a total population of 160. Each household was allotted 1.5 acres of land (Ahmed, 1994, p.6).

The land-use pattern of an average farm at Chandra, as I observed, roughly comprises of 17% (of the land area being used) for homestead, 30% for forest plantation, 53% for agricultural crops (including a few horticultural trees). Plantation is mainly designed and supervised by the local forest office.

In the field, the project is administered by the Chandra Forest Beat Office of Kaliakair Range. Unlike Betagi and Pomora, farmers' own cooperative association to run the project affairs does not exist in Chandra. However, there are some small informal groups among farmers, organized by a voluntary organization (called, Proshika). These groups are not particularly concerned with SF; rather they are formed as a pre-requisite for the credit-services offered by Proshika.

Figure 4.1 Location map of the Chandra project

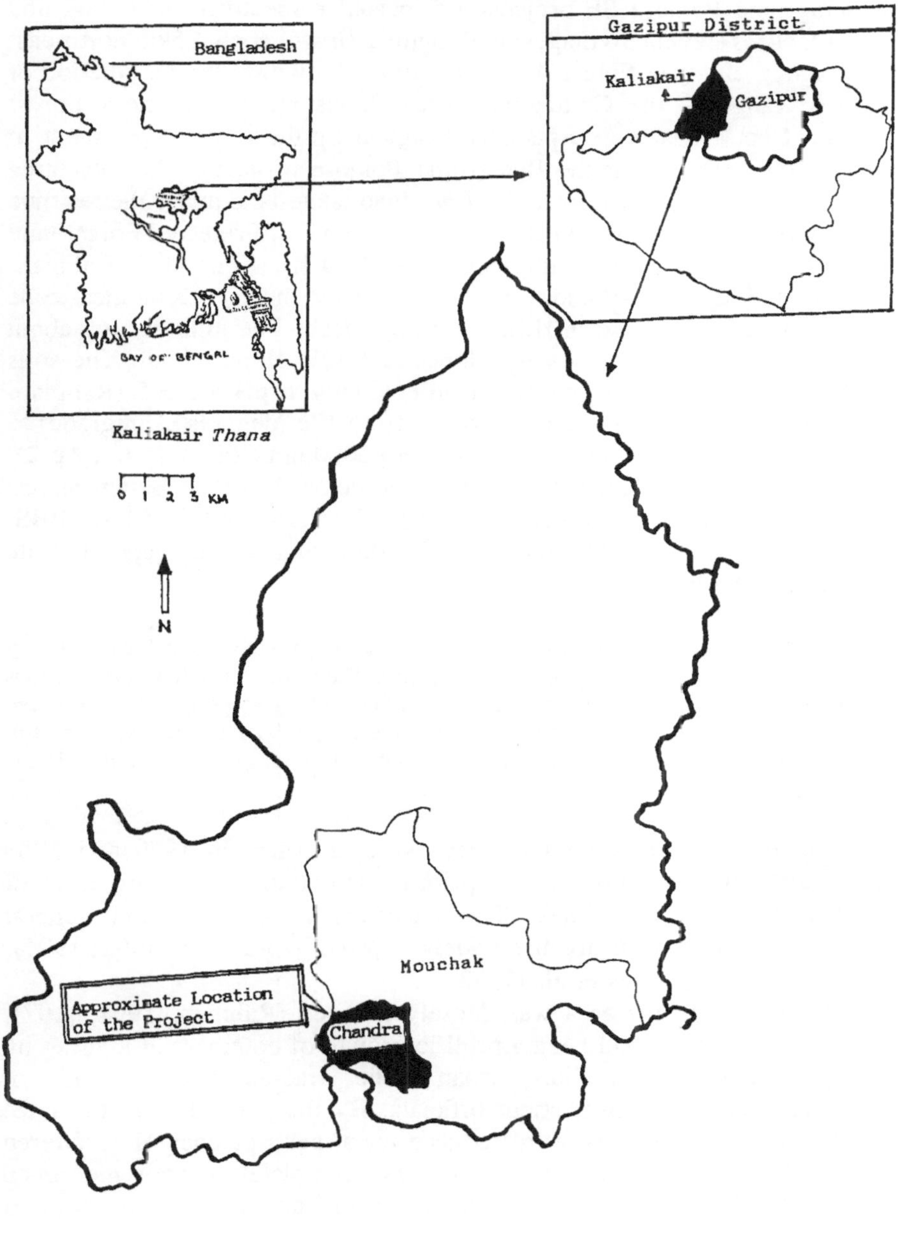

Betagi and Pomora

The Betagi and Pomora SF projects are located respectively in Betagi and Pomora *Mouzas* (revenue-villages) in Rangunia *Thana*, about 25km north-east of Chittagong city (see Figure 4.2). The area falls within the jurisdiction of Rangunia Range of the Chittagong Forest Division. The Pomora Forest Extension Centre has been especially assigned by the forest department to supervise the activities of the Betagi and Pomora projects. The total area under the Betagi project is 190 ha. of *Khas* land (state-land under the revenue department). The Pomora project constitutes 276 ha. of Protected Forest land (under purview of the forest department) and 24 ha. *Khas* land (Rahman, 1992; 1994). The Betagi-Pomora locality belongs to the agro-ecological zone of the "Northern and Eastern Hills" of Bangladesh. The hills occupy about 50% steep slopes, 20% very steep slopes and 30% sloping land. The soils are yellowish brown to strong brown in colour with pH 4.5-5.5 (Rahman, 1992, p.9). The annual temperature varies from 13c (degrees Centigrade) to 33c; and rainfall ranges between 2500mm to 3000mm (Rahman, n.d., p.2). There are 1,610 households (total population being 9,195) in Betagi *mouza* and 3,185 households in Pomora, with a total population of 17,851 (BBS, 1989, p.36). Alim (1988, pp.151-152) summarizes the general "site condition" of the locality:

> The area selected [for the projects] was...land lying absolutely denuded and mostly hilly. The soil lost most of its organic matter. There were very few small patches of plain land in between the denuded earthen hills. Excepting a few scattered bushes there was no vegetation. On the whole the productive potential of the soil was at its rock bottom and the area was really a true sample of waste land in the truest sense of the term.

The Betagi and Pomora projects were launched in 1979 and 1980 respectively. The forest-use scenario prior to the initiation of the projects will be discussed in next chapter. The historical background and general description of the projects are also widely covered by Rahman (1992; 1991), Alim (1988), and Quddus et al. (1992).

Before 1950, this area was densely forested (Rahman, 1994, p.97). Since the 1950s, there had been a prolific growth of commercial logging by an alliance among local elites, urban timber traders, and a section of government and local government officials. By the early 1970s, the once dense forests of the locality were reduced to, at best, patches of "scattered bushes" (Rahman, 1992, p.10) and, at worst, completely barren wide open lands (Alim, 1988, p.147). The deforestation had profound ecological

Figure 4.2 Location map of the Betagi and Pomora projects

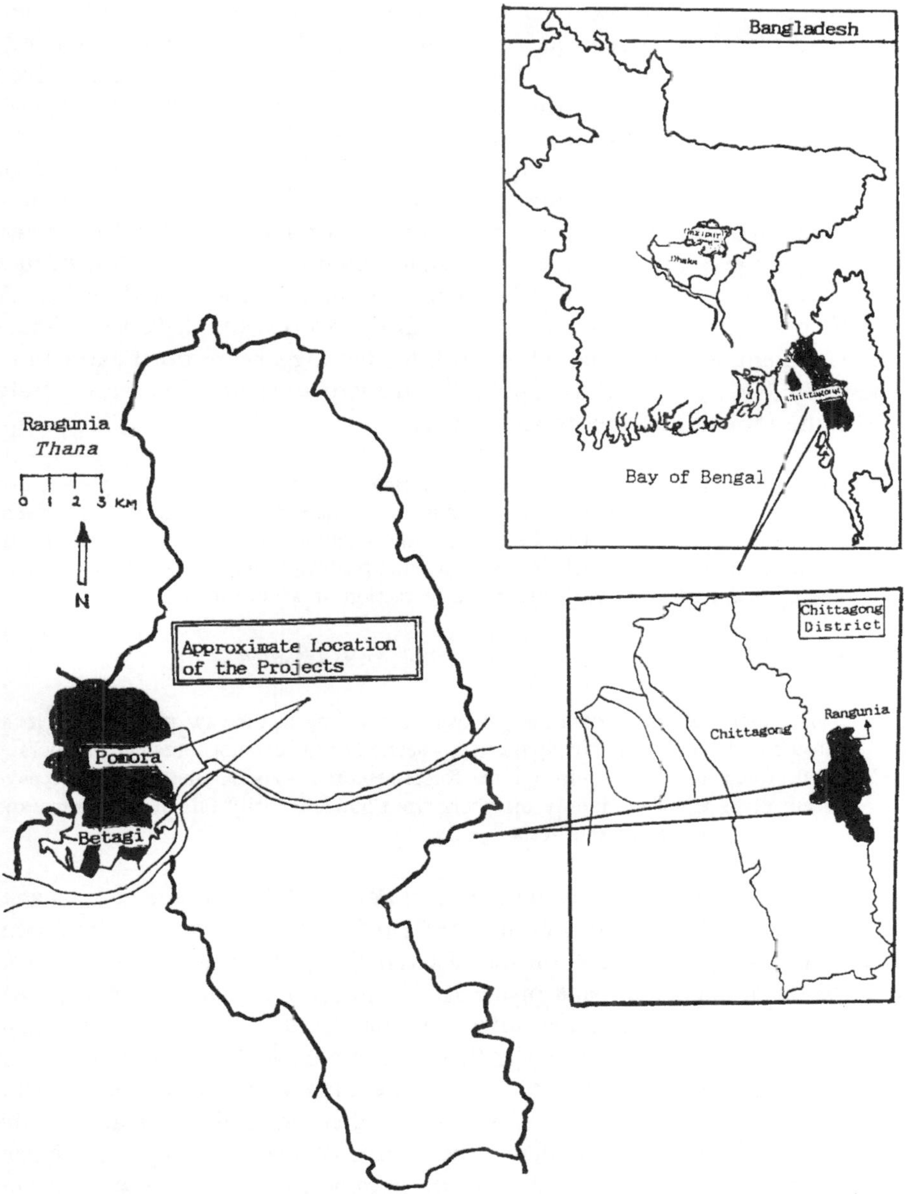

consequences as well. It caused massive soil erosion and degradation both in forest and agricultural lands, and contributed to the reduction of agricultural production in the locality (Bhuiyan, 1982). The state efforts in afforestation and checking the "predatory nature" of deforestation largely failed (Ahmed and Azad, 1987; also see Quddus et al., 1992; Rahman, 1992; Ahamed, 1993; Bhuiyan, 1982; Alim, 1988). "The inaction" of government offices (especially of the forest department), as noted by Ahamed (1993, p.18), was largely attributable to the fact that they "worked hand-in-glove with local influential traders greedy for quick money". With the predominance of commercial interests and the state appropriation of forests, people's traditional rights of and access to forests were curtailed. The local poor and landless people became hostile to the state initiatives; and there had been a sharp rise in the mutual distrust between government officers and people. A section of local poor, in the face of dire poverty and lack of suitable alternative employment, started to work for the loggers in forest extraction. One study (Task Force, 1987, pp.42-43), depicting the situation immediately before the initiation of the projects, notes:

> This area, densely forested within recent times, is now characterized by extensive barren hills....Although these lands are theoretically under government control...,in practice, they were under the unauthorized control of local elites. Landlessness and encroachment are endemic problems and plantation efforts were being thwarted by illegal cutting and extraction of wood for sale.

In the same vein, Ahmed and Azad (1987, p.9) observe:

> The older trees are stolen away mostly for selling at the saw mills....The local influential people, the greedy traders, a section of the corrupt forestry bureaucracy had vested interests in pilfering the forest resources. The landless people, to save themselves and their family members from hunger, easily fall in their traps and wipe out trees from the forests.

"The rampant felling of trees in the Betagi Pomora hills" continued unabated until the late 1970s (Quddus et al., 1992, p.5); and there was an imminent threat of "sweeping attack" on the last remaining patches of state forests, situated on the opposite bank of the Karnafuli river (Ahamed, 1993, p.18). Against this backdrop, some renowned intellectuals, senior government officials, and local philanthropists (notably, Professor Yunus -- Director of the Grameen Bank, Professor Alim -- the then Conservator of Forests, Chittagong Circle, late Mr. Chasi -- the then Special Secretary to the government, Mr. Nurul Quddus -- the then Chairman of Rangunia Central Cooperative Society) envisaged a SF programme for the region with an aim to "rehabilitate the denuded hills with productive trees and the landless with subsistence economy" (Ahmed and Azad, 1987, p.10). It should be noted that the above pioneers of the projects were persons of power and esteem in their respective career, and "without their intervention, the [projects] simply would not have been possible" (Task Force, 1987, p.48). In fact, as Ahamed (Ahamed, 1993, p.18) argued, like local elites, the field government offices were also not very enthusiastic about establishing such people-focussed projects, because "dispensation of land to private individuals was likely to diminish their own authority". But, the pioneers pulled together their power and resources, and were committed to overcome the initial opposition from local vested interests, including the government offices. Under their constant patronage and monitoring, lands were recovered from the de-facto control of local elites. Some landless families were selected from the adjoining villages and were rehabilitated in the SF projects. Each family (farming household) was allotted 4 acres (1.62 ha.) of land on annual renewal basis. In 1987, they were accorded permanent ownership of land. Thus, the pioneers ensured that farmers receive "sufficient cooperation from the government departments and agencies without which the projects could hardly be expected to take-off" (Ahmed and Azad, 1987, p.42).

Currently, 82 and 152 households are participating in Betagi and Pomora respectively. Farmers are expected to follow some conditions set by the government. For example, they are not to work outside the farm, not to sell the land, and to follow the forest department's suggestion (Rahman, 1994, p.98). The projects are managed at two corresponding levels. At the *Thana* level, there is a *Thana* Selection Committee, consisting of representatives from government departments and local government offices. The committee performs coordinating and advisory roles, including the selection of farmers, periodic monitoring of performance, and discussion of problems concerning the projects. At the village level, day-to-day activities of the projects are run by two farmers' cooperatives called the *Bhumiheen Samities* (the landless farmers' associations). The *Samities* arrange meetings to discuss farmers' problems, resolve conflicts among their members, and liaise with the *Thana* committee and other concerned agencies. The *Samity* leaders are elected by the farmers and have to work under the general guidance of the *Thana* committee.

Techniques and Tools of Data Collection

Development studies frequently seem to face a "methodological dilemma";

that is, how to effectively synthesize formal and informal methods, and quantitative and qualitative data. My study, for example, required, on one hand, quantitative data on forest use and SF activities in the context of the specific case-studies (on such variables as income and productivity); and, on the other hand, it needed to address more qualitative issues (such as the transformation of philosophies of forest management, values and attitudes of concerned farmers and state officials, and various social relations and structures which influence forest use). Many researchers share the above concern about how to judiciously balance diverse information and methods, and to suit them to the demands of a particular study (see White, 1992; Rahman, 1994; Ahmed, 1991). As a response to this concern, White (1992, p.10) administered a combination of formal and informal methods in her enquiry into class and gender in rural Bangladesh, and remarked:

> this [concern] is not I think coincidental, or something specific to me, but rather reflects the ambiguous position of development studies. This both prescribes the need for quantitative data (such as details of landholding) and values anthropological methods of a more participatory style.

Accordingly, in line with "methodological pluralism", I deployed a combination of interpretive qualitative techniques such as historical reviews, ethnohistories, observation, key-informant interviews; and quantitative techniques of sampling and questionnaire survey. Both sets of techniques were planned to complement one another.

Sampling and Questionnaire-interviews

Mainly for the purpose of generating quantitative data, survey method with questionnaire and samples was used. However, I was alive to the limitations of structured questionnaires. The common problems associated with questionnaire surveys are non-response, biased response, and artificial findings. They can also be "too demanding" on the researcher (To design and administer a good questionnaire may require an expert researcher) (Bell, 1987; Isokun, 1985; Casley and Lury, 1987). Nevertheless, with all limitations, it is still the widely used tool for easy and quick basic information collection. Bell (1987, p.58) argues:

> questionnaires are a good way of collecting...information quickly and relatively cheaply...as long as the researcher is sufficiently disciplined to abandon questions [and responses] that are superfluous to the main task.

Accordingly, I applied the method with some caution and used other supplementary techniques (such as observation and informal discussion with respondents) to minimize its weaknesses. Two sets of questionnaire (involving both structured and open-ended questions) were designed. The first set was for farmers; and the second, for forest officials. The questionnaire primarily served as a basic list of points and topics, which could then be elaborated by informal exchange of opinion with respondents. Occasionally, after dispensing questionnaires to a number of respondents, group discussions were also arranged. Such informal meetings and group discussions (as complementary tools to structured questionnaire) fostered a friendly atmosphere, provided flexibility of responses, and offered opportunities for the rise and sharing of many interesting issues between the researcher and respondents. Admittedly, the questionnaire for farmers was quite long, and given the wide scope of the study, it could not be helped. However, considering the possibility of respondents' fatigue, I used to divide the discussion into three to four sessions. The time and venues for such meetings were decided by farmers at their convenience. Usually, the introductory meeting was held in the respondent's homestead, and the subsequent sessions used to take place in public-meeting places, such as tea-stalls and *haats* (rural markets).

For a study with time and resource constraints, sample method seemed to be a practical way of quick, economic, (nevertheless, in-depth) analysis. Casley and Lury (1987, p.42) maintain that it

> provides a flexible method that can be adapted to almost every requirement of data collection....[M]any topics particularly those involving detailed transactions of individuals or households require an intensity of interview or observation that cannot conceivably be covered by censuses. A sample is the only mode of inquiry available.

I used "stratified sampling" method. Stratified sampling is argued to be "the almost universal choice for major surveys" (Casley and Lury, 1987, p.57). There were two "sampling groups" in my study: farmers of the case-study projects and forest officials (the "first-class" officers of the forest department).

As regards the first group, I -- with the help of two assistants -- conducted three short pilot surveys in Betagi, Pomora and Chandra (I will discuss about the assistants later). Initially, we stayed 21 days in Betagi-Pomora sites and 13 days in Chandra. The surveys served two major purposes. First, they helped us to rejuvenate our acquaintance and relationship with the localities. Second, a "sample frame", concentrating mainly on the variables of income and population, were prepared for each of the three projects. The frame, in turn, served as the basis for selecting the final samples.

For the pilot survey, farmers' annual income seemed to be the reasonable and practical variable to probe into. Other variables -- for example, land-holding, literacy, occupation -- were not suitable. Because, these variables were uniform and showed virtually no diversity. For example, each household was allotted equal size of land (in Betagi and Pomora, it was 4 acres, and in chandra, 1.4 acres). Farmers were mostly illiterate and were involved in two main occupations -- agriculture (farming) and manual-labour.

138 and 77 households, respectively, in the Pomora and Betagi projects were surveyed to collect data on yearly income; and the "mean" and "standard deviation" were calculated. In Chandra, I followed a slightly different course of action regarding the pilot enquiry. Given the limited population, all the 40 households were covered by a quick census, concentrating on annual income. Shortly afterwards, a sample-survey was conducted covering 10 households. These pilot surveys, besides serving statistical purposes, offered me a general picture about the life and living of the participating farmers in the SF project.

The final samples for questionnaire-survey were selected on the basis of the pilot survey findings, following "stratified random sampling" technique. Depending on the data on income distribution, 5-6 strata were formed and the sample size in each stratum was calculated. The calculations were based on Chacko's (1965) formula:

$n = (ts/d)^2$
where, $d = P \times Y/ 100$
$t = 1.96$ (2)
s = sample standard deviation from the pilot survey
Y = sample mean from the pilot survey
P = margin of error (at 15%)

Accordingly, the sample sizes for Betagi, Pomora and Chandra were calculated to be 24, 40 and 13 respectively; and the samples were randomly drawn (The calculations are shown in the appendix). It may be remembered that in Betagi, Pomora and Chandra projects, respectively, 82, 152 and 40 families (households) are participating.

Once the sample sizes were determined, the questionnaire was pre-tested on 2 households from each of the three projects. "Household" was the focal unit for administering the questionnaire. Regarding the definition of household, I followed Ahamed (1993, p.17), who (in the context of Betagi and Pomora projects) maintain that:

> A household is typically built around a nuclear family, in keeping with the predominant pattern in the project areas. The household members have a joint budget, and by definition, a single unit for production and consumption, each acting as a decision making unit. Normally the oldest male, the father or the eldest brother acts as head of household.

I expected to observe the strategies which household members adopt towards forest use (as a part of their survival strategies), how do households interact with and respond to social relational network, and how their activities are affected by central programmatic interventions including SF. The pre-eminence of household is mainly due to the fact that in the Third World, it is the "most common unit of production as well as consumption" (Casley and Lury, 1987, p.161). The United Nations Statistical Commission (1979, cited in Casley and Lury, 1987, p.173) argues that

> there is universal recognition of the key role of the household sector in the socio-economic development of developing countries. Households account for much of the productive activity and are themselves affected by economic and social changes.

Household appeared to be a suitable unit of data collection and analysis. However, though I concentrated on household for statistical purposes, it did not delimit the focus of the study. Jansen (1986, p.4) also focussed on household "for the purpose of collecting quantitative information", but at the same time, dealt with the "whole range of relationships" beyond the geographic limit of households. Household members have wide social and economic linkages to pursue their survival strategies of production, consumption, use and access to resources. Focussing on these linkages, I expected to understand the surrounding "frame".

We (my assistants and I) interviewed the head of household, usually a male member (father or eldest son), who had two major characteristics. First, he was the (solo or one of the) main earning member(s); and second, he exercised substantial authority over family decisions. This, however, caused a problem. Formally focussing on the male head, meant an exclusion of women. To compensate the limitation, we made it a point to discuss with at least one female member of the household. Generally, I was allowed to talk only with aged women due to the *purdah* rules (see Chapter 7). My female assistant, however, did not have any restriction. We used to meet the selected woman a number of times (Generally an introductory session and 3 to 5 follow-up meetings). Besides, when we were interviewing the male head, women often attended from the background. They listened, and occasionally participated in discussions from inside the house or from behind a purpose-built screen (*dhal*).

The two assistants (one male and one female) were in their 20s, fresh Sociology graduates from the University of Chittagong (Bangladesh), and possessed basic knowledge of research methodology. They knew the localities and local dialects. Returning from villages in the late afternoon, we used to meet together in the *veranda* of the house we resided in (during the field-work). We shared our experiences over supper and cross-checked the results. Any observation made during the day was immediately recorded (usually in a diary; sometimes in a recorder, if we were too tired to write!). The assistants were also briefed about the next day's assignment, especially if there was any particular issue which I wanted them to observe and make note of.

The samples of the second group (forest officials) were selected purposely. More "sophisticated" statistical methods (for example, random sampling) could not be used because of some reasons. First, forest offices spread all over the country; and (given the time and fund constraints) it would have been simply impossible for me to contact the remotely-located officers (if required). Second, many officers showed reservation about sitting for an interview. Third, forest officers maintain busy schedules and frequently remain out of offices on field-tours. It was really difficult to contact them on time. Fourth, it required tremendous persuasion and patience to arrange meetings with officers, particularly the senior officials, amidst the plethora of bureaucratic formalities.

However, my family ties with many senior officials helped me to overcome some of the above problems. Although the samples were selected purposely, officers were chosen from all levels of the administrative hierarchy to make the selection reasonably representative: 1 Chief Conservator of Forests (CCF), 1 Deputy CCF (DCCF), 4 Conservators of Forests (CF), 10 Deputy CF (DCF) or Divisional Forest Officers (DFO), 14 Assistant CF (ACF) or Sub-divisional Forest Officers (SDFO); a total of 30. (The organizational chart of the forest department can be seen in the appendix). Currently, there are 122 "first class officers" in the forest department (Calculated from the Draft Gradation List, vide CCF(E) 2E 239/115-3). Besides, the questionnaire I used was basically a kind of "check-list" of topics (with mostly open-ended questions) and could elicit varied responses and led to the spontaneous rise of many issues (White, 1992, pp.10-11 also found a "flexible check-list of items" to be more effective than a rigidly structured questionnaire). My personal acquaintance with respondents, coupled with this informal approach, helped to relax the stiff, formal atmosphere, commonly found in higher government offices. Once I managed to get an appointment for interview, I found the officers very cooperative and knowledgeable. It took about 4 months to interview the 30 officers.

My intimacy with government officers and staffs was not, however, an unmixed blessing. This discussion will be incomplete if I do not briefly touch on the "other side of the coin". During the initial days of field-work, some farmers and villagers, revealing my connections with the foresters (and other government officers), felt scared and became skeptical about my intentions. I often faced a situation where there was an "eagerness of the respondent to please the interviewer" (Borg, 1981, p.73). I immediately stopped moving with the officials publicly and started to spend maximum time with farmers in their homesteads, farms, and market-places. I also shifted from the widely-known government *bungalow* (rest-house) to an obscure house; and tried to stress my identity of an academic over that of a "son of *Boro Sahib*" (high ranking official). All these measures, coupled with my knowledge of the localities and dialects, helped me to minimize (if not, fully overcome) the initial problems. Similarly, while interviewing the foresters, I had to be constantly cautious about emotional and bias responses. Sometimes, my family-links put me in an uncomfortable situation while raising sensitive issues (for example, bribery and corruption) with the officers. At times, I faced a paradoxical situation in trying to strike a balance between my role as an objective researcher and my image of a "son of a beloved colleague" in the eyes of many senior officials. I am sure that the officers must have felt the same. It could not have been easy on their part to share matters with a researcher "who grew up just in front of their eyes". As Selltiz et al. (1962, p.583) point out, "interviewers are human beings and not machines"; and admittedly, all these problems could not be fully overcome. However, I kept in mind Gavron's (1966, p.159) suggestion: "it is difficult to see how [the problems and biases] can be avoided completely, but awareness of the problem plus self control can help". My awareness of the situation, openness and generosity of the officials, "honestly about the purpose of the exercise", and "integrity in the conduct" (Bell, 1987, p.76), helped to check the problems to a reasonable extent.

Observation

Observation is claimed to be "basic to all scientific enquiries" (Isokun, 1985, p.99; also see Nachmias and Nachmias, 1992). This method is "particularly appropriate when high degree of analytical content is required" (Casley and Lury, 1987, p.64). Bell (1987, p.88) also notes:

> it is a technique that can reveal characteristics of groups or individuals which would have been impossible to discover by other means.... [Formal interviews can] only reveal how people perceive what happens, not what actually happens. Direct observation may be more reliable than what people say in many instances.

Owing to time constraint, "participant observation" in its classic sense, did not seem practicable for my study. Instead, I went for a kind of uncontrolled simple observation to watch and analyze the political-economic relations and structures in a particular site. It enabled me to see the behavior of social actors and processes as they occur in a natural setting. For my study-purpose, observation was a method of data collection "in which the researcher and his collaborators record information as they witness events during the study period" (Lin, 1976, p.205). However, it was more than just "witnessing" or "seeing". It involved "the ability of the observer to read between lines for what could be the latent as opposed to the manifest explanation or meanings of the social activity" (Isokun, 1985, pp.99-100).

I was alive to the common practice of "superficial trips to project areas by the urban professionals and their easily reached environs" (Cernea, 1991, p.514). These cursory research attempts are termed by Chambers (1991, p.518) as "quick-and-dirty appraisal in the form of rural development tourism". Therefore, even within the constraints, I endeavoured to make my observations more than just "snapshots" (Chambers, 1991, p.519). I lived intermittently for about 6 months in the three sites. We (the assistants and I) selected particular places in the localities for meeting the farmers (for example, particular homesteads, markets, local government office premises, selected tea-stalls and plantation-sites). Generally we used to walk, but also rode motor-cycles to move around. We participated in social festivals (especially in *Eid, Maharram*, *Puja* festivals and in a number of wedding ceremonies), and attended farmers' *Samity* meetings. By the end of first month of our residence, we managed to build up a good rapport with local people. The knowledge of local dialect and culture helped in the use of observation method. We used to maintain a field-diary where virtually any observed phenomenon of interest was noted (Wood, 1994, p.451 also widely followed this practice in his extensive field-work on Bangladesh).

Key-informant Interviews, Ethnohistories, Life-histories

The use of key-informants helped in the understanding of complex situations.

For instance, key informants provided valuable information regarding the history of forest use in the localities, underground network of commercial logging, patronage links, and factional politics. Chambers (1991, p.519) noted that conventional studies often tend to "overlook these invisible crucial social facts, [especially] the social and cultural relationship". Key informant interviews drew my attention to some of "these invisible facts". However, caution had to be taken in selecting informants. They were selected only after a thorough survey of major sources of information in the area, and when I had sufficiently familiarized myself with the localities. 11 key-informants were selected from different walks of life, including farmer(s), labourer(s), school teacher(s), local government leader(s), elderly person(s), religious leader(s), and large-land owning elite(s). The selection was influenced by informant's position, reputation, and his(her) knowledge of particular information. In this respect, I primarily followed the "position-reputation method", frequently used by sociologists (for example Quddus, 1978, pp.36-47; Chowdhury et al., 1979, pp.1-18).

Following other studies (Price, 1995; Jansen, 1986; Gordon, 1979), I also made use of ethnohistories and life stories. Ethnohistories are defined as the "histories recalled and recounted by rural people" (Chambers, 1991, p.525). For this study, I recorded general and life histories as narrated by key-informants and villagers in the project areas (as long as I was convinced about the credibility of the narrater). These histories could be of an event, or of living and dead persons, in some ways concerned with forest use and SF in the localities. Mostly, however, for detailed ethnohistories I relied on some particularly reliable key-informants.

All the above methods had other major advantages. For example, these were economic in terms of time and cost, and could be applied with relative simplicity.

Secondary Sources and Historical Reviews

The study extensively used secondary data and historical reviews. Though somewhat "neglected", secondary data analysis has "rich intellectual tradition in the social sciences" (Nachmias and Nachmias, 1992, p.292) and has brought about "an almost revolutionary change in survey research" (Glenn, 1978). On the importance of the use of historical materials, Wilkins (1970, p.52) argues:

> To overlook the contribution historical method can make to the social scientist's work would be to deny him an understanding of those processes which enrich his study of contemporary scene. It would be foolish indeed to attempt to divorce the

> present from the past, as society today is to a very extent a reflection of what has been happening within it over many hundreds of years....[S]ocial scientists, who rightly believe that study of contemporary society without reference to the way in which it has evolved would be only half a study, are likely to rely heavily upon the historical method.

I "relied heavily" on secondary and historical data for some reasons. First, for conceptual-theoretical reasoning, secondary data and historical reviews were the only available sources of information for my study-purposes. They enabled me to handle a wide range of conceptual issues, to share the experience of varied cases, and to delve into history. Second, secondary data also facilitated conceptual-linkage between my study and similar studies in the area, and could be used for comparative purposes (Hymen, 1987; Nachmias and Nachmias, 1992). Third, these sources were economic in terms of cost involved in gathering them, which was a major consideration for me.

Nature, Sources and Procurement of Data

The study used both primary and secondary data. Primary data included quantitative information collected through questionnaire-surveys and interviews, and qualitative data acquired through observations and ethnohistories. The secondary sources included published and unpublished materials, public documentation and reports, subject-specific books and journals, press cuttings, and related television programmes. During the field-work, a good deal of time was spent to consult specific institutions and persons for ideas and information. Micro-data were collected from the records and files of Forest Range and Beat offices, *Thana* and Union council offices, *Thana* Land-revenue, Statistics, Agriculture and Family Planning offices in Rangunia and Kaliakair. The range of people contacted included forestry officials at different hierarchical levels; forestry researchers and consultants; university academics; and farmers, villagers, local government and government officials, positional elites in the case-study localities (A list of selected institutions and persons approached for this study can be found in the appendix). Relevant sources were also exploited in the United Kingdom, especially for theoretical literature and regional case-reviews.

Constraints and Limitations

As hinted in the preceding discussion, the study had a number of limitations. First, the study warranted a systematic investigation into the gradual transformation of forest use. Therefore, a time-series or longitudinal study would have been more befitting. But here, I could only investigate a few particular sites in a given (short) period. This has somewhat delimited the scope of the study. Wood (1994, p.50) also reported similar experience from his survey in Bangladesh: "the fieldwork period was very short, so that we could not reasonably expect to obtain a full picture of the village". However, by extensive use of secondary sources and first-hand data on micro-locations, I hoped to compensate the weakness partially. Second, I had to work amidst major constraints of time, fund and logistic resources. The field-work time was short for conducting a detailed research and no institutional funding was available. Because of the status of "personal" (as opposed to "official") research, I was not officially entitled to any logistic support from public institutions. Third, the description and analyses of the study are incomplete in the sense that "we generated much more data than can be presented or analysed here" (Wood, 1994, p.50). However, I find some consolation in the fact that this seems to be the general tendency of doctoral studies, as noted by Chambers (1991, p.520): "The real or imagined requirements of doctoral research induce students to seek safety and respectability by avoiding shortcuts and finding out more, not less". Fourth, the very topics of forest use and SF suffer from ambiguity and analytical imprecision. For example, the process of forest use has evolved differently in different regions. Though there are some basic similarities, these processes have distinct manifestations of their own. This raises questions about "proper representation" of the case-studies, and makes inference and generalizations difficult. Similarly, SF has become an umbrella term for numerous technical and social interpretations; and despite its popularity, there is no precise definition of SF. Due to the very nature of these subjects and the inherent flexibility and ambiguity contained in them -- any study concerning them is likely to show some degrees of uncertainty and obscurity. Besides, these studies may not likely to be "representative" in strict methodological and statistical terms. To that extent, my study is no exception. However, as discussed at the outset of this chapter, I kept an eye on the question of representation while setting the criteria for choosing the field-cases. However, many researchers do not seem to be particularly concerned about "representation" of empirical cases; and they emphasize more on analytical reasoning and logic of analysis. Worsley (1977, p.120), for example, argues that:

> the general validity of the analysis does not depend on whether the case being analysed is representative of other cases of this kind, but rather upon the plausibility of the logic of the analysis. The generality of the same kind that enabled Sir Ronald Rose to announce the "cause" of malaria when he found the malaria parasite in the salivary gland of a single female mosquito in 1897.

Similarly, Ahmed (1991, p.167) maintains that "generally case studies make no claim to be representative, [rather they] attempt to arrive at a comprehensive understanding of a particular group, institution or situation". Thus, there is a strong body of thought which argues that representation is not an absolute necessity for analytical validity. In her major socio-anthropological research work on Bangladesh, White (1992, p.5, emphasis added) remarked: "Being clear that *there was no such thing as a "representative" village*, I decided fairly arbitrarily the *upazila* [*Thana*] where I would go".

From the methodological perspective, the topics of my research could be compared to a seemingly benign peak of an ice-berg with a huge interior beneath. For example, the study dealt with the history of forest-use with forward and backwards linkages. Though I could empirically investigate only a few small sites in a particular time, I needed to be alive to the process that has evolved all these years (and continues to evolve). Similarly, as discussed in previous chapters, forest use and SF are regulated by varied political-economic relations and structures of the frame. I could cover only a limited number of these variables, but needed to have some idea about the broader frame. Hence, my investigation of a few field-cases is analogous to the visible peak of an ice-berg, where the bulk remains under water. But if one is not aware of the submerged portion, one may end up in a superficial or incomplete understanding of the reality. The methodological challenge for me was to find feasible ways to deal with the surfaced problems, and at the same time, to have some idea about the underground dynamics. Moreover, my research themes themselves were by no means static. Questions I planned to ask culminated into further questions, and the answer revealed, gave rise to new avenues of enquiry. Accordingly, the research techniques and tools had to be re-adjusted to suit to new questions and changed situation. In fact, research of my kind always seems to possess some degrees of uncertainty and flexibility. Issues and concerns frequently change in an enormous (but interesting) process of shaping and reshaping; and the researcher is never truly sure what lies in front of him(her). Perhaps this is why, White (1992, p.5), recollecting her experience of field-work methodology, comments: "The question I was facing was one that dogged me throughout my research: How do you know what you need to find out before you have found it out?"

5 Forest Resource Use in the StudyAreas: Views from the Field

Introduction

This chapter examines the historical changes in the philosophy and practice of forest use in the study-areas. It mirrors Chapter 2 which conceptualized three evolutionary stages of forest use: the pre-commercial and pre-colonial era; the advent and proliferation of commercialisation; and the post-independence era. In line with the general findings of Chapter 2, this chapter traces the characteristic trends which over the years have led to commercialization and alienation in the study-areas. While Chapter 2 reviewed the pattern and evolution of forest use at a more general (and conceptual) level, this discussion concentrates on the particular case-study areas. It focuses on the broader context of the study areas (forests of Chittagong and Dhaka-Mymensingh tracts), with an insight into the localities of the three case-study projects (Chandra, Betagi and Pomora).

The next section reviews the evolution of forest use in the study areas and traces the major historical trends (from the pre-commercial period through current time) and their implications for local people. The third section investigates into the field-case sites in further details. It is a continuation of the preceding section, but it narrows down the focus (from the broader study-areas of Chittagong and Dhaka-Mymensingh) to the boundaries of the field-cases (Betagi, Pomora and Chandra) and relies more on first-hand data. Thus, first, the chapter examines the general (broader) context of the study-areas; and then, more closely focuses on the specifics of the field-cases. The fourth section summarizes the major users of and the nature of competition for forest resources in the study-areas. This chapter, reaffirming the findings of Chapter 2, concludes that there has been a process of commercialization of forestry in the study areas, causing progressive alienation of local communities. An alliance of commercial forces (including urban traders and their local agents, local elites, a group of state and local government officials) dominates forest use, and the poor and weaker local people share a feeling of deprivation and estrangement.

The Changing Trends in Forest Resource Use in the Study-Areas

This section reviews the historical trends in forest use which have alienating effects on local people in the study-areas. It starts with an analysis of the situation in the pre-commercial and pre-colonial period, and subsequently proceeds to discuss the impact of the colonial period (the British rule), and that of the post-independence period (Pakistan and Bangladesh). This section illustrates the gradual transformation of the localized and more informal pattern of forest use to a centralized-commercialized one. It also depicts the process of progressive diminution of local rights and share.

The Early Scenario: Forest Resource Use in the Pre-commercial and Pre-colonial Era

Little systematic information on forest use exists prior to the advent of the British Rule in the Indian subcontinent during the mid 18th century (Forest research Institute (FRI), 1961; Rahman, 1993; Padhi, 1982). Therefore, I had to rely primarily on the limited literature relevant to the study-areas.

Early evidence suggests that forests had a major role to play in flourishing Indian civilizations through the ages. The Mohenjodaro and Harappa civilizations in Sind and Punjab (about 4000-5000 BC), for example, largely depended on forests for their energy and constructional needs (Rahman, 1993, p.21; Dwivedi, 1980, p.2). The Dravidian civilization of 2000 BC also flourished "in consonance with the forests" (FRI, 1961, p.71). Documentations, scripture, seals and paintings from the Indus valley civilizations showed that forests were adored as a "collection of celestial plants and were supposed to be inhibited by divine spirits" (Dwivedi, 1980, p.3). Some analysts have gone as far as to argue that the ancient Indian culture had its very origin in *Aranyas* (forests) and *Ashrams* (religious centres, mostly located in woods) (see FRI, 1961, p.72; Dwivedi, 1980, pp.2-5). Major Indian philosophies and preaching, for example, found expression in the thoughts of *Rishies* and *Gurus* (saints and hermits), who mostly lived in the serenity and obscurity of forests (Dwivedi, 1980, p.3; Anon., 1991a, p.2).

I found a strong body of arguments in the literature on early forest-use situation which asserts that local people had a religious and spiritual affinity to forests (see for example Padhi, 1982; Dwivedi, 1980; FRI, 1961). In favour of their argument, the studies shed light on the fact that there was great reverence for *Vans* (forests) amongst ancient races and religious literature. As early as 1500 BC, *Rigveda* -- the oldest Hindu epic, for example, preached: "Plant, Thus I hail Thee -- The Divine Mother of Mankind" (Rig-X-97-4, cited in Padhi, 1982, p.31). Similarly, the *Puran* considered "plantation of trees and maintaining them were good acts" and suggested that, for "those who do not have sons, trees are there sons" (Dwivedi, 1980, p.7). Later the Aryans, basically a pastoral community and one of the most dominant races of ancient India, revered forests and "maintained all their big institutions in the sylvan surroundings of the forest" (FRI, 1960, p.71).

Besides the argument that forests constituted a part of the religious-cultural heritage of ancient India, some later kingdoms also demonstrated organised attempts in professional forestry management. The *Arthashrastra of Kautilya* (the Economics of Kautilya), written in 321 BC by Chanakya -- the premier of king Chandra Gupta Maurya -- manifested a unique example of the statecraft and resource management in ancient India. It showed that there was a regular forest department during this period, headed by *Kupadhyaksha* (Superintendent of forest products) and assisted by a cadre of *Banopals* or Forest Guards (Rahman, 1993, p.21; Dwivedi, 1980, pp.7-8; Biswas, 1988, p.4; Padhi, 1982, pp.31-32). King Ashoka (273-236 BC) revered forests and had a "well-framed management" of forests (FRI, 1961, p.72). He ordered plantation of fruit trees, medicinal plants and shade trees for public welfare (FRI, 1961, p.72; Dwivedi, 1980, p.7; Bachkheti, 1984, p.1). The *Indika of Megasthenes* (305 BC) noted that India was a country "teemed with" mountains, fruit trees, wildlife and forests (Dwivedi, 1980, p.7). The fall of Maurya dynasty was followed by Kushans, and subsequently, by Guptas (240-673 AD). Guptas also had organised structure for forest management. There were *Gaulmikas* (public revenue collecters) and *Rajmargas* (forest roads) for proper maintenance of forests (Dwivedi, 1980, p.10; Rahman, 1993, p.22).

During 800-1400 AD the Indian sub-continent was divided into many tiny kingdoms which were engaged in ceaseless skirmishes among themselves (Rahman, 1993; Dwivedi, 1980; FRI, 1961). These relentless conflicts warranted large scale forest-felling for defence purposes. On the other hand, for local people forests served as shelters during these crises.

The next important phase in the history of politics as well as forestry was the rule by Mughals (1526-1700 AD). Bengal (of which present Bangladesh is a part) formed a major *Suba* (province) during this time. The Mughals had a "utilizational and aesthetic approach towards forests without any comprehensive outlook" (Dwivedi, 1980, p.12) and "did not have any special interest in the conservation of forests either" (FRI, 1961, p.72). Hence, the state was not particularly keen to restrict local demands for and access to forests. For the rulers forests served the "imperial hunting" and "ornamental" purposes. They used forests as *Sikargah* or game reserve (Rahman, 1993, p.22). The Mughals' aesthetic perspective on forest management was evident by the works of emperor Akbar, arguably the greatest Mughal, who evinced interest in canal-side planting "so as to make it [look] like the canals under the tree in the Paradise" (FRI, 1961, p.72). However, since the Mughals emphasized agriculture over forestry, there had been limited forest-clearing for agricultural purposes (Rahman, 1993; FRI, 1961; Dwivedi, 1980). There are some particular information on forest use in the Dhaka-Mymensingh *Sal* forest tracts during the Mughal period. The dense forests of this area served as ideal refuge for "outlaws" and "squatters". The Mughal emperor Jahangir granted a patch of forests from this tract (called the Atia Pargana) to his "physical-training instructor" Munayem Khan. Similarly, another patch, called Bhowal (close to Chandra) belonged to a family of proprietors locally known as Gazies (Ahmad, 1938, p.12; some information were acquired through personal communication with a descendent of the Gazi family).

On the basis of the preceding discussion, following characteristic features can be observed regarding the nature of forest use in ancient and medieval time.

First, the relationship between forests and neighbouring communities was regulated, at least to some extent, by religion and culture. Worshipping of trees was common-place (Dwivedi, 1980, p.7). This, coupled with spiritual preaching by some rulers, most likely instilled among people some consciousness and interest in forest preservation and propagation. Customs and religion, for example, forbade the use of sickles or axe in certain areas; while in other areas forests were treated as *Devaranya* (gods' grove) from which only twigs and fallen branches could be fetched without causing any damage to trees (Fernandes, 1983). The rulers also fostered the linkage between forests and religious culture. There were forests exclusively "for the study of religion" under some ancient kingdoms (FRI, 1961, p.72). Forests were praised in religious epics. There is a popular belief that during epic days, the Hindu deities -- Krishna and Radha -- had their legendary romance under the serene surroundings of forests in a place called *Gupta Brindaban* (the secret romance), north of Atia forests in the Dhaka-Mymensingh tracts which is not far from my study-site of Chandra (Ahmad, 1938, p.12). As I have found during the field-work, these forests of religious value are still preserved and worshipped by the local Hindu population.

Second, even in the early days there were regulatory state organizations for forest management and forests were not necessarily owned by local communities. However, from the regional literature it seems that these regulatory or policing measures by the rulers were not oblivious of local people's needs; and local people did not find state-actions as a hindrance towards meeting, at least, their subsistence requirements of forest products (see for example Fernandes, 1983; Tiwari, 1986; Dwivedi, 1980). Fernandes (1983, p.11) argued that the "traditional practices in the forest areas were geared to preserving trees while catering to the inhabitant's needs such as firewood, poles,...food". Similarly, Tiwari (1986, p.15) maintained that

> in the pre-British time by and large the rural communities enjoyed unbridled use of forests and wastelands in the vicinity....[U]nder the oriental governments...any one was accustomed, without let or hindrance, to get what he wanted from the forest, to graze his cattle where he liked.

Even the early documents of the British administration seemed to be alive to the fact that people enjoyed a considerable degree of access to forests for meeting their needs. The Board of Revenue of the Imperial British Government of India, for example, noted that "until quite recently...*the people were free to cut what wood they pleased, to graze their cattle, and to exercise common rights without let or hindrance*" (The Imperial Board of Revenue's proceedings of 10.4.1868, No. 2777, cited in Stebbing, 1923, p.16, emphasis added). Thus, it was most likely that the oriental forestry administrations, besides fulfilling their own interest, provided sufficient avenues for catering local needs, and thereby, fostered a peaceful co-existence with local people. The "royal rights" did not overshadow the basic needs of the subjects. For example, under the Hindu rulers while there were "classified forests" for royal requirements and revenues, there were also specially earmarked forests for local people's consumption and pastoral purposes (FRI, 1961; Dwivedi, 1980).

Third, the extent of deforestation was certainly not major during this period, and was mainly caused by wars among rulers, occasional construction and communication purposes and also by (limited) land clearing for agricultural expansion (Fernandes and Kulkarni, 1983; Padhi, 1982). Large-scale commercial exploitation, which eventually was to become the greatest menace, did not seem to be a major contributor to deforestation until this point. Only specified species (of timber-value) were proclaimed by kings as "Royal Trees" for revenue purposes, "otherwise, the forests were open to all and the public obtained their requirements without restriction" (FRI, 1961, p.72). Local forest-based communities did not seem to be involved in deforestation in any significant way. "Frequent changes and political boundaries" and "too many wars which led to destruction of forests" (Padhi, 1982, p.36), evidently, had little to do with general public. Rather, as FRI

(1961) and Rahman (1993) reported, forests served as the "refuge" for people on the face of these crises. Therefore, local people were not the predominant force in commercial exploitation of forest. There were some evidence of forest-based commercial undertakings during medieval India. During the Mughal rule, for example, Dhaka was a ship-building centre, fed by trees from Chittagong Hill Tracts (Khattak, 1979). These undertakings were controlled by urban businessmen, who used forest-based communities, especially the hill tribes, as paid or contractual labourers.

The Advent and Proliferation of Commercialization

Broadly speaking, the process of commercialization of forest use and the consequent alienation of local communities was sparked by the colonial intrusion on the Indian sub-continent. The trends in commercialization were consolidated throughout the British colonial rule and have continued, in substance, till today. For convenience of discussion the colonial era can be divided into two related periods. The first period (1757-1857) was the rule by the "East India Company". This period was characterized by, what Gadgil (1989, p.9) called, "rampant profiteering" by the purely commercially motivated private Company. During this time organized form of government administration hardly existed. The primary motto of the Company rule was to fetch as much earning as possible for their masters. Therefore, the general duties of a government (such as public welfare and services) gave way to maximization of exploitation of a subjugated land and its population. The second phase of the colonial rule (1858-1947) saw the formation of an organized government administration directly under the Imperial British Crown. However, the basic trends in revenue maximization continued unabated. But the exploitation of forests and the state monopolization of forest-ownership were carried out by establishing more organized and constitutional structures (such as the formation of forest department and the promulgation of forestry laws). In other words, the trends in commercialization were not as naked and reckless as those of the Company rule, and were gradually consolidated through more organized and systematic structures of administration.

The British period under the Company rule: the "rampant profiteering" The beginning of British empire in India was marked by the battle of Palassey on June 23, 1757 where a "chartered commercial corporation" (Edwardes, 1967, p.15) -- the British East India Company -- came out victorious. The company placed a "puppet *Nawab*" or king (Hunter, 1967, pp.450-451) to the throne of

Bengal and took de facto control of Bengal with Colonel (afterwards Lord) Clive as the Governor. Shortly afterwards, in 1765, the Company received "formal grant of sovereign rights in Bengal" (Edwardes, 1967, Photograph 1) from the Mughal emperor Shah Alam including the absolute charge of fiscal and revenue administration (Hunter, 1893, p.455). This period marked a distinctive new era in the political economy of forest use and management. This period, eventually followed by a regulatory colonial government under the British Crown, was characterised by an unprecedented monopolization of state authority over forests; maximization of revenue; and a growing proliferation and fortification of the commercial-industrial interests in forestry (Fernandes et al., 1988; Fernandes and Kulkarni, 1983; Padhi, 1982; Nadkarni et al., 1989). Hartmann and Boyce (1983, pp.12-14) summarised the immediate impact of the colonial intrusion:

> After the British East India Company wrested control of Bengal from its Muslim rulers in 1757, the line between trade and outright plunder faded....As Britain developed, Bengal underdeveloped....East Bengal assumed a new role in the emerging international division of labour as a supplier of...raw materials....Resources which could have financed development were instead devoted to subjugating the population.

Here, it may be useful to look at the very motto and philosophy of the early British (Company) government, which may explain the government's continuous persuasion for trade and revenue in all sectors including forestry. Edwardes (1967, p.47), in his extensive study of the British Indian history, noted that the Company administration's "only interest was in profit" and they took recourse to "predatory activities" in fulfilling their interests. In fact, as Sir William Hunter remarked about his own countryman, from the very beginning the Company looked at India nothing more than a fertile ground for extortion and plunder: "The English cherished extravagant ideas of Indian wealth....No funds existed [that could] satisfy their inordinate demands" (Hunter, 1893, p.451). As a price for elevating Mir Jafar as the (puppet) *Nawab* of Bengal, for example, the Company extracted an incredible 1,238,575 sterling in one occasion alone (at 1857 value) (Hunter, 1893, p.451). In a report to the Select Committee on the performance of the East India Company, it was noted that "in thirty years following the acquisition of the Bengal provinces, the nation [Britain] by public and private channels, derived from them alone,...fifty millions sterling (Report to the Select Committee of the House of Commons on the affair of the Company, 1831-32 (734), p. 18, cited in Edwardes, 1967, p.86). The following excerpts from some selected official correspondences of this period will further elicit the contemporary political and administrative philosophies of the state. It may be noted that these correspondences were complied, in their original forms, in an archival collection, known as the Bengal District Records (BDR, 1978). Most of the following excerpts are drawn from the above volume and I have selected particular documents which are concerned with the administration of the broader study-areas of my thesis -- Chittagong and Dhaka (although very little information on Dhaka was available).

The Chief ("Collector") of Chittagong wrote to the Governor in Council at the Fort Williams (the Company headquarters in Calcutta) on 16.1.1761 (BDR, 1978, pp.123-124, Document 2.2, emphasis added) that:

> *The Extending of Commerce, we think very justly Claims our most Serious Attention & the bringing to any perfection [requires] a steady Perseverance.... [W]e come to settle here in order to Establish Trade & Prosperity through the Country.... We are endeavouring to make ourselves acquainted with every thing that may promote a Correspondence & Trade this way* -- one objection to it at present is the Disposition of the Inhabitants who are Yet insensible of a Civil well regulated Government, being no better than a Wild Sett [?] of Mountaners [mountaineers].

In another occasion, the Fort William authorities ordered the Chief of Chittagong on 2.3.1761 (BDR, 1978, p.321, Document 4.9, emphasis added) that:

> you will inform us as soon as possible...*what Duties you think might be levied and brought to the Company's credit or any other Commercial advantages you may be able to procure for our Hon'ble Masters*, that we may send you our directions accordingly.

The Governor in a later correspondence (on 14.6.1761) to the Chittagong Office hoped that: "...by the continual increase which we observe has been made on the Rents of Chittagong, ever since the first conquest of that province..., We have reason to hope that by your diligence they will be further advanced" (BDR, 1978, p.327, Document 4.14). Even at the end of the 19th century, the "revenue-mission" of the government continued unabated. During the mid 1870s, for example, the government noted that: "...land might now be fairly assessed at from four to five times the amount it pays at present as Government revenue" (cited in Hunter, 1876, p.155).

Therefore, it is evident that the government's basic motto was to optimize revenue earning by expanding state control over all prospective sources of revenue. In fact, during the early British "*misrule* in Bengal", as Sir Hunter (1893, p.453, emphasis added) put it, there was "no system of government in Bengal, but merely the tradition that *unlimited sums of money might be extracted from the natives by the terror of the English name*". Moreover, the uncompromising zeal for revenue maximization were often saddled on coercive forces and oppression (BDR, 1978; Hartmann and Boyce, 1983; Hunter, 1893; Edwardes, 1967). For example, the Collector of Chittagong asked permission from the Governor at Fort William for deploying Battalions of *Sepoys* (army) in Chittagong and for sending sufficient weapons on the plea that, "*the Collection as well as Improvement of our Revenues we think greatly depend on the Force that may be kept up here*" (BDR, 1978, p.57, Document 1.6, emphasis added). In similar vein, the Chittagong Chief also opined that "our Revenues in all Respects and every Improvements therein, do & will...depend on the Influence we have in this Country, which influence can only be supported by a proper Force" (BDR, 1978, p.121, Document 2.2). The Governor in his turn also approved the use of force (on 24.6.1761) and wrote: "[in] Company's interest...we think it is highly necessary [that] you should exercise the same authority in your *Cutcherries* [court/territories], as we do in ours" (BDR, 1978, p.327, Document 4.14). Another study also noted that, "besides fulfilling the ordinary duties of Civil Government, the [British] administration had a curious military aspect....The collector was on several occasions indented upon for troops, and more than once was called upon to organize the military defence in his district" (Government of East Pakistan (GOEP), 1970, p.89).

At this point, it may be noted that there was certainly an element of racism and sense of superiority which largely contributed towards shaping the pattern of administration and behaviour of the rulers throughout the British rule. In Chapter 2, I argued that commercial interventions were often saddled on a sense of superiority on the part of intruding forces. Mirroring the observation, Edwardes (1967, pp.32-35) noted that towards the end of the 18th century "the British [in India] were becoming conscious of a sense of racial superiority" and there was "a basic element of contempt for the Indians". Lord Cornwallis (1786-1792 and 1805) the Governor-general of India, for example, was convinced that "every native of Hindustan was corrupt" (cited in Edwardes, 1967, p.33). Similarly, another Governor-general, Lord Wellesly, arriving in India 1798, "brought with him profound sense of racial arrogance" and believed that "imperialism needs the backing of pride, the consciousness of superiority"; therefore, he "had nothing but contempt for Indians" (Edwardes, 1967, p.33). The "racial arrogance" did not remain confined to the highest state officials; in no time it seemed to spread among the Europeans in general. By 1810, for example, as Graham (1813, p.139) noted "every Briton appears to pride himself on being outrageously a John Bull" (also see BDR, 1978; Hunter, 1893). These attitudes could largely explain the characteristics of the colonial rule including the coercive forces used in resource extraction and maximized exploitation.

Chittagong was the first District to come under the British rule (BDR, 1978, p.vii). On 5.1.1761, *Nawab* Mir Quasim ceded this District to the East India Company (GOEP, 1970, p.87). From the very inception, Chittagong was very alluring to the commercial instincts of its new rulers. The state of affairs in forestry was no exception. As early as 16.2.1761, the Chief of Chittagong informed his authorities at the Fort William that:

> the soil [of Chittagong] is very fertile....Rocky Mountains are covered with high wood, a many of them about the Boundaries Yield variety of good Timbers, the Land is capable of producing Quantities of Wheat & Rice, & all other Kinds of grain, Ratten [Ratan], Wax, Oil, Timbers of various Sorts &...Elephants Teeth (BDR, 1978, p.119, Document 2.2).

On 16.2.1761, the Chittagong Office, after comprehending a general idea about the District, confirmed the Governor that: "*we doubt not but the Province of Chittagong will turn out a valuable and important Acquisition to our hon'ble Masters*" (BDR, 1978, pp.58-59, Document 1.7, emphasis added).

Keeping in mind the above observations on the basic philosophy of administration, I now turn to see the characteristic trends as regards the nature and pattern of forest use from the advent of the colonial rule to the early years of the 20th century (In fact, as I will eventually show, some of the trends have well continued up to present time).

This period was characterised, more than anything else, by massive commercial exploitation of forest. Up to the mid 19th century forests of this sub-continent, on the whole, were subjected to "exploitation on gigantic scale" (Rahman, 1993, p.22; Dwivedi, 1980, p.12) through a country-wide commercial network of shipbuilding, communication and construction activities, and foreign export. With the dwindling supplies of British Oak, huge quantities of Indian Teak (*Tectona grandis*) supplemented the requirements of the British Admiralty's Fleet including the Royal Navy (FRI, 1961; Dwivedi, 1980; Bachkheti, 1984; Padhi, 1982). Rapidly growing "military centres" throughout India and the "construction and fortification of public buildings" for expanding state machineries caused heavy felling of the best quality timbers (Bachkheti, 1984; BDR, 1978). In particular, as Tiwari noted, "Railway expansion saw unprecedented assault on nearly all accessible forests" (1986, p.16). The wastage and scale of exploitation also reported to be equally unprecedented. The sub-Himalayan forests of Garhwal and Kumaon, for example, "were all felled in even to desolation, while thousands of trees were felled which were never removed" (Tiwari, 1986, p.12). In addition to timber, trees of ornamental and sundry uses were also felled. FRI (1961, p.72) noted that the highly scented Sandalwood of India, for instance, found its way to different European markets. Following the "Indian Mutiny" of the mid 19th century, the government attached greater priority to building communication network and public offices for smooth supply of logistics and quick military build-up on the face of possible emergencies. This contributed to further deforestation by means of clearance of forest for establishing communication network and providing raw materials for construction purposes (FRI, 1961). There was a series of indents for the best quality Chittagong timbers by the Fort William authorities during the early British rule which were documented in their correspondences (see BDR. 1978 and other archival files of the District Collectorate Library Section and Record Room, Chittagong). Ships loaded with such timbers regularly shuttled between Chittagong and other parts of India especially Calcutta (BDR, 1978; Hunter, 1876). The Fort William authorities notified its Chittagong office to ensure supply of as much timber as may be procurable; "*as the whole will undoubtedly be soon wanted*, either for the new Fort or other Publick [public] business" (BDR, 1978, p.326, Document 4.13, emphasis added). In response to such central (government's) indents, the Chief of Chittagong assured the Governor that: "we have made enquiry regarding the Timber your honour & ca. have indent'd for, & doubt not but we shall be able to comply therewith" (BDR, 1978, pp.316-317, Document 4.4). To cite an example of the amount of timber sought under one typical request for indent, the following list may be noted: "300 Jarril [Jarul] of largest size, 500 small Jarril, 1000 Jarril planks, 250 largest Teak, 300 Teaks of different size, 400 largest Sauls, 1 largest Teak of over 70 feet length" (BDR, 1978, p.314, Enclosure to Document 4.1). Moreover, trees of superb quality and highest timber-value were particularly targeted in his series of exploitation. Regarding the procurement and shipment of trees, for example, the government advised its Chittagong office on 23.6.1761 that: "The best guide to go by is to make choice of these Trees which are free from Knots, of the highest girth & straight...[T]hey [should] run above 30 feet" (BDR, 1978, p.326, Document 4.13). During a period of only six months (ending in February 1876), 52,225 timber-logs and 694,930 bamboos were exported through river-traffic from Chittagong alone (Hunter, 1876, p.194).

The growth of some industries and commercial activities also caused sharp increases in the extraction of forest products, which engaged local people as bonded or contractual labourers. For example, Dhaka and Chittagong were two large ship-building centres, providing regular supplies of ships, boats and schooners for the whole country; and these centres, through their local agents, used a section of local people, especially the tribals, in timber logging. During the 18th century, there was evidence of regular indents for wooden ships by the British administrative headquarters (at Fort William) to be supplied from the Chittagong dockyards (see BDR, 1978). The process seemed to have continued to the end of 19th century. In 1876, Hunter (1876, p.192) noted that ship-building was still "carried on to a considerable extent at Chittagong". Besides, there had been trade of wood-oil and wax and mostly the (tribal) "hillmen" of Chittagong and Chittagong Hill Tracts served as labourers in such trades (Stebbing, 1923; Islam, 1978). There was evidence of a Chittagong-built ship, called the *Snow Buythin* (90 Tons), sailing to Madras with a "Cargoe [cargo] of woodoil" as early as 1787 (BDR, 1978, p.302, Document 3.67).

Some purely commercially motivated businesses, including trade in wild-life, also led to deforestation and negative impact on local people. The *Khedah* (elephant catching) operation, which continued throughout the British rule and well into the Pakistani period (1947-1971), was one such example (Stebbing, 1923; GOEP, 1970). The state's purely commercial perspective on wild-life trade became evident from the words of Governor, who as regards the Elephants, instructed the Chief of Chittagong on 1.2.1761 that: "*If the Elephants... are part of the Company's Revenue*, you will be pleased to receive & dispose of them to the best advantage or if they cannot be sold at your Place you will send them to...any Place where you imagine they may meet with a Market" (BDR, 1978, p.319, Document 4.7, emphasis original). *Kheda* was a purely commercial undertaking. Private contractors were appointed by the government to carry out such operation in return for "royalty" and tribal people were induced (sometimes forced) to act as labourers. It also caused considerable damage to forests, as noted by Stebbing (1923, p.434): "felling work carried out on a large scale to form the stockades [for catching] and the burning of the forest which was resorted to..., resulted in a great waste and deterioration of fine timber forest".

Other commercial undertakings also had profound affect on the pattern of local forest-use. In may be noted that exotic commercial crops, such as cotton, indigo, tea, were "almost entirely an European monopoly" (Edwardes, 1967, p.83). Cotton cultivation in the hills of Chittagong, for example, went apace through the whole period. Tribal women were employed in the hard task of cotton plantation, mainly in planting *Phul Suta* and *Beni Suta* (two particular varieties of raw cotton); and in the 19th century the "raw cotton formed, undoubtedly, the most valuable export from the Hill tracts" (Hunter, 1876, pp.200-203). There had been a gradual increase in cotton export during the last half of the 19th century. In 1858-59, 129.5 tons of cotton were exported from Chittagong, while in 1874-75 it rose to 609.1 tons (Hunter, 1876, p.203). The British government was particularly concerned about the "resistance from hill-men" and there had been series of tensions between such "nomadic hill-borderer" and the "lenient civil government" (GOEP, 1970, pp.88-89). Edwardes (1967, p.84) also observed a "considerable resistance to growing [the exotic variety] of cotton on the part of the cultivators". The government's policy, as GOEP (1970, p.89) noted, "was no doubt prompted largely by commercial considerations, for cotton was grown largely by the hill-men...and it was one of their [government officers'] principal duties to collect cotton and forward it to the factory at Dacca [Dhaka], where there was a large weaving population".

Similarly tea cultivation, which became a "mania" (Hunter, 1876, p.209) for the European planters from the outset of the 19th century, caused considerable forest tracts and wastelands to be converted to territorial tea estates (Hunter, 1876; 1893; Edwardes, 1967; GOEP, 1970). Tea as an industry was pioneered by the government, again, on purely commercial motives. In 1840, the then Collector of Chittagong (Mr. Sconce), started a garden from seeds collected from Assam; and in 1862, another European (Dr. J.B. Barry) first managed to acquire 20,000 acres of public land for tea cultivation (GOEP, 1970). In the following years, Hunter (1876, pp.209-211) noted,

> applications for allotments of wasteland [including forested tracts] poured into the Collectorate...; the mania [got] great impetus [and] the applicants were chiefly Europeans, and often representatives of [European] companies got up for the purpose....An enormous proportion of the total area of wastelands within easy reach of a river, a road or a village path [was ceded to] tea estates.

Furthermore, as Stebbing (1923, p.433) noted, "considerable stretches of forest had been cleared by planters....Planters recklessly devastated the forests in their vicinity to supply their requirements". In the process many village lands were also "taken into the lots and no compensation paid in consequence" (Hunter, 1876, p.210). The results were further alienation of local people from their territories and their growing discontent. Hunter (1876, p.211) noted:

> an antagonistic feeling sprung up between the managers of the waste lots and the natives of different villages....Villagers of all classes often found that the manager either refused to allow grass, bamboos and jungles to be cut at all, or only on payment, and the same as regards grazing cattle....In many instances the villagers tried the effect of a stubborn resistance to the tea-planters.

Another study (GOEP, 1970, pp.229-230, emphasis added), further elaborating on the process of establishing tea gardens and its consequences, wrote:

> *An applicant would put in a petition for any tract he might have prospected and taken a liking to just as if it were the Australian bush; the lot he asked for was put to auction and sold*; and after the planter had built a *bungalow* [house] and planted tea, he learnt for the first time that neighbouring villagers...looked on him as an intruder. *Although the waste lands were Government property, the private estates had no boundaries, and their proprietors had been busy in creating documentary claims over much larger areas than their settlements covered. In some cases...the areas sold included a considerable area of settled land, and land held under various rights and titles,* which thus passed into the hands of the purchasers.... *Villagers were aggrieved by the fact that some managers...refused to allow grass, bamboos, and jungle to be cut...and disputes also arose regarding the right to graze*....Litigation ensued.

Moreover, in order to "improve materially the health of both Europeans and local people in the tea gardens, extensive clearance of jungle and forests" was carried out (GOEP, 1970, p.231). Besides, severe wastage of land was also noted. In 1872, there were 13 gardens with an area of 24,890 acres, of which only 1,200 acres were "truly under tea" (GOEP, 1970, p.230).

Tobacco plantations were also patronized by the government which led to similar effects on forests and wastelands as those of tea estates. The government's policy in the last quarter of the 19th century can be grasped from the following comment of the Deputy Commissioner of Chittagong Hill Tracts (cited in Hunter, 1876, p.204):

> There is no doubt that...the area of [tobacco] cultivation might be largely increased and it is hoped that signs may soon appear of European enterprise, which is already very active in the encouragement of tea-cultivation in Chittagong, taking the direction.

This further intensified the processes of land privatization (under the state's patronage) and of growing dispossession and discontent of local people.

Salt manufacture was another commercial venture which also took considerable toll of forests and local people. The Company had "monopoly" over salt-trade and a part of the salaries of its civil servants came from the profit of the trade (Hunter, 1893, pp.455-456). The salt factories demanded huge firewood for their boilers. The *Mulungees* (local salt workers) were sent to forest to bring firewood almost without any restriction or specification. This led to indiscriminate and wasteful felling. The government's salt-agents were assigned to exploit highest possible revenues from salt-factories. In the early British rule, even the principal district official (i.e. Collector) got personally involved in ensuring regular supply firewood for salt factories, because increased revenue from salt sector (for that matter, from any sources) reflected the efficiency of administration (see for example BDR, 1978, p.263, Document 3.41). As the government documents showed, in many occasions there were records of "oppression" and "violent proceedings" against local workers in salt industries; and in extreme cases, they had to "flee away leaving behind their home lands" (see Documents 3.69, 3.28, 4.41, and 4.57 in BDR, 1978).

Besides the aforesaid central indents for timber and exploitation by different commercial enterprises, this period also witnessed another major intervention on the traditional forest use through the system of "private lease of forests" whereby forests were leased out by the government to commercial contractors. During the 18th century, there was evidence of leasing out patches of Chittagong forests (BDR, 1978). One early document showed that an English gentleman (Captain Forde) was given "an exclusive Privilege of cutting Teak Timbers to the Eastward of Chittagong"; and expressing his opinion on this occasion, the Chief of Chittagong informed Warren Hastings, the Governor, on 26.3.1778, that, such a lease of forests (what he called a "privilege") "will rather be a service than an Inconvenience to the Natives as it will be a mass of Employment to [them]" (BDR, 1978, pp.391-392, Document 4.80). This system of private lease for commercial extraction of forests, as I will eventually discuss, continued well into the next period.

The extension of agriculture, which entailed considerable clearance of forests, received priority over forestry and was encouraged by the government. As Hunter (1893, p.622) observed, "the extension of tillage was considered as the chief care of Government and no regard was paid to the improvident waste of jungle on all sides". The early British government's agricultural policy, again, was prompted by possibilities of commercial production and exploitation. The land legislations related to agriculture, as Edwardes (1967, pp.82-83) noted,

> was to define proprietorial rights as a necessary step towards the efficient collection of revenue....The main agricultural innovations consisted of attempts to exploit, and in some cases, introduce exotic commercial crops.

In 1871, the Collector of Chittagong, for example, reported to his authorities that, "[a] good deal of land...near the hills ...and jungle-land is let on terms very favourable to the cultivators, in order to encourage the extension of tillage" (cited in Hunter, 1876, p.164). In the general context of the whole country, FRI (1961, p.72) noted that, "forests were considered an obstruction for further development of agriculture....The general policy [of the government] was to expand agriculture and to obtain Teak and other timber".

Many forested tracts were handed over to the Europeans through *Patta* (leasehold), on various pleas including establishment of sundry plantations. One of the first things which followed such a transfer of land was a process called "land-clearing" which most commonly involved clearance of forest and eviction of local people. Besides the mainland territories, many off-shore islands (for example Kutubdia and Maheskhali in greater Chittagong, now within Cox's Bazar District) also ceded to private management. "These grants failed to produce any good economic results as expected", because, the incumbents were never truly devoted to their estates and in most cases they sold the land to new comers and migrated to Calcutta (Islam, 1978, p.313). Moreover, in the early years of the Company rule, public officers became so nakedly involved in private trades (mainly in salt, timber and various commercial crops) that the Company devised, with little success though, some regulatory measures to restrict "such commerce" (BDR, 1978, Documents 2.29, 3.30, 4.46). On the nature of such trade and extortion, Hunter (1893, p.456) noted:

> The company servants had the privilege of carrying on their private trade, throughout Bengal, free from inland dues and all imposts....All the officers, civil and military alike, were tainted with the common corruption. Their legal salaries were paltry,...but they had been permitted to augment them, sometimes a hundred-fold, by means of private trade and gifts from Native[s].

Edwardes (1967, p.87) also noted that civil and military officers under the Company rule found "their habit better adapted for commercial pursuits [and] in the course of time, carrying on a successful commerce, many became possessors of large capital and returned to [Britain]".

Thus, the period between the 18th and the mid 19th centuries witnessed a state-patronized process of heavy exploitation of resources. This era, in fact, set in motion fervid attempts towards forest commercialization, which, as I will eventually discuss, continued through to present time. The next period, which was not essentially much different, saw the consolidation of these commercial processes under more organized structures of administration and governance.

The British period under the Crown: consolidation of the state forestry administration This section focuses on forest policies and practices during the latter half of the 19th century until the independence of India in 1947. By this time, in 1858, the administration of India was transferred from the East India

Company to the hands of an organized colonial government under the British Crown. The transfer, which was a partial response to "the plundering of the Company" (Edwardes, 1967, p.94), aimed at creating "a better government" (Hunter, 1893, p.495). "In a modernizing zeal...the British consolidated their power" by attempting to create "a rational and efficient administration" (Edwardes, 1967, p.149). Through these reform attempts, "the spirit of governance" was fortified and under the new government, as Edwardes (1967, p.176) argued, "the sprit [grew more] cold, bureaucratic, optimistic, and racially arrogant". As regards forestry management, this period saw the consolidation of commercialization of forest use, the formation of state forestry organisations and policies, more organised attempts in exploitation and revenue maximization, state monopolisation of forest ownership, regeneration activities, and systematic restriction of local rights and access. The events and experiences of this period are too wide to cover in this brief discussion. I will concentrate on major historical dates and incidents which had been instrumental in shaping the patterns of Indian forestry with particular reference to my study areas.

It is interesting to note that, following the unprecedented commercial exploitation, the very first steps towards regular conservancy of forests in India were also prompted by commercial motives. In 1800, for example, a Commission was appointed by the government to enquire into the availability of Teak in Malabar forests for commercial exploitation (Padhi, 1982, p.37; FRI, 1961, pp.72-73). In 1805, another government forest committee proclaimed "Royalty Rights" over Teak timbers in the south (FRI, 1961, p.73). Following which, in 1806, while wondering about "the question of regular supplies of timber to the Navy", the post of first Conservator of Forests in India was created and "his work was to arrange the exploitation of forests" (Dwivedi, 1980, p.12). The Conservator himself "established a timber monopoly throughout Malabar and Travancore and arranged for plentiful supply of timber" (FRI, 1961, p.73).

In 1855, Lord Dalhousie, the Governor-general, promulgated the first outline of forest conservancy in India in the form of a memorandum (Memorandum of the Government of India, dated 3.8.1855), which as Stebbing (1923) termed, was "The Charter of Indian Forests". Again, the "ruling principle" centred around commercial exploitation, although some realization of the "maintenance of forests" seemed to have occurred: "Teak timber should be retained as State property...trees which had reached exploitable size they should be removed in conformance with the future maintenance of forests" (cited in Rahman, 1993, p.22). The so-called "scientific forestry" had its footing by this time whose predominant works included the state appropriation of forests, maximization of organized exploitation, building logistics and communication networks, artificial "mono-culture" plantation, and hatching the growth of the future "invincible" bureaucracy. There had been a sway of "scientific forestry" operations in many parts of India. A "timber agency" was established in 1852-53 with a depot near Sialkot "to explore the forests of Western Himalayan Range" (FRI, 1961, p.73; Padhi, 1982, p.38). Prior to this, in the 1820s another government report on the forests of Himalayan foothills concluded that, "forest wealth was inexhaustible and that the exploitation of the forests could continue unabated" (cited in FRI, 1961, p.73). The "mono-culture" of exotic species in artificial plantations started and was imposed on many sites without any consultation with local people. By the early 1860s, exotic tree species, such as *Acacia* and *Eucalyptus* were introduced in Madras (Padhi, 1982, p.38). Bengal forests, it may be noted, had not been particularly targeted for "conservation through scientific forestry" until this time (FRI, 1961; Padhi, 1982; Dwivedi, 1982).

The year 1864 witnessed a major step towards the consolidation of state authority over forestry management. The "foundation of scientific forestry was firmly laid down" (Rahman, 1993, p.22) with the launch of the forest department under the Government of India and with the appointment of a professionally trained German forester, Sir Dietrich Brandis, as the first Inspector General of Forests (Padhi, 1982). The formation of a forest department was originally conceived in the Governor-general's despatch of 1.11.1864 to the Secretary of State (cited in FRI, 1961, p.76, emphasis added) in which he commented :

> the idea of allowing proprietary right in forests to individuals should be abandoned, as such rights were almost certain to lead to destruction of forests concerned. It was considered necessary that all *Government forests should be strictly set apart and made inalienable*. It was conceded that provision would have to be made for the continued exercise of certain private rights already existing, but *it would be a good policy to extinguish even those rights...wherever possible*.

Therefore, from its very birth, the forest department had been expected to foster proprietary rights of the state over nearly all major forest areas and also to ensure a clear seclusion of local people from "government forests". The despatch also made room for the appointment of the Inspector General of Forests and for a comprehensive forest act enabling the government to demarcate "reserves" and thereby, paving the way towards "extinguishing the local rights". Accordingly, the Government Forests Act 1865 (VII of 1865), the first attempt at forest legislation in British India, was promulgated. It made a series of restrictions on local rights. One provision of the law, for example, prescribed (cited in Stebbing, 1923, p.9):

> The preservation of all growing trees, shrubs and plants, within Government Forests...by prohibiting the collecting and removing of leaves, fruits, grass, wood-oil, resin, wax, honey, elephants' tusks...or any natural produce of such Forests; by prohibiting the ingress into and the passage through such forests...by prohibiting cultivation...and the grazing of cattle.

It also made government officers invincible against any public protest by such provision as: "No suit or other proceeding shall be commenced against any person for anything done in pursuance of this Act" (Clause 16 of the Act, cited in Stebbing, 1923, p.11). This Act also made provision for constitution of the "reserved" and "protected" forests; and "forest officers...endeavoured to put the Rules in force with uncompromising rigour amongst [the local] population" (Stebbing, 1923, p.14), causing rapidly growing popular discontent and alienation. As regards the nature of "reserved" forests, Hunter (1893, p.625, emphasis added) noted:

> [these were] managed as the property of state, *with a single eye to their conservancy and development as a source of...wealth*. Their limits [were] demarcated after survey;...*cultivation by hill tribes [was] prohibited: cattle were excluded from grazing*.

Even the Board of Revenue of the Indian government (Board's proceedings of 16.4.1868, No. 2777, cited in Stebbing, 1923, pp.15-17, emphasis added) itself, commenting on the nature and impact of the 1865 Act, noted:

> *while...useless for good, the Act opens a wide and dangerous field for oppression and extortion....It will be in the power of the pettiest "peon" in the [forest] Department to harass the people at his own pleasure....The people, in the exercise of their prescriptive rights, would be placed at the mercy of ill-paid and almost uncontrolled officials*. The Forest Officers naturally desire to have ample powers, but the Board would observe that there are other things to be considered besides the value and preservation of timber. The grazing of cattle in forest tracts is one of the chief grievances, but it is not too much to say that this was the saving of a very large proportion of the cattle of the country....*The fact is that a very large proportion of the so-called "forests"..., in truth, merely jungles with villages scattered through them; rights of way...and privileges existing..., and it is impossible to introduce into such tracts the stringent system of conservancy..., without exciting much popular discontent, and incurring serious risk of oppression....Government cannot declare that to be their property which is not clearly their own*.

Forest Conservancy in Bengal started in 1864 with the appointment of Mr. Anderson as the Conservator of Forest for the Lower Provinces (FRI, 1961; Dwivedi, 1980; Rahman, 1993). On assumption of the office, his first assignment was the exploitation of Sikkim Forests and establishment of a timber depot for extraction and trade (Stebbing, 1923). The "conservancy" in Bengal was justified (and given priority) by the government in consideration of the amount of prospective timber extraction and revenue earning. By the mid 1860s, the "Home authorities were fully alive to the fact that the introduction of Forest Conservancy,...so far as conserving and *perpetuating the timber supplies, had become a question of the first importance*" (Stebbing, 1923, p.375, emphasis added). In fact, the process of reckoning forests as prospective commercial commodity, as already discussed, started long before and it assumed an organized form through government intervention on forest management. Therefore, it was no surprise that a report, as early as 1832, in the *Journal of Asiatic Society of Bengal* (No.7, July, 1832, p.305) commented: "It is further probable that, as the forests have never been thoroughly examined by any *intelligent European*, we are still ignorant of many *valuable productions which they contain* (cited in Stebbing, 1923, p.400, emphasis added).

While there had been progressively stringent regulations on local access and rights, the system of leasing out forests to private traders for commercial exploitation continued with equal vigour. On the nature of the system of lease, a report by a British official in the latter half of the 19th century read: "It was only necessary for a contractor [mostly European]...to obtain a *parwana* or order from the Civil authorities to cut timber, and then he started work; and as every Gond carried out an axe, the forests were soon filled with fallen logs" (Report of the Superintendent of Forests, Central Provinces, around 1861, cited in FRI, 1961, p.75). Until nearly the end of the 19th century, the government's view was to "open forests to private enterprises" (Bachkheti, 1986, p.5). In the process, the greatest forested tract of Bengal -- Sundarban -- was also exclusively leased out to a private commercial agency (the Port Canning Company) in 1866. Such "monopoly" resulted in "brutal oppression" of the local forest-based people by the way of "collection of fees" (Hunter, 1876, p.304). Later, however, the lease was resumed; but the damage had already been done. In 1873, the government assigned a Deputy Conservator of Forests to reckon the forest of Sundarban which was considered an "inexhaustible jungle" (Hunter, 1893, p.624). The government's interest in the Sundarban was primarily because of the view that, it "forms a very important article of export" (Hunter, 1876, p.304). The Deputy Conservator's report chiefly confined to offering proposals for establishing "toll stations" (collection-points for government revenue) and issuing licences for timber-extraction (Hunter, 1876, p.304). The officer also opined to police government forests strictly to make the people "aware that the forests now are worked by Government"; and further prescribed that,"if a few men are caught felling against orders, and punished, it would have effect of stopping others" (Report of Mr. A.L. Home, Deputy Conservator of Forests, cited in Hunter, 1876, pp.311-312). Therefore, the government's basic orientation and approach to forest management were hardly different from those of the previous system of private contracts. Besides, although monopoly of a singular private agency was withdrawn, nevertheless, private business on different forest products went on full suing. The Calcutta-based timber merchants had their agents in Bengal who ensured regular export of trees from different parts of Bengal forests (Hunter, 1876). Hunter (1893, p.624), for example, noted that "Calcutta had, from its foundations, drawn its supply of firewood" from Sundarban.

The Chittagong Forest Division was one of the earliest Divisions to be formed in Bengal; but there had been a "curious lack" of regular records about these forests (Stebbing, 1923, p.403). Chittagong remained to be termed as a "backward division" in the government papers until as late as 1955 (see for example Ghani, 1955; Cowan, 1923). This was partly because "the great mass of forests of Chittagong and Hill Tracts" was considered "an unknown country peopled...by aboriginal tribes [and] a queer nomadic population", even until the 19th century (Stebbing, 1923, pp.403 and 441). Second, from the very beginning of "regular attempts in forest conservancy", there had been growing tension between local people and the government (Cowan, 1923, pp.16-19). An early "working plan" of the government, for example, noted that "the antipathy of the people in the [Chittagong] Collectorate and the time occupied in dealing with forest offenses have been contributing factors in retarding development [of forest conservancy]" (Cowan, 1923, p.18; also see Ghani, 1955, p.17). It was the prospect of earning huge revenue that induced the government to establish forest management in Chittagong. Around 1865-66, the revenue collected by the Civil authorities led the government to decide that the "forests of this district will be worth the expense of conservancy;...it is probable that on careful inspection valuable productions will be found" (Hunter, 1876, p.408). Accordingly, the Government of India (Despatch No.93-F, dated Fort William 1.3.1867) and the Secretary of State (Despatch No.Rev./For./14, dated 8.5.1867) opined that Chittagong forests should be brought under the forest department's control "as soon as it can be managed" and foresters' "authority should be extended" (cited in Stebbing, 1876, pp.408-410). It may be noted that the original Chittagong Division was a huge administrative charge, consisting of the greater Chittagong Hill Tracts and Cox's Bazar Districts. At present there are as many as 8 Forest Divisions over the area that formed the original Chittagong Division.

Prior to the take over by the forest department in the late 19th century, Chittagong forests were under the purview of the revenue department as the *Khas Mahal* Forests (Khan, 1980). In the *Khas Mahal* forests, it seemed that the rules were somewhat flexible and local people could meet their demands to some degree. From the launch of the forest department until the first decade of this century, forest policy of this region centred around "creating Reserves and collecting revenues" (Khan, 1980, p.71; also see Cowan, 1923, p.18; Ghani, 1955, p.18). In 1871, the charge of forest toll stations of the Chittagong Division were taken over from the tribal "hill-chiefs" by the forest department. Thereby, the indigenous authority of hill communities were over-ruled by the state and "a more rigorous method of collection was at once installed, with a consequent great rise in revenue" (Cowan, 1923, p.19; see Table 5.1).

Table 5.1 Government revenue earning from the Chittagong Forest Division (1862-1875)

Annual Average	1862-1871	1872-73	1873-74	1874-75
Rupees	11280	45429	57589	102446

Source: Cowan, 1923, p.19. Note: The Rupee figures do not appear to be corrected for price fluctuations.

Massive "reservation" work started from the 1870s onward with a proportional reduction of local rights. By a notification in the Calcutta Gazette (dated 1.2.1871), nearly the whole Chittagong Hill Tracts district (that is, 5,670 out of 6,882 sq. miles) was declared a "government forest" (under section 2 of the Act VII of 1865) (Khan, 1980; Cowan, 1923; Ghani, 1955; Baten, 1969). In 1875, the Conservator of Forests of Bengal, Sir William Schilich, visited Bengal and his visit was reported to have "marked the beginning of a golden era for management of Chittagong Forests" (Khan, 1980, p.72). The following excerpt of the Conservator's report (cited in Cowan, 1923, pp.19-20), following that "epoch-making" visit, shows his view about the indigenous practices and local rights:

> The system of forestry in force in the Hill Tracts until a few months ago may be shortly described as follows: The hill-men had the free run of the forests for their

> domestic requirements....Individuals who desired to export forest produce had to take out a so-called permit to cut....The effects of the above system have been detrimental to the forests as the free run of them and the almost unrestricted cuttings told very severely on the growing stock of the timber....The great enemies of forest conservancy were...the practice of *jhuming* [and] the wasteful manner in which timber has been cut....In short, there can be no doubt that the forests were on the road to ruin.

According to his prescription for rescuing forests "from the road to ruin", the following years witnessed a constant rise in levies, tolls and in the expansion of "reserves". *Jhuming* (an indigenous form of shifting cultivation, practised by tribal and indigenous communities) and all forms of extraction and access by local people were prohibited in reserved forests. Levies, for example, on Jarul timber (which tribal people extensively used for preparing dug-outs) were enhanced by as high as 50% (Ghani, 1955, p.19). In the year 1875 alone, 570 sq. miles of area was declared reserved forest which included the whole land-mass of major river-side valleys in Chittagong (namely Sungoo, Maini and Kasalong). The newly formed "reserves" were primarily inhabited by hill tribes and aborigines. In 1884, the total area of reserved forest rose to 1,345 sq. miles (Ghani, 1955). Mono-culture trials of exotic species (with an almost total disregard to local choice or demand) were also carried out, side by side the process of reservation. In 1871, Teak seeds, obtained form Burma, were planted at Sitapahar and Pahartali of Chittagong (Cowan, 1923, p.19). From the very inception, these plantations faced public opposition and distrust. One government document noted, for example, that "villagers made a point of doing mischief at night to the young plants which they considered likely to grow up and shut out their pasturage" (Ghani, 1955, p.18). The ever-expanding "reserves" also necessitated huge expenditure (Cowan, 1923, p.22) and consequently, more importance was attached to revenue maximization. Meanwhile in 1878, a new forest act (Act VII of 1878) came into operation with further emphasis on forming reserved and protected forests. Since in protected forests some form of existing people's rights were (at least) recorded (although not settled), the government did not seem to compromise with its ever-growing monopolistic proprietorship; and therefore, protected forests "were gradually converted into reserves" (Stebbing, 1923, p.470). During 1889-90 there were 56,000 sq. miles of protected forests in British India. By the end of the century reserved forests were extended to 81,400 sq. miles, while protected forests were reduced to 8,800 sq. miles (Stebbing, 1923, p.470). A process called the Forest Settlement Work (which has continued to be a major duty of foresters until today) was very active throughout the Indian public forests during the period between 1871 and

1900. Its main purpose was to demarcate and settle governmental proprietorship and royal rights over community or private rights. Mr. Ribbentrop (who later became the Inspector General of Forests), while describing the objective of settlement works, wrote: "[its aim is] to enquire and record to what extent the proprietary rights of the State are limited by existing adverse rights of private persons or communities" (cited in Stebbing, 1923, p.476). He did not want "*the forest [to] remain burdened with rights of pasture or wood cutting*" and maintained that "*a forest can be said to fulfil its highest function when it produces, in a permanent fashion,...the best possible return to the proprietor*" (cited in Stebbing, 1923, pp.476-477, emphasis added). Moreover, chapter II of the Act of 1878 made provision for "village forests" which were supposed to respond partially to local needs. But, even as regards the village forests, the Act "demanded that the forests should first be created [as] reserved forests"; and this provoked public resentment and suspicion (Stebbing, 1923, p.471).

Accordingly, reservation works in Chittagong forests went on apace in line with other parts of India. Communications and construction activities were carried out; logistic and manpower supplies were enhanced; patrolling and policing were intensified; and many "inspection paths" and *bungalows* (rest houses) were built (Cowan, 1923; Ghani, 1955). However, these had little effect in appeasing the growing bitterness between the government and local communities. After the reservation of Ramgarh-Sitakund tract in Chittagong in 1893, for example, incidents of so-called "forest offenses" increased by 300% (Cowan, 1923, p.22). Again, these "offenses", in a kind of cyclic order, provided the government with the coveted plea and justification for further reservation. The demarcation of government territories and collection of tolls constituted the predominant tasks of forest staff (Cowan, 1923, p.22). In 1879, there were four toll stations (Chandraghona, Syllock, Ichamoti, Matamuhuri) in Chittagong Division and by 1881, additional stations were erected at Halda, Idgaon and Rezu (Ghani, 1955, p.19). In the process, public discontent boiled up rapidly and consequently, the government was obliged to make some room for compromises and ventilation of tension. For example, 120 cases (involving 1,059 persons) of "forest offenses" (concerning *jhuming* by the locals) in 1905-6, led Maini reserved forest (an area of 337 sq. miles) to be declared "disforested" (deforested) (Khan, 1980, p.76). Similarly, the government had to relinquish the proposal of reserving another huge tract in southern Chittagong, called Naaf Sundarbans, in the same year (Ghani, 1955, p.20).

In 1894, British India's first forest policy was formulated. It "gave preference to agriculture over forestry" and proposed that "demand for cultivable land can be, to some extent, met by clearing forest areas" (Hussain, 1992, p.18). Understandably, it gave renewed impetus to the process of "land-clearing" that had long been active in Bengal, causing considerable damage to forested tracts. It also made it clear that "Royalty for the Government must be collected for various facilities enjoyed by people" (cited in Rahman, 1993, p.24; also see Wadud, 1989, p.5). These facilities included limited "concessions" for pasturage and fuel wood collection. Rahman (1993, p.24) argued that "the main aim [of the policy] was to collect revenue and to satisfy the local population by granting so-called rights and concessions".

The first "working plan" (the departmental work and operation schedule) for Chittagong forests was drawn by Cowan (1923). Subsequent working plans (for example Ghani, 1955; Baten, 1969; Khan, 1980) had been stereo-type planning practices, each mirroring the preceding plan. Two explicit objectives of all the plans were to maximise economic return from forests and to ensure uninterrupted supply of raw materials to forest industries. Although there had been promises of catering for local demands, these were mostly confined to papers only. All the plans and many departmental records (see for example GOEP, 1964; 1964a) continued to identify local people as heinous enemies of forest, primarily for such activities as *jhuming* and pasturage which local communities had been practising for generations. Government documents repeatedly declared that "there are no right and concessions of any sort in the reserved forests of this division" (Ghani, 1955, p.11; Baten, 1969, p.16; Khan, 1980, p.37). In the same vein, the working plans of Dhaka-Mymensingh forests were equally vocal against the "mischievous" local people especially "squatters" and "encroachers" who, according to the government, "considered forest as no man's land" and to whom forest "offers a permanent attraction for encroachment and illegal clearance" (Chowdhury, 1957, p.11; also see Ahmad, 1938, p.3). Accordingly, the plans suggested a complete denial of all "rights and concession" in the "reserves" and prescribed strictly regulated limited grazing on particular areas only (Ahmad, 1938; Chowdhury, 1957).

Meanwhile, a revised extensive forest act (The Forest Act, 1927: Act No.XVI of 1927) was enunciated with a radical increase in the power of the government to form reserved and protected forests. It also covered a host of regulatory and punitive measures concerning cattle grazing, forest settlement, local rights, and the control of timber and forest products (Alam, 1992). Reservation works continued with great vigour. The forests which used to meet some local demands were also engulfed in the process. In 1930, for example, the *Khas Mahal* forests of Chittagong were taken over by the forest department from the civil administration and were made into either reserved or protected forests (Baten, 1969, p.33). During the second World War there was another surge of exploitation in the already dwindled forest mainly to cater for the mammoth demands of the government's war-time communication and construction activities. "The forest department undertook to organise the extraction and delivery of all timber required;...and the huge demand...during the years 1942-45...resulted in all accessible parts of the forests being heavily depleted" (Khan, 1980, p.80; also see GOEP, 1961, p.1-V). After the war, forests were indented for "special works for improvement under the post-war Reconstruction Schemes" by a departmental plan called the Dent's Scheme (Baten, 1969, pp.33-34). The "coupes" or blocks of forests were sold out to private contractors who extracted the products by "clear-felling" (Baten, 1969, pp.33-34). Here again, local people's share of the forests was sacrificed for commercial interests. Private and community forests were also "ruthlessly exploited during the war-time" (Dwivedi, 1980, p.15).

In the 1930s, private forests under the control of *zamindars* (land-holders) and other private proprietors also ceded to state control. This gave impetus to the process of horizontal expansion of state authority (and the subsequent proprietorship by the state). There were many privately owned forests throughout Bengal and the early attempts to regulate working of these forests met with public opposition. The government could not proceed with its intension of bringing these forests to its management "for want of public support" (Dwivedi, 1980, p.14). From the mid 1930s to the early 1940s, more laws were passed enabling the government to take over the control of these forests. In the process, the forests of Dhaka-Mymensingh tract were appropriated by the government in the mid 1930s. They were later declared reserved or protected forests, limiting or prohibiting local access and share (Ahmad, 1938). Similarly in Bihar, for example, the "Private Forest Act was passed according to which any private forest could be taken over by the Government" (Dwivedi, 1980, p.15).

In sum, the colonial period was marked by state monopolisation of forest ownership, intrusion of commercial interests into local economy, and a systematic denial of local rights. The trends in forest commercialization were sparked by the early Company rule which was characterised by "rampant profiteering" and maximized exploitation. Since the mid 18th century, more organized forms of government and state activities were noted. However, the basic trends in commercialization and alienation continued unabated. This later phase of organised governmental activities also witnessed the introduction of bureaucratic forest management whose main purposes being revenue maximization, restricting local rights, further expansion of state control and proprietorship, and more stringent rule enforcement.

Amidst these predominantly commercial considerations, as Hartmann and Boyce (1983, p.12) argued, by the time "when the sun set on the British Empire in India, eastern Bengal [Bangladesh] had been reduced to an impoverished agricultural hinterland". Similarly, Roy (1987, p.25) also maintained that at the end of the colonial rule, "the present Bangladesh area receded into background, [serving as] a hinterland of the industrial nucleus of Calcutta".

The Post-Independence Era: Continuity and Consolidation in The Pursuit of Commercialization

As far as forest management and local people's access were concerned, the post-independence period, in essence, was not distinct from the philosophies and practices of the colonial rule. In fact, as the following sections will elaborate, the trends in commercialization and alienation not only continued, but in some instances, were further intensified.

The Pakistan period The independence of India and the formation of Pakistan in 1947 brought about little change in the nature of forest use and management. The Pakistani period (1947-1971) was a continuation and outcome of the colonial rule and exhibited similar characteristics. Revenue-orientation of forest policies, isolation of government officers from people, emphasis on maximum economic return from forests, governmental patronage for forest-based industries, maximum exploitation, and the expansion of state proprietorship over forests -- were the main features of forestry during this period. Roy (1987, pp.25 and 39) explained:

> Pakistan was an ardent follower of the industrial path; so in development planning emphasis was always given to maximising overall growth through accelerated industrialisation....The forestry practices was not different from the colonial period....Up to 1960, the activity of the sector was mainly limited to realising revenue for forest produce extracted from Govt. forest at a speed greater than that during the colonial period, as more forest produce was required to support the urban economy....After 1960...the forestry sector...took the strategy to utilize any existing resource that could be found as base for western, capital-intensive, export-oriented industries.

Therefore, local demands received little attention. Instead, a number of capital-intensive forest-based industries were set up during this period including the Newsprint Mill in Khulna, Paper Mill in North Bengal and the Rayon Complex at Chittagong. Besides, there was mushroom growth of hardboard, particle board, plywood and match industries during the 1960s (Roy, 1987). Industries in Pakistan period, as a rule, received over-whelming state support and protection, fed west-Pakistan markets, and most of the benefits (accrued from these industries) were funnelled to their entrepreneurs who were mostly west-Pakistani capitalists (Roy, 1987, p.26; Alam and Rahman, 1993; it may be noted that the province of "East-Pakistan" constituted present Bangladesh). Until the late 1950s, the development of forest industries remained almost entirely confined to a group of private capitalists, implying a purely commercial focus of the regime (Zabala, 1990, p.12). In 1960 however, a major initiative was launched with the establishment of a state-owned and publicly-managed forestry enterprise called the "East Pakistan Forest Industries Corporation" (after 1971, known as the Bangladesh Forest Industries Corporation). Its main purpose was "to initiate the use of mechanised extraction methods in accessible hill forests areas" (Zabala, 1990, p.12; also see Das, 1982, p.10). But, despite the great state protection and subsidized supply of inputs, from its very birth, the Corporation had been running, with rare exceptions, at loss. In 1969-70 the loss was Taka -3.01 million which rose to Taka -17.81m in 1978-79 (Bangladesh Forest Research Institute (BFRI), 1982, pp.71-72). Moreover, the Corporation seemed to have contributed to further depletion of forests; because initially it was assigned the "challenging task" of "clearing the best natural forest" of the Chittagong Hill Tracts with modern equipments to feed the industries (Roy, 1987, pp.42-43). Huge quantities of "hill bamboos" (which were commonly used by poor people and called "poor men's timber") were mechanically extracted "to feed the Karnafuli Paper Mills" (GOEP, 1961, p.1-V).

On the replenishment and plantation side, there had been two trends during the Pakistan period. Until 1960, the government followed colonial footsteps in toto by concentrating on policing people and raised only 13,163 ha. plantations in the 14 years period between 1947 and 1960 mainly with Teak (Roy, 1987, p.39). During the 1960s, though large-scale plantation schemes were designed, these were primarily focused to feed the growing industrial needs. Besides, as Roy (1987, p.43) observed:

> The existing natural forest was considered as low-yielding and unprofitable [and] as a result...the best natural...forest areas of the country...were replaced by the artificial Teak forest....In the 1960s, 51,197 ha. of natural forest were converted into artificial plantation mostly with Teak.

In fact, exotic Teak plantation was favoured over indigenous species and local choice, and was advocated all-through the colonial period on predominantly commercial considerations of "high quality timber and nominal planting cost" (Griffith, 1941, p.129). In I955, in the same vein, the government noted that the "Teak is almost certain to be continued as the most important species in future forest plantation programme" (GOEP, 1955, p.4); although as I have earlier discussed, there had been public discontent right from the very first Teak plantation in Chittagong as early as 1871.

The Pakistan period witnessed the formation of two forest policies. Though apparently devised to cater the need of a newly independent nation, the Forest Policy of 1955 depicted all characteristic manifestations of the colonial forest administration including the expansion of state territories, "scientific" extraction of timbers, fortification of the bureaucracy by increased training and manpower, and managing all forests through rigid departmental plans (see for example Hussain, 1992, p.18). In 1962, a second policy was launched (The Forest Policy, 1962), which stated (cited in Hussain, 1992, p.18, emphasis added):

> *The management of forests to be intensified to make it a commercial concern*, utilization of forest produce was to be improved,...regeneration speeded up to keep pace with increased harvesting, irrigated plantations primarily to produce industrial wood to be included...and *timber harvesting in Chittagong and Sundarbans was to be accelerated.*

The local rights and demands remained as ignored as before.

The Bangladesh period Bangladesh surfaced as a sovereign state on the global map in 1971 following a historic civil war of independence. In the first decade of independence, there were two economic plans in Bangladesh: the First Five Year Plan (1973-78) and the Two Year Plan (1978-80). Philosophically, these plans "aimed at economic rehabilitation of the war-damaged economy as well as acceleration of economic growth mainly through public sector" (GOB, 1990, p.II-1). But practically, in the forest sector the "development resources were allotted mainly for raising plantations with a view to meeting the future demand of raw materials for forest industries and...the potential of rural forestry sector, the most important forest resource of the country, was totally ignored" (Roy, 1987, pp.44-45). The government's support for homestead and rural forestry remained nominal. The only notable service from the government was offered in the form of establishing some nurseries, most of which were located in urban areas, benefiting a section of urban population (Roy, 1987, p.44). 47 new nurseries were set up during the First Five Year Plan period (making a total of only 59 nurseries for the whole country) and the so-called (bureaucratized) "extension services" were extended to 6 districts (Anon., 1991a, p.10). Rural local demands and choice remained largely obscured in these initiatives.

As discussed in Chapter 1, even without or with very little assistance from the government, rural homestead and community forests have continuously catered for the predominantly major portion of the country's demands for forest products (GOB, 1990, p.V.E-1; 1992, pp.45-49; ADAB, 1992, p.8; Anon., 1991, p.3; Task Force, 1991, p.125). These forests (as compared to state-forests), notwithstanding their invaluable contribution, remained neglected in the country's macro plans until the early 1980s.

The latest forest policy was announced in 1979. Again, this was "a two-page manifesto-type statement" with obscure and "generalised directions", "mostly focusing on the forest department" (Anon., n.d., pp.5 and 18). Its suggestions included "horizontal expansion of the forest area" under the government which was to be "carefully preserved and scientifically managed" by a (centralized) "cadre of forest officers", "setting up new forest based industries", "optimum extraction forest produce", and protection of forests from the (so-called) "encroachers" (see for details GOB, 1979). Rural forestry and local people received no major attention, except as a cursory and vague call for a "mass motivational drive for tree planting" (GOB, 1979). In fact, the policy "expressed the views of the traditional foresters, overlooking the overall development strategy" (Roy, 1987, p.45) and was hardly adequate for addressing the current needs and crises of Bangladesh forestry (Task Force, 1991, p.219; Anon., n.d., p.18).

The overwhelming emphasis on revenue maximization continued unabated over the years. In citing an example of the trends in revenue acceleration, Figure 5.1 depicts the earnings of the Chittagong Forest Division over the last 50 years.

Bangladesh forestry showed little change from the traditional colonial-industrial approach to forestry. It concentrated on "managing vast tracts of publicly owned forest land by a bureaucracy, keeping people out" (Roy, 1987, p.44). With the exception of slight reshuffling in the mid 1980s through the launch of SF programmes, the colonial approach more or less continues to be the dominant trend in the country's forestry sector until today. It has been said that "sustained yield" is the main "obligation" of the public forest management of Bangladesh (Khan, 1980; Zabala, 1990). Under the umbrella of "sustained yield", the objectives of forest management (for different categories of forests) are as follows: (a) for the Hill Forests the "mainstay" of forest management is to "convert irregular forests by valuable and fast growing species" and "to derive maximum economic benefit under the principles of sustained yield"; (b) for Inland *Sal* Forests the objectives are "to

Figure 5.1 Public revenue earning from the Chittagong Forest Division: 1928-1977 (in million Rupees/Taka)

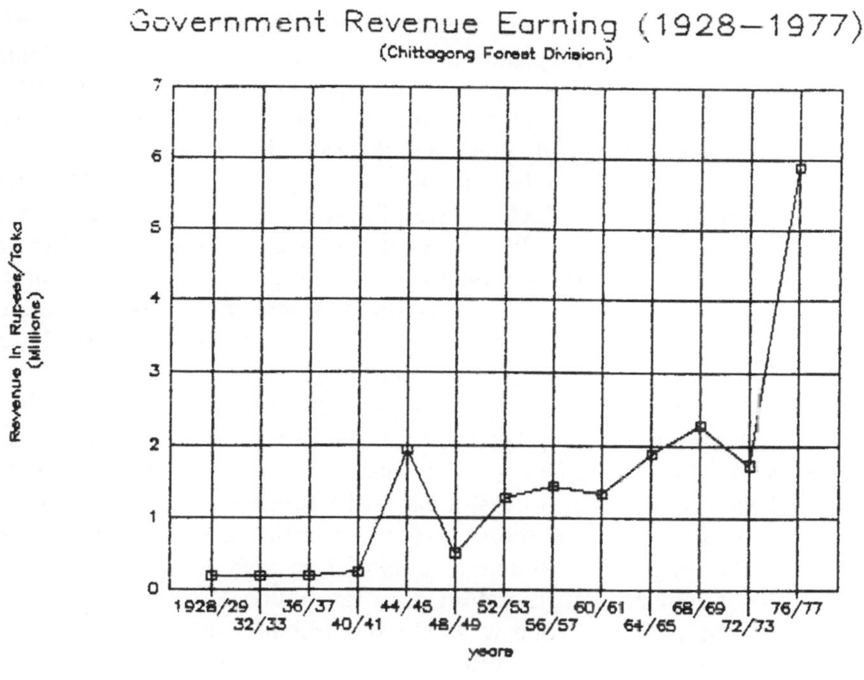

Data sources and notes: Information for the period between 1928-29 and 1963-64 was collected from Baten, 1969, pp.120-121; and between 1966-67 and 1977-78, from Khan, 1980, p.89. The (revenue earning) figures do not appear to corrected for currency fluctuations.

bring the forest under scientific management" and "to create recreational facilities in these forests for town dwellers" (Zabala 1990:16-17). There has nearly been a complete lack of any significant provision for rural people's involvement in forestry in pursuing the so-called "sustained yield" principle.

The Fourth Five Year Plan (1990-95) also noted that the performance of public forestry, to say the least, has been poor. The achievements, especially as regards SF projects, lagged far behind the targets. The "unclassed state forest" lands were primarily used for catering to the industrial needs. In many government plantations the survival rate of trees was poor. The stereo-type principle for plantations (which the government inherited from the colonial rule) was: "clear-felling with artificial regeneration". The "clear-felled" areas were mostly planted with Teak without regard to "species-site suitability" or "local choice" (GOB, 1990, pp.V.E2-V.E3). The very method of "clear-felling", which is often followed by "burn off" operation, entails some clear maladies. It can cause damage to top-soil and micro-organisms, make the site more vulnerable to biotic interference and natural calamities, delimit bio-diversity and seed-sources, and encourage soil erosion (M.K. Hossain, Institute of Forestry and M.K. Alam, Bangladesh Forest Research Institute, personal communication).

Industrial needs claimed priority in the forestry sector over local demands. "Public industrial plantations using fast growing species" were emphasised by the government (Magno and Ali, n.d., p.1). The paper and match factories "have eaten up vast areas under forest in the Chittagong Hill Tracts and in the mangrove forest in the Sundarban area" (Ahmad, 1990, pp.15-16). The long-term extraction rights of forest products to paper mills and the Forest Industries Corporation, at a rate much lower than stumpage rates (for instance, Gewa wood was sold at a ridiculously low rate of Taka 0.06 per cft. to the Khulna Newsprint Mills) were uniformly maintained, regardless of the setbacks in the forestry sector (Task Force, 1991, p.220). According to the latest "working plan", Chittagong forests' primary consumers have continued to be rural and urban saw mills (15% of the country's saw-mills are located in Chittagong), city furniture shops, and sundry wood-based industries (Choudhury, 1993, p.29). Needless to point out that local people have no significant share in such a list of consumption. Furthermore, non-timber forest products (such as bamboo, sungrass, wax, honey), which are of great utility to the poor, have received little importance and have been branded under a degrading caption of "minor forest products" (Roy, 1987). Notwithstanding their contribution to local as well as national economy, these so-called "minor products" of general public importance have continued to be neglected in government documents, and little systematic attempts have been put forth to augment their production (Khan and Khan, 1994; GOB, 1992).

As regards forest legislations, no major qualitative change was made in the colonial Forest Act of 1927 to update and mould it to the distinctive demands of an independent sovereign nation. Although Bangladeshi officers were vocal about changing the act, they primarily confined themselves in suggesting more "stringent punitive measures" and "further power" of policing for them (see GOB, 1977). Little deviation from the colonial time has been noted in the culture and attitude of the bureaucracy regarding people, as echoed in the following comment of a senior forest official (Bhuiyan, 1977, cited in Wadud, 1989, p.17, emphasis added) :

> In an underdeveloped country like ours *the regard of the people towards any law is governed by its harshness in application....The Forest Act being too old, the penal provisions contained in it are too lenient....Because of this lenient application...rules do not have a compelling and demoralising effect on the minds of the offenders.*

Similarly, another officer suggested that the punitive measures should be "considerably increased" and "forest official should be vested with magisterial powers to deal with forest offenses" (Chowdhury, 1977). Accordingly, "some cosmetic forms of" amendments of the Forest Act of 1927 were made in 1989 (Anon., n.d., p.19; also see Ahmed and Bhuiyan, 1993, p.4) with provisions of more stringent measures for policing local people and further restrictions on public access (Alam, 1992).

Overview

Thus far, I have reviewed the evolution of forest use in the country with special reference to the broader study areas of Chittagong and Dhaka-Mymensingh forests. The review focused on the processes of commercialization of forestry and the consequent alienation of local communities. It also revealed that, although these processes were initiated by the colonial rule, they have continued with full vigour through the post-independence period. Bangladesh forestry is still pursuing the colonial heritage in the forms of bureaucratic-custodian management of forests, restrictive and punitive laws, planned isolation from public, para-military orientation of state officials, state-patronage for industries, and commercial exploitation and revenue maximization.

Now, in an attempt to focus more closely, the next section examines the changing patterns of forest use in the specific context of field-case sites (Betagi, Pomora and Chandra). It is basically a continuation of the above discussion, but it is more probing of the field realities and reliant on more first-hand data. Very limited secondary sources of information were available on the field-cases; and I primarily relied on personal observation, questionnaires, key-informant interviews, ethnohistories and a series of informal discussions with local people from different walks of life.

Evolution of Forest Use in the Field-case Sites

This section uses historical and contemporary evidences to shed light on the processes of forest commercialization and public alienation, which I have identified in the preceding review of forest use in the broader study-areas. As virtually no secondary information is available on the field-sites prior to the British administration, this section concentrates on the evolution of forest use since the colonial era. In particular, it focuses on such processes as state appropriation of local forests, progressive diminution of local rights and access, growing hostility between state officials (and its commercial allies) and local people, maximized exploitation and revenue earnings by the state, and the intrusion of commercial forces in the localities under state patronage.

Chandra

As noted before, very little record on Chandra forests exists, except occasional mentions in a number of government documents of the colonial time. Chandra forest is said to be a conterminous patch of the broader Bhowal forests in the Dhaka Mymensingh *Sal* forest tract. This tract was clothed with dense forests, predominantly (about 95%) of *Sal (Sheora robusta)* trees, locally known as *Gajari or Gazalli* (Ahmad, 1938, p.5). The growth of the trees was so rapid that these forests were considered inexhaustible. The inaccessible forests of this locality served as shelter grounds for outlaws and other migrants who had been driven out of their homes for such reasons as dire poverty, political unrest, natural calamities, and criminal activities. Prior to the 1940s, these forests were private properties of *zamindars* and other influential families (Chowdhury, 1957). *Zamindars* earned some revenue from these forests through their revenue officers, called *Naibs*. They were not particularly interested in the forests, except as an alternative source of revenue (Ahmad, 1938). The so-called "offence of encroachment" by local villagers, as later condemned by the government, did not seem to have bothered the *zamindars* much. A later government document also noted that "they [the private proprietors or *zamindars]* acknowledged no responsibility for protection [of forests] so that encroachment...passed unpunished" (Ahmad, 1938, p.12). Therefore, as reported by many elderly villagers whom I interviewed, it seems most likely that the poor and forest-based local people were able to use forests for their day-to-day requirements, and were not particularly targeted by any stringent prohibitive regulation from the *zamindars*. Moreover, this forest tract was inhabited by a considerable number of tribal population consisting of *Garos, Kooch, Hajongs* and *Mandhais* tribes (Ahmad, 1938; Chowdhury, 1957). An off-shoot of the *Kooch* tribe, locally known as the *Barman* family, still lives in Chandra (Ahmed, 1994, p.7) and constitutes a major section of my respondents. Dhaka-Mymensingh forests were characteristically different in many ways from other forests. These forests are "generally intermixed with cultivation and habitations of local tribal people" (Chowdhury, 1957, p.1). Many of these tribes have lived here for generations (Ahmed, 1994). As my informants reported, they have practised shifting cultivation for ages and seemed to have inflicted no major damage on forests (as compared to the deforestation and plunder of trees caused by commercial logging in recent times). But, the governmental documents consistently condemned the practice of shifting cultivation and identified it as "the greatest enemy of forest" (see for example Ahmad, 1938; Chowdhury, 1957). One document goes so far as to accuse that "due to the practice of shifting cultivation by the local tribes more than 80 percent of the forest has been affected" (Chowdhury, 1957, p.12). However, in informal interviews with me, a number of forest researchers of the Bangladesh Forest Research Institute opined that the damage caused by shifting cultivation was "much less than what the official propaganda have suggested"; and shifting cultivation as an agent of deforestation, "was of no significant bearing as compared to that of illegal commercial logging" (personal communication). Even the government's Forestry Master Plan experts note that, although *jhuming* "is universally condemned as undesirable...but it is often forgotten that it was a remarkable innovation for primitive cultures responsible for revolutionary transition from food-gathering to food production" (FMP, 1992, p.44). The tribal (*Chakma*) chief of Chittagong hill tracts, *Raja* (king) Devashish Roy (cited in Gain and Moral, 1994, p.29) maintains:

> Ironically enough, those who speak or write negatively about *Jhum* cultivation...usually seem to have little or no practical knowledge about *Jhuming*. Had they looked closely enough, they would have realised that this age-old mode of cultivation is not really the evil that is made out to be,...and that *Jhumias* too are neither primitive nor nomadic.

In fact, the limited deforestation in the *zamindari* period was primarily caused by timber merchants (and not by local people) who misused the lease issued by *zamindars*. In many instances, "forests were freely leased out to the timber merchants" (Ahmad, 1938, p.13). The main routes of commercial logging by these merchants were through the rivers, especially the Turag, Bangsha and Banar rivers (Ahmad, 1938, p.9). Besides, some *zamindars*, in the face of financial hardship, took resort to forest extraction. Parts of Bhowal forests, for example, suffered massive extraction by the successors to *Raja* Rajendra Narayan Roy "to provide the necessary funds to put the estate on its leg" (Ahmad, 1938, p.13). However, these activities did not seem to affect the life and living of general villagers. They were habituated to live among forests and to cultivate the adjoining lands (locally called, *Baid* lands). As noted above, they had little restriction on obtaining the daily requirements from their neighbouring forests. As my evidence suggests, their subsistence demands had little negative impact on forests and they did not seem to have played any major role in commercial extraction. Though forests were under private (mainly *zamindari*) ownership, in many instances local people did not even know whom the forests belonged to. In other words, the private ownership status of forests (vis-a-vis state control) was most likely not a major hinderance to local people's access to forests for meeting their subsistence requirements. Indeed, the fact that villagers had a reasonable degree of access to forests was even recognized by some government officials. For example, at the advent of reserving these forests (during the early 1930s), the Forest Settlement Officer himself noted that the people "enjoyed unfettered rights under the management of the *zamindars*" (cited in Ahmad, 1938, p.3).

In line with the sway of state appropriation of private forests since the early 1930s in many parts of the Bengal (see Dwivedi, 1980), Dhaka-Mymensingh forest tract ceded to government management. The first patch of these forests, Atia forests, was taken over by the government (under section 38 of the Indian Forest Act, 1927) as early as 1925 (Ahmad, 1938). Later a series of reservations was made; and by the mid 1950s, nearly the whole tract came under government's purview. In many occasions, the process of reservation met with public opposition. The Forest Settlement Officer, for example, when trying to enquire into local rights on these forests around the year 1925, "an organised opposition started against forest settlement as it interfered with the unfettered right" (Ahmad, 1938, p.3). The government pleaded that "private [forests] were taken up for scientific management by the forest department" (Chowdhury, 1957, p.25). In 1951, the Dhaka-Mymensingh Division was split into two separate Divisions and a new Mymensingh Forest Division was launched, following which massive "vesting and acquisition of forests took place" (Chowdhury, 1957, p.25). The Chandra SF project area partially covers two *mouzas* (revenue villages) of Kaliakair *Thana*, namely Kalampur and Chandra. On 14.11.1951, a total of 850.83 acres of land at Chandra were acquired by the government under section 7(a) of the East Bengal Private Forest Act of 1949 (Notification No. 11,397) and 11.8.1952 (Notification No. 9,490). Similarly, 393.58 acres at Kalampur were acquired on 14.11.1951 (Notification No. 11,397) and 11.8.1952 (Notification No. 9,490) (source: Notification File, Forest Range Office, Kaliakair).

Prior to the take over by the government, Chandra locality was under a *zamindar*, locally known as the *zamindar* of Baliadi. He was hardly interested in forests. Some local influential families, on payment of a cess called *Khazna*, enjoyed de facto control over forests. Some of them were engaged in commercial extraction and marketing of timber. They usually used the Dhaka-Tangail road and the Turag river for despatching timber to urban centres (primarily to furniture shops and timber depots). As my informants reported, the poor locals were hardly barred from entering forests for leaves, barks, twigs, branches, and (occasionally for) poles or small logs. Some villagers, such as Abul Basit (aged 56), used to act as paid labourer (*Kamla*) for timber traders. Basit recalled:

> I used to work as a helmsman in big carrier boats, plying in Turag [the nearest river]. My *Compani* [employer] used to buy *gazalli* [*Sal*] from local *sowdagars* [merchants]....Many labourers, especially the strong [able bodied] *Barmans* used to bring the trees from forests to the *ghat* [river-station] either on their shoulder or on *van-gari* [rickshaw van].

Since the state intervention, a massive policing and eviction operation was carried out to evict local people, who by that time had become "encroachers" on the land where they had been living for ages. The evicted people assumed a kind of nomadic life by resettling in different parts of forests before being chased again. In the process, they also lost their *baid* lands adjacent to forests where they used to cultivate paddy. Many local villagers recalled that uncertainty became a part of their life and consequently, they started to grow desperate. It was at this point that some of the local poor (especially the evicted or chased villagers) took recourse to causing deliberate damage to forests. The local Chairman of Mouchak Union council told me that since the evicted villagers were pushed into hardship and into an unsettled life, they joined hands with commercial loggers who by that time were very active in these forests. Loggers used to employ these people (at very low wages and exploitative terms) to extract timber. The people, in the absence of any better alternative, found it easy to "bring a head-load of *gazalli* and get cash-money from the *Munshi* [the agent or clerk of the timber trader], waiting for them at a nearby point with a truck" (Chairman, Mouchak Union, personal communication).

Immediately after the state acquisition, the indigenous practice of cultivation "was put to an end" (Chowdhury, 1957, p.25). Instead the *Taungya* system was patronized by the forest department, whereby a selected group of villagers were allowed to reside in forests and practise "a method of establishing forest crops in temporary combination with agricultural crops" (FMP, 1992, p.45; also see Chowdhury, 1957, p.84). The primary purposes behind the formation of these departmentally patronized "forest villages" (through *Taungya* villagers) were twofold. First, to ensure "sufficient and continuous supply of [free or cheap] labour" for raising government plantations and "other works of improvement" by the forest department (GOEP, n.d., p.13); and second, to neutralize the pressure of public opposition and discontent by engaging a section of the local poor against the rest majority. These "forest villagers" added fuel to the bitterness between local people and the forest department. In many occasions, as reported by my informants, they served as "accomplices, spies or undercover informants" in governmental coercive or regulatory actions on local people (for example, during eviction operations and periodic raids). Besides, as a result of the state intervention there seemed to be a process of displacement and dispossession, as noted by one government working plan document (Ahmad, 1938, p.5):

> with the advent of the Forest Department in Bhowal, most of the squatters [local people] have left the forest area....The old cultivated areas are now found as blank areas overgrown with weeds almost in every block.

Subsequently, the local right of grazing was also strictly regulated. Increasing numbers of toll stations and *ghats* (river stations) were leased out to private merchants for realizing the transit fees and revenues of forest products. The process of departmental extraction through private contractors was initiated. Blocks of forests were demarcated for extraction (a process called "coupe-operation") and were "clear-felled" patch after patch (Ahmad, 1938; Chowdhury, 1957). There was a systematic rise in government revenue. For example, in the Bhowal forests, which is very near to Chandra, the revenue earning was Rupees 5,870 during 1933-34, and in 1935-36, it shot up to Rs. 105,328, and by 1948-49 it rose as high as Rs. 738,496 (Ahmad, 1938, p.15; Chowdhury, 1957, p.28). "A network of forest roads" were departmentally constructed to facilitate further "extraction" (Chowdhury, 1957, p.88). Thus, throughout the Pakistan period rigorous rule enforcement went apace on local people on the one hand, and massive deforestation by commercial extraction went on simultaneously on the other. While local rights diminished, governmental extraction and revenue increased. Consequently, by the late 1970s, the forests dwindled considerably, and more importantly, mutual distrust between the government and local people cropped up, resulting in desperate activities from both sides. A number of serious assaults were made on foresters, and the government launched eviction operations in retaliation. Plantation activities started to fail as a regular phenomenon (as noted by the Chairman of Mouchak Union and Kaliakair Range Office staff, personal communication).

When the war of independence started in 1971, the final series of assault on these forests was launched which continued unabated until the early 1980s, resulting into the barren look that the area possesses today. In fact, the whole Inland *Sal* Forest was affected by the war (Zabala, 1990, p.14). Following the war, an armed gang of highly organized timber traders intensified commercial logging to an unprecedented extent. Local influential families joined the loggers "with a clear alliance with government offices" (Chairman of Mouchak Union, personal comminication). Immediately after the country's independence, the government imposed a ban on timber felling for 10 years (Zabala, 1990, p.14) which hardly produced any significant result (Zabala, 1990, p.14; Byron, 1984, p.1; GOB, 1990, p.ve-2). Systematic deforestation was carried out and, as reported by my informants, there had been widespread corruption and connivance on the part of government offices (especially the police, land revenue and forest departments). One informant remembered:

> The embittered local people witnessed the destruction of the neighbouring forests right in front of their eyes. Then they started to damage [government] plantations....Replanting activities simply could not match the extent of devastation that had been in progress [through] commercial logging.

The forest department carried out periodic eviction operations to "wipe out" the "squatters" and "encroachers". On many occasions, these actions were recorded in the personal notes of the duty officers; and the officers also reported about the mounting local resistance they faced during such operations (see for example the remarks of the duty-officers in the Visitor-Book of the Forest *Bungalow* at Chandra). By the late 1970s, as one key-informant remembered, the area became "literally devoid of any significant block of tree"; and "on these wide-open lands, massive migration of people [both from the adjoining villages and remote places] took place". Most of the large chunks of (the barren) land was first occupied by local elites (including local-government leaders, large land-owning families, political leaders, timber traders and other businessmen). The elites were in "close understanding" with government offices. In their turn, elites started to act as patrons and let these lands (under their de-facto control) to the poor locals and migrants in return for some forms of clientage and servile support (I will discuss the clientage supports in Chapter 6). A major proportion of these new settlers consisted of people who were pushed out of their original habitat by natural calamities such as flood, draught, devastation by rivers and also by socio-economic forces of poverty, political turmoil and unemployment. These new settlements also allured some left-wing political parties who organized the poor settlers against any governmental or private attempts of eviction. The Bangladesh Workers Party, for example, organized a *Khet-Mozur Samity* (land-labourers association) among the settlers, more or less, on a communistic agenda; and as I eventually realised, many of my respondents shared this political identity.

Thus by the 1980s, the de-facto control of the area ceded to influential elites and the land was covered with settlers. The forest department tried in vain to raise some plantations, mainly to signal their proprietorship over the land. In the locality the last attempt of raising plantation was made, covering 50 acres of land, in 1975. But by 1979, "there was not the slightest sign any tree left....Even the *Mothas* [tree-trunks] were uprooted...and we [foresters] tried no more" (Forest Range Officer, Kaliakair, personal communication). Today, the Inland *Sal* forest tract, as a whole, exhibits one of the worst case of deforestation. 50% of the so-called "forest" have been cleared of and occupied by settlers (Byron, 1984, p.35). "Illegal exploitation", as reported by the government itself, has resulted in "less than 30%" tree-cover in *Sal* forests (GOB, 1990, p.ve-2). In the same vein, another study (ARDP, n.d., p.1) noted that:

> the greatest transformation of forest lands has taken place in the *Sal* forests of Dhaka-Mymensingh region. There, about 250,000 acres are classified as public forest land, but less than a third of that area has actual tree cover.

Betagi-Pomora

Again, little recorded information is available on Betagi-Pomora area prior to the induction of the SF projects in the late 1970s. Consequently, I drew mostly on observation, key-informants and ethnohistories.

In the 18th century, the area was densely forested. Lands, including forests, were under *zamindars* and the government used to fetch revenue from them. As early as the 1760s, there had been some forms of state intervention on these forests. As historical evidence shows, cotton trade was quite active at that time (see BDR, 1978). The government used to cut trees in these hills by employing paid-labourers (*Coolies*), brought from the plain-lands. Such governmental extraction schemes were faced by local people's resistance, especially by the *Kookie* tribe. The *Kookies* were a nomadic tribe of Lushai hills in Chittagong hill tracts (GOEP, 1970). On several occasions, the army was indented for protecting the government wood-cutters and extraction squad (see BDR, 1978, p.382, Document 4.73). Similarly, the government periodically sent agents (*Sezawal*) to collect revenue from *zamindars* of Rangunia and very often, armed forces used to accompany them to resist public opposition. For example, the Collector of Chittagong, asking escort for such a revenue-collecting agent (*Sezawal*), wrote to the military Commanding Officer of Chittagong (Major Ellerker) on 11.10.1791 that (BDR, 1978, p.267, Document 3.44, emphasis added):

> The *Waddadar*...is about to send a *Sezawal* into Ranganeah [Rangunia] Province for the realizing of the Revenues in those parts. I request you will give Instructions to the Party detached there *to protect him from any Violence or Obsuructions that may be offer'd to him from the People*.

Revenue from Rangunia was of considerable value to the Chittagong Collectorate (see for example BDR, 1978, p.148, Document 2.11). These evidences show that the initial attempts of state intervention into the locality were prompted by the considerations of commerce and revenue, and from the beginning, these attempts met with public discontent.

After these occasional documents of the early British rule, no official record could be found on the situation until the 1930s. However, it was reported by my key-informants that local *zamindars* (as I showed in earlier instances) were never particularly interested in forests, except for occasional extraction for family purposes or meeting exigencies (for example, *zamindars* used to cut trees to meet expenses of socio-religious festivals). Some influential families, in understanding with *zamindars*, were in actual control of forests. No major organized logging was reported prior to the 1950s. Similar to the case of Chandra, in this locality too, neighbouring villagers had reasonably wide access to forests, as far as their daily household requirements and pasturage were concerned. In some instances, villagers, in return for their rights in the forests, had to render some personal (clientage) services to local elites (such as working in their fields, providing manual labour, and helping in household chores). One key-informant, Chakrabarti (aged 77), remembered:

> *zamindars* never particularly cared about forests. Forests were primarily treated as "jungles" and pastoral grounds for villagers' livestock resources....*Garib-guinna* [the poor and weak villagers], especially the neighbouring people, had [almost] unrestrained access to forests for fetching firewood, leaves, *chaan* ["sungrass" --

> thatching material] and for rearing their herds.

In parts of forests in the locality, some indigenous communities, especially a section of the orthodox Hindu population, used to worship forests and uphold their sacredness. There was a large Hindu temple and community centre in the locality, called the Jagatpur Ashram, which also held control over a vast area of land (including forests). The temple is still there, and on my visit, one of the elderly priests recalled:

> In the old days, the managers of this estate did not particularly think of realizing *Khazna* [revenue] from forests....People were never barred to enter [the forests] unless there had been report of any large [scale] felling with a mind to selling [i.e., commercial exploitation].

Similarly, at least five *Mazars* (grave-yards of saints and religious seminaries) of Muslim priests are still seen in the area; all situated in (the recently denuded) forests. My informants reported that these saints lived amidst these forests and preached religion to neighbouring villagers. Forests were their most-loved rendezvous for practising and teaching religion. Therefore, it seems that forests and religion, like in many parts of the Indian sub-continent, had been closely related; and saints attempted to convey the message of this relationship to the locals. Interviews with some elderly villagers suggested that the preaching had, at least to some extent, influenced the conservation and propagation of forests.

Besides, it seems likely that commercialization of forest, entailing its indiscriminate exploitation as a "commodity" (see Chapter 2) did not become active in the locality until the mid 20th century. There was limited felling of the commercially important trees. This was done, as already noted, mainly by *zamindars* for their own use, or for limited exportation to urban centres. But, the mushrooming of timber traders, as seen today, was certainly not the case.

Most families, even the poor agriculture labourer class, had (forest-based) livestock as auxiliary means of income. Livestock rearing was a comparatively easy task in the area. Villagers took their herds to forests, let them free to graze for the whole day and brought them back at the sunset. There was no major restriction on local rights of grazing (and thereby, earning an additional income).

The area had dense forests even up to the 1950s and wild life were also abundant. For example, even within close quarters of local forest office, a touring officer saw a leopard roaming about (source: Remark of Mr. Y.S. Ahmad, Conservator of Forests, Chittagong, on 24.4.1949 in the Visitor's Book of Chiringa Forest *Bungalow*). Similarly, I met a man, whose father hunted a rogue tiger in the very area (around 1953-54) where the Pomora SF project is now situated. Massive land acquisition by the government started from the early 1930s. Within five years time, an overwhelming land-mass of Rangunia came under the purview of the forest department and was declared protected forests. These newly acquired lands included Betagi (year of acquisition: 1934), Jungle Ghatchek (1935), Sarafbhata (1935), Ghatchek (1934), Nogaon (1934), Baniakhola (1934), Pukiakhola (1934), Dahmirchara (1934), Andharghona (1934), Dingallunga (1934), Tinchodya (1934), Chenkhali (1934), Saidbari (1934), and Daskin Ichamati (1934) (source: Forest Land Records file, Range Office, Rangunia). Figures 5.2 and 5.3 present a comparative picture of the forest department's territorial boundary in Betagi, Pomora and surrounding areas in two points over the history: during the early years of state acquisition (1930s) and as the situation stands today (1990s). These newly acquired tracts, as reported by my informants, not only covered forested or tree-covered areas, but also included village pastoral grounds, market places, crop-fields and some homesteads. The Chairman of Pomora Union showed me around some of these "forests". I found that parts of the so-called "forests" were in fact open grounds which contained *madrassa* (religious seminaries), *haats* (markets), homesteads and a number of ponds.

Following the take over by the government, immediately local access and rights were drastically curtailed. One key-informant remembered that "one fine morning" they were told by "some men with huge mustaches and *khaki* [brown-coloured] trousers" that villagers could no longer enter the forest without permission from the *tila ghar* (literally meaning, "the house on the hill" -- the local forest office at Chiringa). Immediately afterwards, some villagers were arrested by foresters and taken to the *tila ghar*. It was at this time that the spawn of corruption and bureaucratic oppression slowly began to creep into the day-to-day village life, ultimately to grow into today's predator. Many incidence of corruption by policemen and foresters were reported by my respondents. Subsequently, in the face of mounting opposition from local people, the government had to surrender some of the areas (by declaring "disforested" zone). The "disforested" (deforested?) areas included Jungle Parua, Jungle Ghat Chek, Nogaon, Chengkhali, Tinchudia, Baniakhola and Andharghona (source: Office records of Pomora Beat Office).

Although massive land acquisition took place, little productive utilisation of land was observed. Virtually, no organized planting efforts were recorded in Betagi-Pomora upto the late 1960s on the newly acquired lands; and foresters' duties were confined to policing their territories.

Furthermore, on the plea of "scientific management of forest" and "systematic plantation", a large part of the local forests were sold on public

5.2 A sketch map showing the Forest Department's (approximate) territory in Betagi-Pomora and surrounding localities in the initial years of state acquisition (in the 1930s)

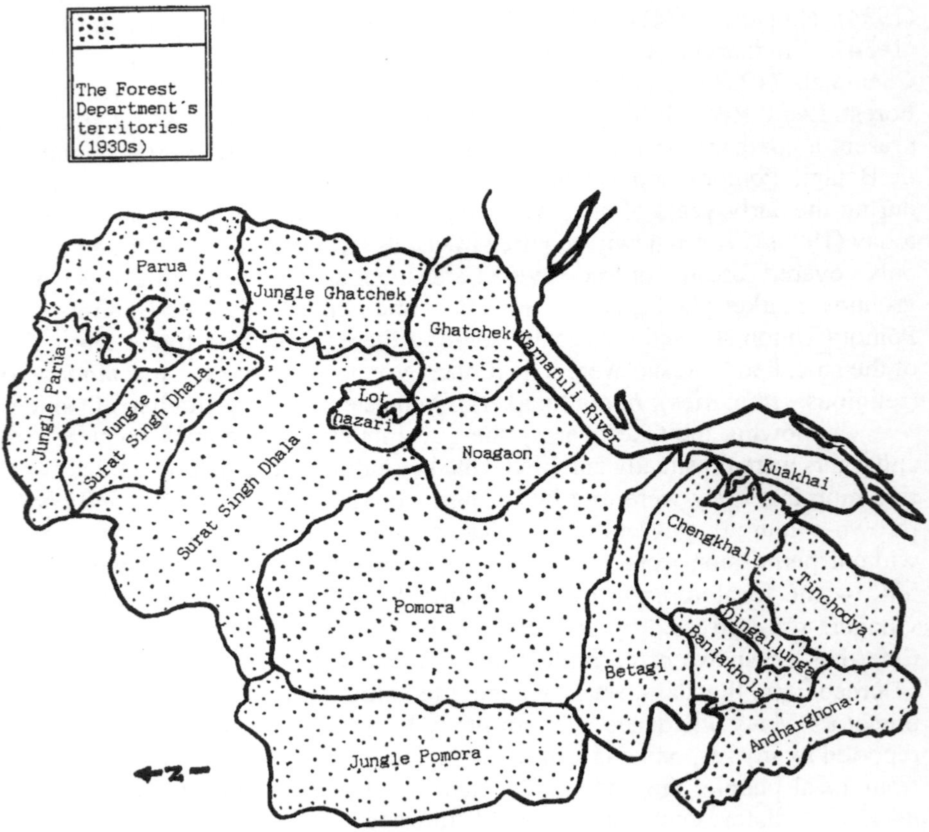

Data sources and notes: This map is drawn on the basis of the office records of the Pomora Forest Beat office and the Rangunia Range office, and also on the information collected from the key-informants. The map is not exact to scale and boundaries.

5.3 A sketch map showing the territories under actual control of the Forest Department in Betagi-Pomora and surrounding localities at present (in the 1990s)

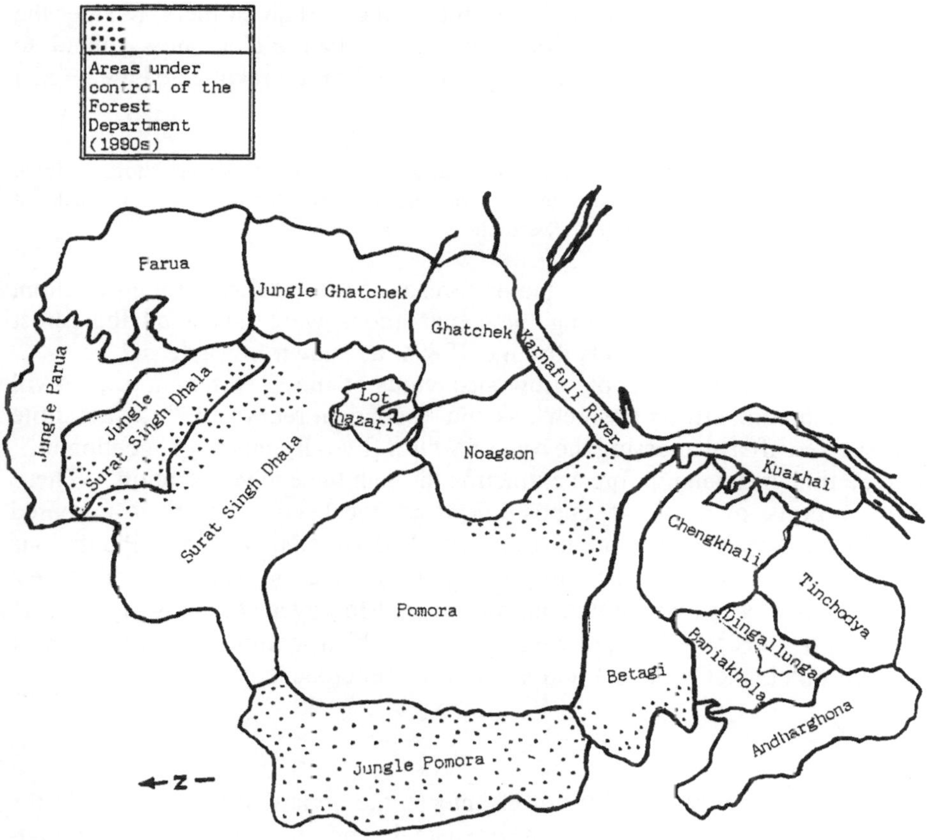

Data sources and notes: This map is drawn on the basis of the office records of the Pomora Forest Beat office and the Rangunia Range office, and also on the information collected from the key-informants. The map is not exact to scale and boundaries.

auction during the late 1950s and early 1960s. Private traders were granted permission to "clear-fell" forests through the "coupe operation". It was at this point that urban timber traders were invited to join the scene and commercial logging had its legitimate beginning (as reported by key-informants and as confirmed by the Chairmen of Pomora and Betagi Unions). Traders, in turn, relied on local elites for timber extraction. In the process, a powerful local petty trader community emerged as agents of their urban partners. Besides the vigilant foresters and policemen, local and urban traders now joined to "protect" "their" forests from the people. One informant recollected the impact of coupe operation:

> Coupe operation continued for about four years....All big trees were chopped down and many of them had never been brought to market owing to lack of communication; many trees were put to rot that way.

Thus, after clearing the major forests of the locality through auction, coupe and subsequent burning, new plantations were raised by the forest department in the late 1960s through 1980s, as shown in Table 5.2.

However, local people grew suspicious of the government's goodwill in the plantation efforts ("which", as one respondent remarked, "was not quite unexpected after witnessing the role of *Sahibs* [literally meaning "gentlemen"; here, public officials] in forest extraction through the coupe operation"). There were reports of constant opposition from local villagers; they uprooted seedlings, damaged plantation, even attacked isolated foresters. Plantations started to fail and repeated re-planting was done (in vain) by the forest department, as shown in the above table. Moreover, the government, in retaliation, intensified its policing operation which included apprehension of a large number of villagers and eviction of "encroachers".

Overview

Mirroring the general findings of Chapter 2, the review of the situation in the broader study areas of Dhaka-Mymensingh and the field-sites of Betagi-Pomora and Chandra revealed similar trends in commercialization of forests and the resultant public alienation. The recent scenario, as I will elaborate in the next section, is hardly different; because this has been more or less a continuation and outcome of these trends. In line with previous discussions, the next section examines the process of competition more closely in the micro context of field-case sites. Considering its complexity, the manipulation of "losers" (the local poor and weak people), and its far-reaching implications for people and environment, I view the process of competition as a "game".

The findings of the next section will also help to draw my conclusions more confidently about the nature of commercialization and alienation.

Table 5.2 Public plantation work at Betagi-Pomora (under the Pomora Forest Beat Office)

Year	Acres	*Mouzas* (revenue villages)
1968	388.95	Jungle Pomora
1969	419.00	-ditto-
1970	167.00	Surat Singh Dhala
1971	61.00	-ditto-
1975	50.00	Pomora
1976	100.00	Surat Singh Dhala
1977	N/A	Betagi
1978	100.00	Surat Singh Dhala
1979	150.00	-ditto-
1980	350.00	-ditto-
1981	150.00	-ditto- (replanting on the 1970's plantation)
1982	50.00	Jungle Surat Singh Dhala
1983	115.00	-ditto-
1984	50.00	-ditto-
1988	50.00	Surat Singh Dhala (replanting on the 1970 and 1981 plantations)
1989	50.00	-ditto- (replanting on the 1970 and 1981 plantations)

Data Source: Office records of Pomora Forest Beat office, Rangunia Range, Chittagong.

The Players in the Game: The Users and Nature of Forest Use

Principally, the forms of forest use in the field-case localities include grazing, timber logging, fuel wood extraction for commercial purposes, domestic fuel collection, and public afforestation activities. Commercial use of forest supersedes all others. It includes illegal commercial logging by urban merchants (through their local agents) and catering the needs for urban and local furniture shops, saw-mills, tea stalls and brick kilns. There are, to cite an example of the volume of such commercial use, nearly 15 brick kilns, 27 saw-mills and 25 furniture shops and numerous tea stalls in Rangunia *Thana* alone (source: Office records, Forest Range Office and Thana *Nirbahi* (executive) Office, Rangunia). Similarly, in Kaliakair Range there may be as many as 40 brick-fields and 12 saw-mills (as informed by the Beat Officer, Chandra).

It may be relevant here to focus in more detail on the extent of commercial forest use in the locality. For burning 100,000 bricks, 1,500 mounds (1 maund = 37.32 kg) of fuel wood is used (source: *Samoya*, a Bangladesh Television documentary, broadcasted on 23.5.1994). As my key-informants informed, each of the Rangunia brick kilns generally use 3,000 mounds of fuel wood in burning 1 "round" of brick (a "round" consists of 300,000 bricks). At least 4 "rounds" of bricks are burnt in each brick manufacturing season of roughly 4 months. On the basis of this estimate, for 15 kilns, approximately (3000x4x15) 180,000 mounds of wood are seasonally burnt in Rangunia alone. In fact, brick kilns are one of the largest consumers of fuel wood in the country which absorb about 23% of total fuel wood consumption (Hussain, 1992, p.15). Another major consumer of wood is the tobacco kilns. There are, as reported by local villagers, about 150 kilns in Rangunia. The tobacco produced from these kilns, feed the collecting-depots of three leading commercial tobacco companies of the country. I visited a number of kilns and was informed that these kilns remained operational from February to April each year. On an average, each kiln absorbs 600 mounds of wood. Therefore, it is estimated that roughly (150x600) 90,000 mounds of wood is burnt during each season in Rangunia alone. Similarly, numerous tea-stalls in the locality also consume substantial quantities of fuel wood. One of my every-day venues for meeting with the Betagi-Pomora farmers was a tea-stall, locally known as *Burir-Dokan* ("the old lady's shop"), with a seating capacity of 25 persons. Even this small tea-stall consumes about 75-90 mounds wood every month (as reported by its proprietor). Tea-stalls are very popular among villagers of all class and creed, and can be found almost everywhere in the locality. A tea-stall is not just a beverage shop; it variedly serves as a public meeting place, a small grocery shop, a venue for socio-political discussion, a source of news and information, and a leisure and pastime place. And, what is more relevant to me, is the fact that, tea-stalls are a major consumer of illegally collected wood, and therefore, remain as a constant source of encouragement for the illegal fuel wood trade. Besides, piles of sundry logs are openly visible in the *verandah* of local saw-mills. The main supply of wood for saw-mills comes from the illegal logging network. After nearly a complete destruction of major forests in Betagi-Pomora area, now Chittagong hill tract has become the principal target of the loggers.

The forest department's use of land in the localities mainly consists of plantation works. It carries out annual planting, but even foresters themselves agree that failure is a frequent phenomenon. They attribute the failure to public opposition, whereas local people have a different explanation. The Chairman of Pomora Union argued:

> The forest department forbids my people [villagers] to burn forest, but they themselves practise "slash and burn" every year....Poor quality saplings are planted, no care is taken....Maximum plants die in utter rain and sun. They scold us..., make us the scapegoat, plant trees again in the next season....You know, planting means money...the more you plant, the more [you qualify for] increased budget allocation..., the equation is very simple.

The illegal logging network, the most dominant user of forests, is organized, well-linked and extensive. Loggers have an underground administration which, due to corruption and connivance of government bureaucracies, nearly runs parallel with the local public administration (see Figure 5.4). The loggers have regular "pick-up points" for collecting timber; modern vehicles -- mostly jeep, trucks and marine-boats (boats are run by locally altered and upgraded irrigation engines); paid cadre of employees consisting of informers, labourers and middlemen to liaise with government offices (mainly with the police and forest offices); coded pass for carrying out underground logging; and a network of contacts with urban and local politics and administration. Logging-routes in Chandra forests are through the Turag river and Dhaka-Tangail highway. Loggers have underground offices at "points" such as Fulbaria, Mouchak and Chandra Bazar. Similarly, in Betagi-Pomora locality loggers have a wider network with regular "pick-up points" in Malirhat, Shilak, Chiringa, Dhoparghat, Shantirhat, Gochara, Mariamnagar and Ghatchek -- to name only a few. Bolirhat and Lamburhat, at the border of Chittagong city, are the two largest depots of illegally collected timber where even the police or foresters hesitate to raid. Timber merchants from all over the country gather there and these timbers find their way to different

5.4 A simplified schematic presentation of the commercial logging network (as observed in the study-area)

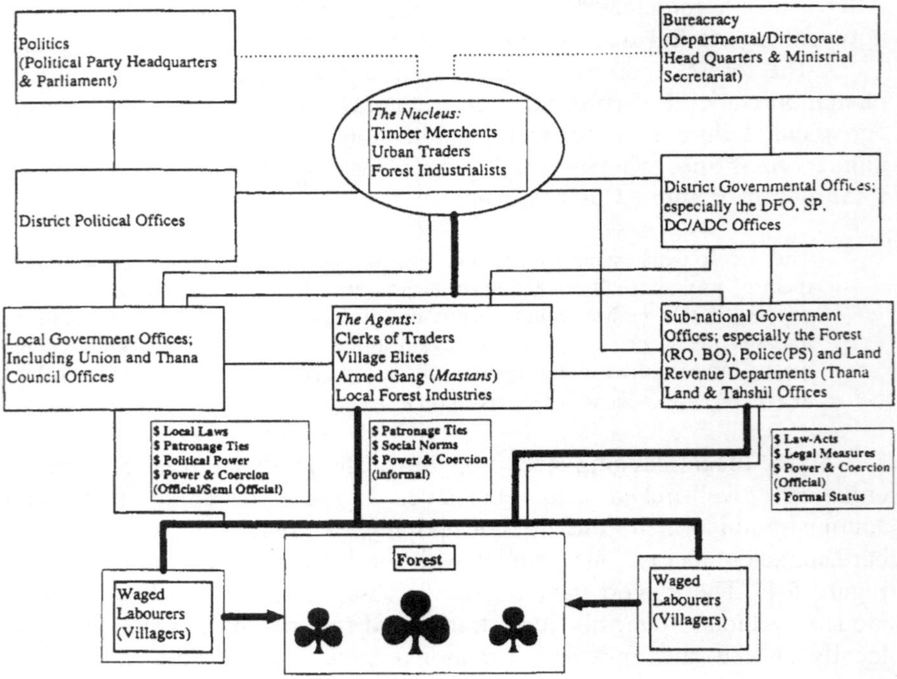

Notes: Based on personal observation, ethnohistories and key-informant interviews. For an index to this figure, please see the next page (p.193).

Index to figure 5.4	
▬▬▬	Principal line of extraction and exploitation (regular): The main line of commercial logging.
———	Supportive line (regular): These agencies act as partners and accomplices to the loggers. They maintain regular contact with the nucleus of logging network, and facilitate smooth operation of the network.
···········	Contingency line (Occasional): If there is an emergency or extra-ordinary situation which can not be resolved by the local management of the logging network (the nucleus), this line is used to reach the apex of national politics and bureaucracy. At this stage, the senior public officials and politicians get directly involved with the functions of the network.
$	Authority-base: These are the means or instruments of authority exercised by different agencies involved in the logging network.
Abbreviations	FD = Forest Department; DFO = Divisional Forest Officer; FRO = Forest Range Officer; SP = Superintendent of Police; BO = Beat Officer; ADC = Additional Deputy Commissioner (Revenue); PS = Police Station; DC = Deputy Commissioner.

parts of the country. The most commonly used routes for timber trafficking are the Karnafuli river (with its several tributaries) and the Chittagong-Kaptai highway. Loggers use a coded pass, called "Token", unofficially issued by the police and forest departments for transportation of the illegally extracted timber. The system of "token" is efficient and it is nearly impossible to prove the massive corruption involved in issuing such undercover passes. Thus, the processing and marketing of illicit forest products are surprisingly well-organized and secured. I have also collected an extensive profile of the merchants engaged in logging. They seemed to have some common characteristics: their power is rooted in landed properties; they have wide network of horizontal and vertical linkages in bureaucracy and politics at the macro and micro levels; they are mostly in their mid 40s; they enjoy some forms of political, social or religious, formal or informal institutional positions in the locality; and have intimate "understanding" with government offices especially the police, forest and land-revenue offices. Some of them are political leaders, currently holding or previously held chairmanship or membership of respective Unions; and in most cases, they are close kin to each other. They have fairly sound knowledge of forest and police laws (including their loopholes and limitations). Under different disguises, I observed the logging network very closely (On my informal discussions with the Chairmen, Pomora and Mouchak Unions and foresters of Pomora Extension Centre, they corroborated the above observations). I have come to the conclusion that illegal commercial logging is the most dominant form of forest use in the localities and the primary factor contributing to deforestation. It has also given rise to a cemented alliance among local elites, government and local government officers, urban traders and other commercial forces. My observations as regards the predominance of logging network in the political economy of forest use are in close similarity with the findings of another in-depth village study (BRAC, 1986, p.49) which notes:

> Large timber merchants are cutting at rapid rate with unofficial cooperation of the [government] officers....This cooperation is in sharp contrast to the harassment and punishment experienced by poor and destitute people when they cut head-loads of firewood...to sell for survival, as they have always done....Possibly no other activity demonstrates so clearly the imbalance of power as the way in which different [people] are able or unable to exploit this common resource.

SF farmers also opined that commercial logging and the government's connivance were the major causes of recent deforestation in the localities (Table 5.3).

The domestic and subsistence usages by general villagers are nominal and consist of fuel, fodder and thatching goods collection and livestock rearing. On many occasions, even while trying to meet these ostensible requirements, local people, especially women and the poor, are reported to be harassed by the police, foresters and local elites. In fact, as my experience suggests, the roles of the poor and weak section of villagers in forest use are threefold. First, some of them act as paid labourers in commercial logging. They also face all the odds and risks involved in the operation. Sometimes they are apprehended by the government and then eventually freed by their patrons (loggers). The second group, who previously depended on forests for their livelihood to a large extent, has now nearly stopped entering forests and washed off their hands from the "game". It seems that they have failed to come to terms with loggers, concerned elites and government officials. As an alternative strategy, these people have reduced dependence on forests by changing their primary occupation (mainly) to daily-labour and share-

Table 5.3 Major causes of recent deforestation in the localities (as reported by respondent farmers)

Causes	Betagi-Pomora locality (n=64)						Chandra locality (n=13)					
	1		2		3		1		2		3	
	N	%	N	%	N	%	N	%	N	%	N	%
Commercial logging and timber trade	42	66	20	31	0	0	8	62	3	23	1	8
Govt.'s corruption and connivance	10	16	32	50	9	14	3	23	3	23	1	8
poor villagers' demand	9	14	6	9	7	10	1	8	2	15	0	0
abuse, unrestricted use	3	5	6	9	0	0	1	8	0	0	2	15

*Notes: * This table and the following tables (5.4 and 5.5) are based on the field-work. * N = Number * The percentage figures are rounded up to the nearest whole number. * For the Betagi-Pomora locality, n = 64 (24+40). These two projects are situated in the same locality in Rangunia Thana. * Degrees of severity of the cause (as assessed by respondents): 1 = most active cause, 2 = very active cause, 3 = active cause.*

cropping. The third party, which includes the most poor and desperately needy villagers, enters forests for domestic and small scale commercial purposes, get arrested, face jail, come back, and return to forests again. Their abject poverty compels them to return to forests despite the continuous harassment and manipulation by commercial forces. Thus, these three sections of local people are the weak and insignificant players in the competition for forest use. The "game" is often too expensive for them to win, as one of my respondents points out:

Who are we in the *Lakh Taka karbar*? ["one hundred thousand taka business"]...;

> you can buy the ink of the *sahibs* [officials] if you have money...and once you have the "token", you have the forest at your foot.

Conclusion: "Sure Exclusion of Locals"

This chapter examined the field-realities in light of the conceptual and general pattern of forest use elaborated in Chapter 2. It investigated the process of forest use at two related levels: at the broader context of the study-areas (Chittagong and Dhaka-Mymensingh forests); and at the more localised setting of the field-cases (Chandra, Betagi and Pomora). Reaffirming the findings of Chapter 2, the empirical investigations in this chapter allude to the predominance of commercial considerations in forest use which has caused a systematic deprival of local rights. The process of commercialization (and the consequent public alienation) was sparked by the colonial intrusion on Bengal forests, but over the years it has propelled somewhat independently, even after political independence of the nation.

The far-reaching consequences of the process of commercialization and bureaucratization of forest use, and its relevance in explaining some of today's major forestry problems, are astutely summed up by Shiva et al. (1983, p.49, emphasis added):

> the crisis in forests...is rooted in the organisation of forestry [in the Indian sub-continent]....Although the rural people had been using forest resources for centuries before organised forest management was introduced,...the forest stock remained healthy under community management. Indeed, the rapid deterioration of the forest stocks...can be traced both to the reservation of large forest areas for monopolistic use of the...Government and to sudden and drastic reduction both of those forest lands open for community use -- the civil or revenue forests [*khas* forest lands] and those forests which belonged to local village councils. *An irreversible trend in forest management and land-use was initiated, the most significant social impact being the slow but sure exclusion of local communities from the use and management of forest resources. The result was the alienation of the whole rural population from the forests* which, in its eyes, now came under the exclusive control of such "external" authorities as forest officers and contractors.

Thus, the state actions, primarily propelled by "commercial and industrial interests", had been "at the root of the malaise" in forestry which not only led to the "process of alienation" of local people (Fernandes, 1983, p.1), but also facilitated the deforestation process (Gadgil, 1983). The advent of organized state-management in the 18th century, as I reviewed in this chapter (and in Chapter 2), resulted in a

> ...shift to revenue and monetary economy, commercialisation, and consequent change over from community [use] of forests to state control....Though [the government] claimed that their legislation aimed at preserving forests from people who did not know what was in their interest, in reality ..revenue was the main interest (Fernandes, 1983, p.13).

In light of Chapter 2, I will now briefly recapitulate the aforesaid process of alienation. The state-monopolization of authority and proprietorship over forests led to public alienation primarily in three ways: loss of local rights and control; distortion in community regulations and traditional patterns of forest use; and exploitation, conflict and dispossession (see Chapter 2). As I have shown, the very motive behind the state-appropriation and bureaucratization of forests through reservation was prompted by commercial interests; and it, almost as a rule, entailed the denial of local rights and access. The following remark of Stebbing (1923, p.415, emphasis added), who held charge of Chittagong Forest Division sometime in the early 20th century, amply brings home the philosophy of "scientific forest conservancy" as regards local rights:

> *Local requirements must become subordinate to the state of the forests and the exigencies of conservancy*....If the interests of conservancy render it necessary that certain restrictions be put..., local supplies must be regulated accordingly....This will, of course, at first lead to dissatisfaction and complaints. The state of things is unfortunate. *The present generation has to suffer for the errors and improvidence of the past....Conservancy will require the strongest support of the Government to enable it to resist the pressure [from local people]* which will from time to time be put against the Forest Department in the shape of all kinds of [public] complaints.

He (Stebbing, 1923, pp.465-466, emphasis added), echoing the voice of government, further maintained that

> *[the] prescriptive rights of user are often incompatible with the scientific and economic working of a forest*....Government as the guardian of all public interests, *must insist on the regulation of these rights* so as to render it possible a good management of the reserved forests.

Accordingly, reservation works, which have continued through to present time, resulted in systematic restriction of local rights and access. Furthermore, local people were made to work under exploitative terms and contracts, leading to further aggravation of the situation (Joshi, 1983, p.27). The state-led development works in forests in the forms of communication and logistics build-up often "resulted in the penetration of economically viable outsiders in the forest regions who, in order to make quick profits, have started exploiting the forest dwellers, displacing them from their land and making them bonded labourers" (Joshi, 1983, pp.27-28). Propelled by the old principle of "divide and rule", a departmentally patronized group of villagers, loyal to the government, were formed in the name of "forest villagers" to subdue and distract local opposition (by using a section of local people against others) and also to ensure cheap labour for plantation works. Revenues from forests mounted up, but very insignificant amount of such earnings were ploughed back to the improvement of forests. Socio-religious traditions and community regulations of resource use were largely disrupted and over-ruled by the government proprietorship. Parts of Bengal forests, Stebbing (1923, p.465) noted, previously considered as the "property of the people" were "inherited by British government by the right of conquest". Local people were consistently blamed for forest destruction in major government documents (for example Cowan, 1923; Ghani, 1955; Baten, 1969; Khan, 1980; Ahmad, 1938; Chowdhury, 1955; Stebbing, 1923); whereas, ironically, they seemed to be the major suffers from the commercial intrusion on forestry. ADAB (1992, p.8) argued that through the process of reservation, for example, "communities' right on forest were usurped and taken over by the state in order to serve the urban, commercial and colonial interest...denying people's access to forest resources". The Bangladesh Agroforestry Plan (1990-95), a coordinated national action-research agenda developed by the government, similarly noted that (BARC, 1991, p.5)

> the fundamental cause [of deforestation] is the breakdown in the traditional mechanisms that controlled people's access to and protection of the forests. The root of the problem can be traced to restrictive forestry laws that severely limited individual and community rights to the forest, and favored the flow of forest products to urban, commercial and industrial users.

Today the above trends in alienation culminated into a situation where a powerful commercial alliance has turn out to be the "winning" force in the competition for forest use in the study-areas. The alliance consists of a section of government and local government officials, local elites, urban traders and their local agents, and local businessmen. The "losers" are the poor and weak villagers who partially meet their subsistence needs from the neighbouring forests, especially shepherds, fuel-wood collectors (mostly women) and thatching material collectors. The colonial legacy has hardly changed. Even nearly half century after the official cessation of the colonial rule, the reservation act, for example, "still remains with full force denying people's access to forest" (ADAB, 1992, p.8). Illegal commercial logging, carried out by the above-discussed commercial alliance, is the prime contributor to deforestation and the most dominant form of forest use. The social and institutional base of this logging network, as has been explained before, is nearly invincible. It has, at least, three immediate implications. First, the network engages a section of the local poor in timber logging at exploitative wages and conditions. Second, by its predominance and control, it further reduces even the nominal share of the local poor who enter forests for subsistence requirements. Third, it takes heavy toll of the environment -- the most visible manifestations of which are extensive deforestation and soil erosion in the localities. Recently, some senior government officials are increasingly becoming alive to the centrality of commercial logging operation in forest destruction and peoples' alienation (see for example Choudhury, 1993, p.60; Hussain, 1992, p.8). Similarly, of late, there has been a growing public opinion against the menace of illicit logging which is manifested in the country's leading news-media (see for example The Azadi 17.6.1994, The Dainik Bangla 26.7.1994, The Ittefaq 7.7.1994, The Courier 29.4.1994, The Janakantha 3.3.1401 [Bengali yr.]). The local poor in the vicinity of forests generally share a feeling of deprivation and isolation. Majority of my respondents opined that they should have right of access to forests for meeting their subsistence needs (Tables 5.4 and 5.5).

Table 5.4 Farmers' opinion about their right of access to public forests (Q. Do you think you should have right of access to public forest lands?)

Sites	Affirmative		Negative		No-comments	
	No.	%	No.	%	No.	%
Betagi	7	29	8	33	9	38
Pomora	18	45	10	25	12	30
Chandra	11	85	2	15	0	0

*Notes: * Affirmative = farmers think that they should have the right of access to the neighbouring public forests); Negative = farmers who do not think that they should have such rights).*

Table 5.5 Farmers' opinion about the nature and type of rights on public forests (as perceived by respondents)

Nature of rights	Betagi		Pomora		Chandra	
	No.	%	No.	%	No.	%
To procure firewood, fodder and thatching materials	4	17	12	30	9	69
To fetch timber	1	4	2	5	0	0
For pastoral purposes	2	8	4	10	2	15

Currently, the local poor are largely deprived of such rights and their inner feeling of deprivation is strong, although they do not express it publicly for obvious reasons. One local labourer ventilated his desperation to me, in confidence, idly sipping a cup of tea:

> everybody is having his share [then he explained, in specific details, the corruption and illegally acquired wealth of the concerned local public officials, namely the police, land revenue and forest officials]. But I cannot have a single *Sagun* [Teak] pole for mending my hut....I went to the [government] nursery for seedlings..., the *Mali* [plantation worker] wanted 40 Taka for each seedling....I have arranged to enter the forest in one of these nights. I have settled it with...*Sowdagar* ["merchant"; he mentioned the name of a well-known timber trader]. He [the merchant] said, he would send me to the ...forest [name of a specific forest in the area] with his other men....I can bring as many poles as I can carry on my shoulder and [in return] I have to assist his team in bringing *gach* [timber] for him.

In the above context of public alienation, growing hostility against the governmental efforts in afforestation, and the inability of the government to manage their territories singlehandedly -- there has been an apparent realization among the upper echelon of state officials about the necessity for a policy-shift in forestry. Parker et al. (1989, p.9, emphasis added) note:

> foresters [have begun] to realize that more attention must be paid to local needs....Until now, *most of the riches produced from the forests have been captured by distant elites and have therefore been of little benefit to the communities surrounding the forests.*

As discussed in Chapter 3, these realizations received extensive support from the international donor quarters. Accordingly, during the end of 1970s, the government's realization coupled with the wide international advocacy, set the stage for the experimentation with a number of participatory forestry projects (commonly coined as SF) in Bangladesh. The next two chapters will examine the performance, achievements and the surrounding frame of three selected SF projects (the case-studies of my study).

6 Social Forestry at Work: The Frame

Introduction

The previous chapter traced the major historical policy changes and current practices which have led to the progressive commercialization of forest use and the alienation of local people from forests in the study-areas. Apparently, in an attempt to compensate the sores of public alienation, the Bangladesh government has committed itself to bring about "a dramatic change" (GOB, 1990, p.ve-5) in the forestry sector since the 1980s (see Task Force, 1987, pp.1-3; Hussain, 1992, pp.4-5; GOB, 1992, pp.26-27; Alim, 1988, p.17; ADB, 1989, pp.6-7). The "change" can be seen in the form of introducing participatory approaches to forest management, more popularly coined as SF. This chapter and the next one (which is basically a continuation of the current discussion) focus on the functioning and performance of three SF projects (the case-studies) in Bangladesh. Mirroring the observations of Chapter 3, this chapter shows that SF projects function within the surrounding frame, and are profoundly influenced by such variables as social relations and power structure; role of local government, national government and other institutions; land tenurial arrangements; market opportunities; role of public forestry organization, policies, and donor agencies. It discusses the implications of these variables for the performance of SF and argues that they are not generally conducive to fulfilling SF's ambitious objectives.

In line with other south Asian countries, there has been great enthusiasm for SF in Bangladesh (Chapter 1). The government has attached the highest priority to SF and it has become the dominant strategy in the country's forestry sector (Task Force, 1987; GOB, 1992; FMP, 1992). Although the goals of SF vary according to the nature of particular projects and regional peculiarities, they most commonly include empowering "the weaker sections" of local communities through active participation in forest use and management, promoting a cooperative structure to articulate the interests and voices of local people, improving their standard of living, meeting local forestry needs, and promoting self-sufficiency and social equity among local communities (Magno and Ali, n.d., pp.9-10; Rahman, 1992, pp.6-7; Task Force, 1987, pp.1-17; GOB, 1992, p.27; Byron, 1984, pp.7-11; Alim, 1988, p.51; Salehuddin and Rebugio, 1990). The last national plan (Fourth Five Year Plan) has set similar goals for SF (see GOB, 1990, p.ve-7). Furthermore, the planners often particularly emphasize the goals of "empowerment" and "social equity" (Rebugio, 1985; Salehuddin and Rebugio, 1990; Srivastava and Pant, 1979). My case-studies (Betagi, Pomora and Chandra) have also borne similar goals (see for details Rahman, 1992, p.10; 1991, pp.11-13; ARDP, 1989, pp.2-3; Alim, 1988, p.148; Ahmed and Azad, 1987, pp.10-12). In order to achieve some of these goals, especially the goals of fostering social equity, self-sufficiency and cooperation, SF has also (directly or implicitly) aimed at empowering local people to free themselves from some of the social structures of dependency and exploitation. For example, the planners of one of my case-study projects (Betagi) assert (Proceedings of the meeting on Community Forestry, held in the office of Assistant Commissioner, Rangunia, dated 4.6.1979, p.2, Translation):

> [the Betagi project] will rehabilitate "naked" [barren, deforested] hill with "naked" [landless poor] people. It will show the potentials and strength of landless people...farmers will join together [like] one family and strive towards autarky....[They] will also help themselves to face all opposition from the claims of *jotdars* [large land-owning local elites].

Thus, one can see that the expectations from SF have been quite high. However, the unfavourable social milieu (the frame), within which rural development projects including SF operate, and which constrains the fulfilment of their lofty goals, has often been ignored or been paid little attention by the protagonists and planners of SF. The functioning of SF can be visualized as a system of filtration: SF's goals and inputs are filtered through the regulating and competing frame, ultimately to produce the real results. This chapter examines the implications of some selected variables of the frame for the performance of SF in the field.

Social Relations and Power Structure

In the study areas, social and economic transactions involving SF farmers and affluent villagers largely occur through a pervasive patronage network. Patrons are commonly called *Murubbi* (which could variedly connote an "elderly and experienced person", "guide", "counsellor", "protector", "notable or respected individual") in the study-areas. The term *Kamla* (literally meaning a worker, labourer) represents a wide range of clients including share-croppers, different forms of contractual or hired labourers and workers, family servants and household workers. In this section, it is impossible to cover the varied manifestations of social relations (especially the patronage network) and power structures. I will focus only on the dominant aspects of such dependency relations in the study-areas.

Before proceeding to discuss the implications of the patronage network and power structures, I shall briefly note the nature and characteristics of patrons. Patrons constitute a range of rural elites who, with regional variations of wealth and social status, generally are the incumbents of formal or informal power-positions. In following Bottomore (1964, p.14), "elites" are defined here as those "functional groups which have high status for whatever reason in a society". They are groups (including small leadership groups) with "considerable social influence" (Bottomore, 1964, p.13), who can "collectively manipulate the masses" (Ahmed, 1980, p.19).

In the study-areas, I observed the following general characteristics of patrons (elites). First, they enjoy large-scale legal and/or de facto possession of land. Second, patrons have wide vertical and horizontal connections in the local bureaucracy and politics. Most of them are directly or indirectly related to local politics, and thereby, to the local government positions. Third, they are in possession of arms and hired cadre of "musclemen" (*Mastans*). Fourth, they are backed up by the support of a large family-lineage, and/or clan and client groups. Fifth, patrons are involved in diversified occupational and socio-economic activities. These characteristic sources of power have also been widely observed in many other regions of Bangladesh (see for example Ahmed, 1989, pp.22-23).

There are some major factors which foster and sustain patron-client relations in the study-areas. First, the grossly unequal pattern of ownership and access to land immediately puts the controllers and owners of land in an elitist "superordinate class" (Kerkvliet, 1990, p.114). Although SF farmers have received lands from the government, these plots are situated in the hilly terrain unfit for agricultural purposes. Therefore, a majority of farmers are still dependent on patrons for work as agricultural labourer and share-cropper in their (patron') fields. Second, there is an acute scarcity of employment opportunities in the localities. Patrons are in control of the limited employment sources including agriculture, timber and other businesses. The income from SF is not sufficient for covering the yearly expenses of the farmers. I found that 13 (54%), 27 (68%) and 8 (62%) farmers in Betagi, Pomora and Chandra respectively were in deficit in meeting their household expenditure during 1993 (for details on farmers' income and expenditure, see Chapter 7). Therefore, farmers compete among themselves to tap the limited employment sources under the disposal of the elites to supplement their income. Third, farmers also need political and physical protection against local elites and their *mastans*, and against other crises such as natural calamities and harassment by petty government officials. The protection of a patron safeguards the farmer from possible oppression by other elites. On the other side, patrons can "extract labor, food, rents and interests from the farmers" (Scott, 1985, p.xvi). Through the patronage links, patrons can ensure a loyal work-force which they can use in times of labour scarcity. They can also demand assistance in household chores and various other unpaid services. More importantly, patrons can find a political support base in their respective clients during electoral politics and during the times of competition and contention with fellow elites. My observations on these catalytic factors for the establishment and consolidation of patronage network are closely similar to the findings of many studies on rural Bangladesh (for example Jansen, 1986, pp.165-181; 1990, pp.25-26; Rahman, 1994, pp.221-224).

To show how the patronage network functions and influences SF projects, I shall consider the power structure of Pomora village which centres around a family-lineage called the Hafez *Paribar* (family). The late Hafez Chowdhury was the Chairman of Pomora Union Council for nearly 20 years in the Pakistan period. At that time, the family mostly relied on landed property as their income source, and had good connections including social visits with local public officials. As regards the status-hierarchy, being *Chowdhuries*, the family enjoyed (and still enjoys) villagers' esteem and recognition. After independence of the country in 1971, the local political leadership continued to be confined to this family; all the four Union Council chairmen hailed from this lineage-group, including the current chairman. However, recently there have been some changes in the nature of their power-sources and application of authority. In the face of increased political and social unrest throughout Bangladesh, the family has attached greater emphasis to the coercive forces of arms and *mastans*. In order to meet the additional costs of maintaining these forces, they have diversified their income sources. Besides income from agriculture and lands, they have indulged in various commercial undertakings like the timber trade, becoming contractors for various public works, and urban-based small and medium-scale businesses. Similarly, they have diverse political affiliations as well. The four chairmen are members of three political parties (2 of *Awami League*, 1 of *Jatiya Party*, 1 of *Bangladesh Nationalist Party* -- the three principal political parties in Bangladesh). Ideologically, these parties are poles apart. But members of this lineage have successfully exploited their respective affiliation with these parties in retaining their power and hegemony for generations. Whenever the government changed, they switched their loyalty, and managed to have one of their family-members included in the local government structures of the new ruling party. They also have intimate connections with the *Thana* and District level officials of different government departments. Some officials pay regular informal social visits to the Hafez family. A number of young members of the family received their education at the Chittagong city-based colleges. This has enabled them to establish links with the district offices of the government bureaucracy. These links were further consolidated through varied commercial businesses, when the family interacted more with the bureaucracy.

The Hafez family has profound influence on the functioning of SF projects and on the overall pattern of forest use in the locality. The principal patronage supports include monetary help, advice and counsel, providing security and occasional employment opportunities, and lobbying in the government offices (see Table 6.1).

Table 6.1 Nature of the most commonly solicited patronage supports (as reported by farmers)

Nature of support	Betagi		Pomora		Chandra	
	No.	%	No.	%	No.	%
Monetary	4	17	5	13	4	30
Advice and counsel	5	21	4	10	1	8
Approaching government offices	12	50	21	52	7	54
Providing security and employment	3	12	10	25	1	8

*General notes: * Tables in this chapter, except otherwise stated, are based on the field-work data. * Betagi: n = 24, Pomora: n = 40, Chandra: n = 13. * The percentage figures are rounded up to the nearest whole.*

This table also confirms the importance of lobbying local level government offices in all three study-areas.

The provision of security and employment was the second most common patronage support in Pomora. The main employment opportunity for clients (farmers) comes in the form of hired labour in commercial timber logging. Major patrons in the locality, including the Hafez family, are actively engaged in illegal logging with their urban allies, such as the city-based timber and furniture contractors. They employ poor villagers, including SF farmers, in timber logging. For poor villagers and SF farmers, this is an important source of income, given the limited prospects of employment especially in the off-peak agricultural seasons.

Many of my respondents (or their forefathers) have been affiliated with the Hafez family for generations as family servants, workers, share-croppers and political party workers. During land distribution and selection of farmers for the SF project, the family recommended many of their long-acquainted clients to get the lands (the farming plots). It may be noted that the Union Chairman was an influential member in the *Thana* Selection Committee, which was responsible for selecting farmers for SF. In the process, some fairly rich people, with the "blessing" of the Hafez family, also entered the project as (fake) "landless". For example, Azam, who controls nearly 10 acres of land, managed to get a plot in the project through the local Chairman by virtue of his long affiliation with the family. Azam's father served as a clerk in the Union Council under a chairman who was a member of Hafez family, and Azam claims kinship with them. Today, Azam has assumed a dominant position in the Pomora SF project and heads one of the contending factions of farmers. He, in return of the favour, serves as a "vote bank" (Rahman, 1994, p.382) for his patron -- the local chairman -- at elections. On occasions of relief distribution, the chairman patronizes a very select group of people (his clients) which includes Azam and his factions.

The situation of Betagi and Chandra mirrors the experience of Pomora. The patrons exert varying influence and control over SF projects. As in Pomora, the selection of project farmers was largely facilitated through patronage networks. Three Union Council officials (Members) selected 70 farmers in Betagi alone, out of a total of 82 (source: Proceedings of the *Thana* Selection Committee's meeting on 4.6.1979).

Many of the SF farmers continue to share-crop their patrons' land as an auxiliary income-source. Although, given the great scarcity of employment in the localities, share-cropping provides some benefits to farmers mainly in terms of an opportunity for additional income and of partially reducing risks in various contingencies. The terms and conditions of share-cropping arrangements are unequal and heavily favour the patrons. The usual pattern of sharing in the study-areas is that the farmer provides all "input" expenses of cultivation including seeds, fertilizer, insecticides, agricultural equipments and other costs of repair and improvement which may be necessary in any emergency. The costs of harvesting are also borne by the farmer. At the harvest, the crop is divided in two halves between the patron and the farmer. Even the by-products of crop, primarily the paddy straw, is also equally divided. Besides, farmers often need to borrow money from patrons at usurious interest rates. A sharecropping farmer is also expected to be available for unpaid services at his patron's house (such services commonly include cooking, cleaning, washing, gardening and other household chores especially during socio-religious festivals). In more general terms, Jansen (1986, Chapter 5) considers such process of "extracting diverse advantages" from share-cropping farmers by landowners as one of "surplus appropriation". Similarly, Cooper (1983) terms sharecropping as the "dependency web" for "Bengal peasantry". Momen (1993, p.70) also views it as a "structure of subordination" where farmers generally have hardly any "power to bargain" with landlords and are obliged to work "under onerous terms". From his general review of sharecropping and mortgaging pattern in Bangladesh, Rahman (1994, p.215) concludes that:

> sharecropping [is] a predominant form of exploitation which in effect contribute to the consolidation of the class of rich [landowners or patrons]. Share-croppers are [in fact] patron-tied labourers [who] are also trapped in perpetual indebtedness.

Thus, the advantages which patrons extract through the sharecropping mechanism outweigh the marginal benefits which farmers get. Involvement in SF has neither freed the farmers from such unequal exchanges and obligations, nor has it reduced their vulnerability to any significant extent.

The main patrons (with their respective factions) are contending among themselves to establish control over SF projects. This has caused strife inside the projects and has led to the emergence of corresponding factions among SF farmers. Each patron utilizes his allied group (or faction) in the project to support him in factional contentions and struggles in the village power structure. Patrons consciously attempt to impede the growth of any collective activity or institution among SF farmers.

Besides exploiting farmers as support-base and contract labourers, patrons also capture some of the benefits specially designed for SF. For example, in Betagi and Chandra a substantial portion of "inputs" (for example seedlings and fertilizer) which the government provided to SF farmers free as an incentive, were siphoned off by the patrons. In Betagi, rich patrons took up the innovative agroforestry technology specially developed by forest scientists at the Bangladesh Forest Research Institute for SF farmers. Patrons acquired the knowledge from their clients in the project, and successfully applied the technology in their own plantations. This silent transfer of technology was facilitated by mutual understanding. Patrons were generally interested in testing new agricultural technology which could potentially augment production (for example, as the local Agriculture Extension officers told me, the rich landowners were particularly enthusiastic for trying the

High-Yielding Variety of paddy in their land when this was first introduced in the locality). The clients also had an opportunity to offer this service free of charge to their interested patrons to be in their good books. In fact, in the "inequitable social order" of Bangladesh such "bubble up" (Burch, 1988, p.85) of benefits (arising from rural development interventions) has been widely observed (see for example Hartmann and Boyce, 1983; Rahman, 1994; Siddiquee, 1993). Hinting at the unequal patronage network, Heaver (1982, cited in Burch, 1988, p.85) also noted that in rural areas where "the poor live next to, are tenants of, owe money to and work for...prosperous larger landholder [patrons]; such leakage of benefits...is inevitable.

The government's effort to evict fake "landless" farmers from the Betagi and Pomora projects largely failed due to patronage-protection and factional politics. In 1992, the latest attempt at eviction failed. This time poor landless farmers opposed the government's action and supported their rich neighbours. After a series of interviews and informal talks, I realized that this was yet another manifestation of patronage politics. Patrons did not come into the foreground this time; instead they manipulated their clients to frustrate the government's bid, and thereby safeguarded their integrity and authority.

SF farmers call their respective patrons by fictive terms like *Chacha* (uncle), *Bapjan* (father), and *Borda/Bhaijan* (elder brother). These fictive connections reinforce patron-client contracts (Mashreque and Amin, 1993, pp.22-23) and clients accrue "financial and political benefits" (Sarkar, 1980, p.55). Patrons thoroughly relish such explicit demonstration of subordination which signals increasing prestige and status for them. Besides, "through reinforcing bonds with poor", patrons can effectively "command a large labour force" (Mashreque and Amin, 1993, p.23) in the locality, which can overcome problems of labour supply at periods of peak activity.

On days of socio-religious festivals (for example *Eid* for the Muslims, various *Puja* festivals for the Hindu population), the farmers visit their patrons (including the fictive rich kin). The patronage network becomes expressive and invigorated during these festive occasions through ritual exchange of goods and services (Mashreque and Amin, 1993, p.26).

Involvement in the timber trade has brought patrons (elites), government officials and urban entrepreneurs into close contact, resulting in an intensification of the relational network of patronage and extraction. In the process, there has been an alliance, to use Wood's (1994, p.84) words, between "rich peasants and petty bourgeois" which commands resource distribution in the country-side; and the members of this alliance "have become agents of the state". Farmers appeal to their respective patrons for resolution of factional conflicts among themselves or with other elites. They also seek help from patrons who have regular contact with local officers to approach government offices which otherwise remain inaccessible to farmers.

There are hierarchical differentiations among SF farmers. Although labelled as a "landless group, participating in SF for collective improvement" (Proceeding book of the Betagi *Bhumiheen Samity*, pp.34-35), farmers are by no means a uniform community. In fact, differentiation and competition within the peasant class have been meticulously surveyed by many studies on Asia. Two most interesting works are of Scott (1985) and Kerkvliet (1990), which have focused on the complex details of day-to-day class relations and resistance in two villages of Malaysia and Philippines respectively. Kerkvliet (1990, p.76), for example, notes:

> Society is...not simple. There can be considerable competition among people of the same class. For instance, landless workers scramble over each other to get scarce work, small business people take customers from rivals, peasants upstream steal water from others whose fields are downstream,...buyers try to deprive competitors of grain sales.

In my study-areas, some farmers claim superior social status, based on their lineage or family-heritage, economic position and their linkage with powerful patrons. They like to build relationships with outside villagers of equal family background. In the initial years of the projects in Betagi-Pomora, poverty and land-hunger brought farmers into a temporary unified platform. But as they have improved their financial position, farmers are tending to show more consciousness towards differential social status and recognition. The status and attitudinal differences among farmers are subtle but strong. Some farmers never feel comfortable to be identified with those with whom they have lived side by side in the projects for years. These farmers are trying to increase their status and social recognition, often degrading (or ignoring) fellow farmers, and thereby, causing rupture in group-solidarity. I met a Betagi farmer (a *Chowdhury*) employing a hired-labourer to work in his garden, though I was told that he was a daily labourer himself prior to joining the SF project. Hiring a labourer reflects increased status for the household. The institution of marriage is another area where the working of status-differentiation becomes clear. With increased solvency, farmers are marring their sons/daughters to wealthier and higher-status families. At least 15 of the Betagi-farmers have established marital connections with families of higher social status; and thereby, they seem to have upgraded their status as well. This pattern of status-differentiation has profound implications for farmers' group solidarity, and social mobility and recognition. Here, status is largely contingent on the economic position of farmers. With increased status, a farmer can move around in the upper echelon of the villagers more comfortably, though it also alienates him from his fellow-farmers in the projects. Furthermore, as I have already noted, farmers compete among themselves for patronage favours, especially for additional employment opportunities and protection. However, the weaker farmers tend to lose out in this competition. A patron prioritizes his clients while distributing scarce benefits. He would typically prefer kin, close acquaintance and relatively better-off farmers for a number of reasons. First, helping the kin is a social obligation; it also enhances his benevolent image and social recognition. Second, as one patron told me:

> if you need to put oil on one's hair, it is always better to go for someone who has already got some oil on his hair, [because] he will require less oil.

Trying to help the poorer farmers requires more effort which patrons are not interested in making. Third, a patron does not want to lose a stronger farmer fearing that he might join the rival group and contribute to strengthen his (the patron's) opponent's position.

During informal discussions, I have noted that farmers are not naive or ignorant about the society they live in. In fact, as some other studies have also observed (for example Kerkvliet, 1990; Scott, 1985; Jansen, 1986), they were well-informed of the processes of exploitation which surround their life and living, of "the considerable inequalities and of where everyone is positioned" (Kerkvliet, 1990, p.14). But they still maintain their allegiance to power holders and patrons for reasons discussed above. The logic, however, was most astutely summarized to me by a SF farmer (who received primary education and some formal training from the forest department) in a local tea-stall, where we used to meet together in the evenings and chat for hours:

> you can't fight with a *Kumir* [crocodile] when in water....Besides if you don't have a boat, the [crocodile], if he is in a good mood, can even help you cross the river on his back!...and more importantly, if you are a friend of the [crocodile], smaller lizards won't harm you.

Farmers have a subtle way of showing their resistance and indignation against the various forms of exploitation they face day-to-day. They go for "low profile techniques" what Scott (1985, p.xvi) terms as the "everyday forms of resistance" (also see Kerkvliet, 1990). Such "everyday resistance" in my study-ares commonly assumes the forms of foot dragging, false compliance, back-biting, character assassination, jokes and puns, and minor pilferage. Farmers seldom attempt to make more stubborn and organized forms of collective resistance. Because, first, farmers lack the institutional base for overt unified defiance. Second, the strife and divisions among farmers reduce the chance of collective action. Third, the risk of getting in the open confrontation with patrons are too high for farmers. "*Murubbis* are everywhere", as one farmer told me and the punishment, mainly in the forms of losing additional income sources and physical and mental harassment, can be severe. Fourth, the limited patronage benefits what farmers receive are "sufficient to ensure that [they] choose to conform in most respects to the standard of seemly behavior that is defined and imposed by the village elite" (Scott, 1985, p.25). Fifth, as I have repeatedly noted, farmers are in ceaseless struggle among themselves to compete for the scarce resources from patrons. "Poverty, itself, therefore, inhibits [them] to openly confronting the rich [patrons]" (Kerkvliet, 1990, p.176). Thus, farmers have little recourse beyond vocabulary, gossips, idiom, and discussion among themselves (against the patrons); "simply because they are too vulnerable to do more and are divided" (Kerkvliet, 1990, p.198).

The social relations including patronage links are not infallible or static; rather they are in a continuous process of shaping and reshaping according to changing situation. For example, the degree of intimacy in a patronage relationship fluctuates substantially over the time. A patron may choose to select new clients from farmers, or farmers can also switch sides responding to better prospects. I noticed a situation in Pomora when a patron (a local government official from the Hafez family) favoured two new clients (among SF farmers), who had become quite active in the latest Union council elections for his (the patron's) opponent party. The farmer's tenacity and enthusiasm during the election period attracted the attention of this particular patron; and he, by promise of a substantial reward, induced the two farmers to switch allegiance. Farmers, especially the poorer section, however, cannot avail themselves of this opportunity so easily, primarily because the risks and initial repercussion of changing sides are sometimes too strong for them to sustain. Clapham (1982) and Kerkvliet (1990), among others, have also noted the "fluid character" of political clientelism. In the context of a Philippine village, Kerkvliet (1990, p.9) for example, writes:

> The alliance and factions, however, are unstable. If individuals think they can gain more from a rival faction, they will very likely switch sides, making an ally of a group that was previously a competitor....This instability contributes to, and is a reflection of, constant manipulation both of the people by political elites, primarily through patronage, and of the competing politicians by voters.

Besides, in some parts of my study-areas there are emerging signs of market principles of demand and supply influencing the transactions between elites and poor, and thereby, occasionally disrupting the traditional patronage considerations. I will elaborate this issue in the discussion on the implications of market for SF.

The above discussion shows that these dependency relations and structures of domination are deeply ingrained in the social fabric of the study-areas. They have profound implications for forest resource use and SF. From the above experience, four broad conclusions can be drawn. First, SF intervention has brought about little change in the nature, patterns and implications of the dependency structures such as patronage. SF farmers are still under heavy influence of elites and they have to perform within limits set by the patronage network. SF has done little to free farmers from the manipulation by the network. Second, patronage relations have partially benefitted the better-off sections of farmers in terms of increased status and opportunities for additional income, although patrons have extracted many more benefits (in the forms of unpaid labour, servile works and political support) from this unequal exchange. The poorer farmers, however, have lost out in competition with the better-off farmers to tap benefits from patronage sources. Consequently, they have been pushed into further poverty and marginalization. Third, patronage network, status-differentiation, and power relations delimit the growth of cooperation and collaborative initiatives in the study-areas, and thereby, frustrate one of the major goals of SF intervention. The factional politics, inter group conflicts and divisions among farmers, which are fuelled by above social relations and structures, disrupt group solidarity and reduce the chance of inter-farmer cooperation. Remote from such romantic observations as "all the Betagi families live like a joint family" (Alim, 1988, p.157), I have noted that farmers are actively engaged in factional politics, modelled on the broad pattern of power politics in the villages. Besides, there is a growing pattern of status-differentiation among the project farmers with malign implications for group solidarity and social mobility of the excluded. Fourth, the social relations and structures enable elites to consolidate their power and position; and serve as a leverage to manipulate the functioning of SF projects. These structures also provide elites with an avenue to appropriate a substantial share of the benefits from SF. In general terms, the structures of manipulation and domination including patronage, power relations and status-differentiation (with all their manifestations) are a part of the deeply ingrained inequality of the society (see Chapter 3).

The Role of Local Organizations and Public Offices

SF farmers have to deal primarily with the respective Union Councils and a number of government offices at *Thana* (especially the forest, police, land-revenue, and agriculture offices). In the Betagi-Pomora projects, the cooperation and assistance of the Union Councils and concerned government offices have gradually waned with the passage of time. There are three explanations for this declining support. First, the initial enthusiasm of local offices was geared up by personal initiative and interest of some centrally-placed senior officials and intellectuals -- the project pioneers (see Chapter 4). Now, in the absence of such vigilance, local offices have gone back to their usual naive behaviour in providing service to the projects. SF is no longer in their priority list. Second, farmers' economic upliftment and increased status have instigated a sense of envy among other sections of local people. Third, farmers as a group is seen as a rival power-base by some local government leaders/officials. I interviewed the chairmen of the Betagi, Pomora and Chandra Unions. Two of them maintain their respective client groups among SF farmers (as discussed in the previous section). Nevertheless, they believed that there should not be a farmers' cooperative institution (*Samity*) to manage affairs of the projects; instead they suggested that the projects should be managed by them (that is, by the Union Council administration). The third chairman commented:

> [the SF farmers] do not deserve what they have been given....Four acres of land for each farmer!....Oh! its too much. You can't expect a dog to digest *ghee* [pure butter-oil].

The Union Councils hardly pay any attention to the projects and the Council officers seldom visit SF areas (see Table 6.2). The role of local government institutions in supporting SF activities has been poor. The Forestry Master Plan (FMP, 1992, p.4) also noted the *Thana* administration was entrusted with SF activities (mainly strip plantations); "but it was a dismal failure" (also see CFD, 1994).

In Chandra, the forest department is more active because the project is in its nascent stage and is still receiving central attention. Apart from the forest department, institutional support from other public offices is insignificant (Considering its central role in SF, I will elaborate the forest department's performance in a coming section). I found that many *Thana* officials were simply unaware of the existence of SF projects, although their offices, according to governmental papers, were supposed to be assisting such rural development initiatives. One *Thana* Agriculture Officer told me that since the agriculture office did not have "formal links with SF", therefore his office was not "obliged to [render] any service" to SF projects. By "formal links" he implied compulsive written instructions from his central line-authority to cooperate with SF projects. During the initial years of the Betagi and Pomora projects, majority of the District and *Thana* officers used to attend the project meetings, seemingly because high-ranking officials including a Central Secretary (one of the project-pioneers) used to be present in these meetings (for example the Proceedings of Thana Committee meeting of 10.11.1978 showed that as many as 20 government and local government officers were present, including the *Thana* Agriculture Officer). Alim (1988, p.147) also noted that senior officers' presence was the main cause of local officers attendance and prompt action. Similarly, when I visited one Family Planning and Welfare office, they could hardly recognize the SF project I was referring to, although the project was within their walking distance and a number of their village-workers live near the project area. In general, visits of government officials to SF projects are few and far between (see Table 6.2).

Table 6.2 Inspection and visit of the concerned officials to the projects (as reported by farmers)

Frequency of visit	Projects	LG		FD		FP		AgD/ DAE		Others	
		N	%	N	%	N	%	N	%	N	%
Regularly	B & P	0	0	0	0	0	0	0	0	0	0
	Chandra	1	8	4	31	0	0	0	0	0	0
Occasional -ly	B & P	16	25	12	19	5	8	5	8	5	8
	Chandra	6	46	8	61	4	31	3	23	6	46
Rarely	B & P	24	38	6	41	10	15	16	25	35	55
	Chandra	4	31	0	0	7	54	6	46	2	15
Never	B & P	16	25	16	25	30	47	24	37	16	25
	Chandra	0	0	0	0	2	15	3	23	0	0
No Comment	B & P	8	12	10	16	19	30	19	30	8	12
	Chandra	2	15	1	8	0	0	1	8	1	8

*Notes: * B & P = Betagi and Pomora, where n = 64 (24+40). These two projects are situated in the same locality in Rangunia. Chandra (n = 13) is in a different location (Kaliakair). * N = Number. * LG = Local Government (here LG refers to the political representatives especially Union Council Chairmen and members. FD = Forest Department, FP = Family Planning, AgD = Department of Agriculture, DAE = Department of Agricultural Extension. * "Others" include land revenue, police, and rural development officials.*

Although corruption and public harassment seem to be the common norm of behaviour of majority of the public offices, I heard grievances against the police more than any other offices. The Officer in Charge of the local police station was made a member of the Betagi-Pomora *Thana* Selection Committee. Initially, the police was forced by the central authorities to help farmers in their struggle against local influential land-holders (see Proceedings of Thana Selection Committee meeting on 27.3.1981). But those days are history now. I heard dreadful reports of harassment and physical injury caused to farmers by the police. Any reasonably serious observer can immediately see the network of intense corruption by the police. Extortion and intimidation have become the way of life for police in the study areas. On 2.3.1988, on one occasion, the Rangunia police openly collaborated with a local elite against a SF farmer in exchange of a handsome bribe. The farmer was arrested on a false case of fire-arms possession (vide 19 A and F clauses of the Arms Act). He was brutally tortured in the police custody before being sent to court. On 19.3.1992, after four years of untold miseries, (with the help of fellow farmers and kind gesture of some senior officials and a solicitor), he was acquitted from the Special Tribunal Judge's Court in Chittagong. Although all the allegations against him were unsubstantiated, the damage had already been done. The cost incurred in the case has not only left him indebted until now, but also he is constantly haunted by memory of the incidence. He burst into tears while telling me the story and showing the court-documents. This case also received media-attention (see for example The Daily Nayabangla, 10.6.1991), but with hardly any impact on the behaviour of local police. In sum, my experience suggests that government offices, with rare exceptions, will only render even the minimum expected services under constant vigilance and force from their superior line of command. Otherwise, routines become exceptions, and public services cost money and harassment.

Besides their involvement with local government and public offices, some farmers have occasional exposure to radical left-wing political parties.

The membership of such parties, at least temporarily, fortify their position in their struggle against governmental actions and local elites. For example, some of the Chandra and Pomora farmers shared identity with *Khet Mazur Samity* ("land-labourers association" -- a leftist local political organization) and they showed political resistance against government or local attempts at eviction. This political link also runs on a dyadic reciprocal process. For the political parties, SF farmers act as their much-coveted support-base; and for farmers such a membership is a source of some help during crises. But formally no attempt was made to incorporate these political supports into the project activities (which could have a positive effect on promoting farmers' interests). These links remain informal and often, dormant.

Farmers have virtually no involvement in other voluntary associations (for example, in a youth club) in the localities. It seems that voluntarism, as compared to prospective material interests, has little influence on fostering cooperation.

Land Tenurial Arrangements

In context of the review of land tenurial issues in Chapter 3, this section focuses on some of these issues and their implications in the field. The issues include tenurial security and its impact on farm productivity and group solidarity; land and tree tenure, and its impact on women; the government's role in determining land tenurial and related policies; and the social justification of providing land-ownership to particular groups of rural people at the expense of excluding others.

Mirroring the observation of many studies (see Chapter 3), tenurial security of land is the most-coveted demand of the farmers in my study-areas. The Pomora and Chandra farmers are most vocal for securing their land rights because currently they use public land on annually renewable lease-contracts. The Betagi farmers, on the other hand, have received permanent title to land in 1987. In the former cases of temporary or contractual tenure of land, the contract document imposes a plethora of terms and conditions, with clearly unpalatable impact on farmers. For example, the document given to the Pomora farmers (by the forest department) includes a condition (Condition No. 2 of the Permit issued by the Chittagong Forest Division) which reads:

> The land covered by this permit is part of Government...forests and is therefore inalienable and is liable to resumption whenever such a course is found necessary by the Divisional Forest Officer and no claim whatsoever shall be entertained in respect of such site.

Furthermore, it restricts keeping livestock on the plot (Condition No.10), which is a highly impractical condition. Consequently, farmers share a feeling of uncertainty and grievance over the insecurity of land-tenure in Chandra and Pomora.

However, on the other hand, rather contrary to the popular belief (Task Force, 1987, p.91; Bentley et al., 1987, p.233; Alim, 1988, p.154; Quddus et al., 1992, pp.111 and 114), I have found that tenurial security alone does not necessarily bring about increased farm production or group solidarity (see for details Khan et al., 1996). Betagi is a vivid case in point. In Betagi, the longing and demand for permanent ownership of land was by far the most propelling factor behind farmers' enthusiasm in the initial years, until they managed to achieve the permanent title of the land (in 1987). They were given verbal assurance by the project pioneers that permanent title to land might be accorded to them, provided they could show satisfactory progress in respect of land and plantation development (A. Rahman, Bangladesh Forest Research Institute, personal communication; Khan et al., 1996; Alim, 1988). Motivated by the hope of secured tenure, farmers showed spectacular development in terms of plantation activities, income earning and group solidarity until 1987 (Rahman, 1987; 1991). In 1987, farmers were offered permanent ownership of land with the expectation that tenurial security would further accelerate the community spirit and farm production. But during the post-tenurial period (1987 -- todate) there has been a gradual decline in their enthusiasm. Some elements of stagnancy have been observed in the project (Khan et al., 1996). The community feelings have drastically eroded. Presently, farmers are showing more inclination to off-farm activities and individualistic pursuits. Farmers' engagement in farm activities at Betagi has gone down to an average of 9 labour/hour per day from that of 16 labour/hour in 1984, as reported by Rahman (1987). Several factors can explain this declining trend. First, although farmers received permanent title to land, there have been constant conflicts over territorial boundaries (demarcation) of such lands (plots), between farmers and elites, and also among the farmers themselves. In many plots, even after 7 years of formal ownership, farmers have not been able to establish their control over the land. The much-needed institutional support of the government (for example the collaboration of the land settlement and land revenue offices) has gradually declined. After formal division of plots and individual ownership, farmers' collective spirit has also eroded. A farmer who does not have a problem, is no more interested in helping his troubled neighbour in lodging a collective complaint to the government, or in forming a cooperative resistance against contending elites. This has also given greater opportunity to a section of corrupt public officials (especially in the land revenue and police departments) in manipulating the situation to their vested ends. They are, on the one hand, helping elites to misappropriate (or at least, to create trouble over) the land and, on the other hand, asking huge bribe from farmers. Some respondents reported that government offices in alliance with local elites want to prolong the conflict over land tenure, because such tensions are a perennial source of income for them. Thus, as one can clearly see, the central government's (macro policy) decision of providing farmers with land-title has not been adequately secured and backed up by local level offices (at the micro level). Second, farmers do not seem to find SF or afforestation schemes to be the most economically rewarding alternative land-use. "Long term forest cropping", for example, does not attract farmers because of the long gestation period and the danger of plunder of the commercially valuable trees. In many instances, armed group of *mastans* have forcibly chopped down trees from SF projects at night. Very few of these incidents are reported to the government. One farmer explains:

> informing the police or forest offices only creates further trouble. They will ask you "what happened?", "who did it?" and "when?"....Then [they] will go the person who did it, but will do nothing to him....They will start taking money from both sides [from the offender and the victim]. It's like bringing wolf to safeguard *Kura* [chicken].

Instead of planting trees, farmers have other utilities of the lands allotted to them. These utilities are somewhat reclusive. In Betagi, for example, land-ownership has increased farmers' social status and recognition. By virtue of land ownership, they can arrange greater loans from informal sources, especially from the *Mohazans* (informal money lenders; generally rural traders) and can establish marital connections (for their sons and daughters) with better-off families. I have observed a number of cases in Betagi, where farmers have been able to attract *bhala* ("good" or "established") bride-grooms (such as urban factory workers, petty traders) for their daughters by unofficially (verbally) assuring a share of their lands. I have met a number of such new-comers who are residing in the project plot with original owners (farmers). In some places the new comers have even taken de-facto control of the land. One farmer told me that his new son-in-law, whom he had allowed to reside with him, pushed out his own son and took complete control of the plot. But the farmer could not report the incident to project authorities, because he illegally invited the son-in-law to share the plot at the first place. Thus, the increased social position through permanent title to land has brought about some vices as well.

Therefore, land ownership is more of a symbol of prestige, status and increased social connections for farmers, than an avenue to increased plantation and production. Farm productivity is not directly dependent on permanent title to land, rather it depends on many other factors such as institutional support from government offices, greater economic feasibility of tree-planting operation, security of harvest, and assurance of benefits. Trees on SF lands entail different implications on their own. A number of farmers reported that some valuable trees on their lands were demanded by elites (who used to control the lands prior to the SF projects) on the ground that the trees were planted by them (the elites). Many instances of attack on such trees were reported. Tree tenure, therefore, is another major consideration for SF. The government does not have any policy decision in this regard. The local public offices have kept silent on tenurial issues, partly because they can exploit the present ambiguity to their own interests. Some fruit trees have also become an apple of discord between SF farmers and villagers (outside SF area). Prior to the advent of the projects, villagers used to have a share of fruit and medicinal trees. Now, with individual ownership and control, such common usage has been put to an end. Nevertheless, villagers at large think that such "special purpose" trees (such as medicinal or fruit species) are "for all", irrespective of the legal position of the land on which they grow.

Unlike the general "tendency of most land tenure reforms which expands men's access to land" at the cost of "decreasing or informalizing women's right of ownership" (Rocheleau, 1987, p.84), the Betagi women have the privileged position of enjoying co-ownership of the plots (along with their husbands). But this has little visible impact on their day-to-day affairs and decisions. I found that a good number of women (the land co-owners) either did not know at all of their share in the official ownership; or even when they knew, they were mostly unaware of the rights and privileges associated with such ownership. Many women whom I interviewed were often oblivious of their share of ownership and did not have any understanding of the ownership document ("deed"). The "deed" being a complicatedly phrased legal paper, was far beyond farmers' (and women's) understandings. The use of land as a lever to increased social status or informal loan, as I discussed earlier, is largely confined to men (the farmers). Women's co-ownership of land has not contributed to any significant increase in their power in household decision-making, except some involvement in marriage decisions. Besides, land ownership has done little in linking women to "formal sectors where contracts are signed, land is bought or sold, technical assistance is given" (Rocheleau, 1987, p.85). The Betagi women, like majority of women folk in other study-sites, still work mostly within informal household spheres. It seems that, providing permanent tenure of land alone would bring little change in their position, without substantial corresponding changes in other related structures, for example, more education for women, increased access to cash income and other productive means, increased power in household decisions, and some flexibility in their work-responsibilities and socio-religious taboos.

The individual allotment of land to SF farmers has instigated discontent of local elites and common villagers alike. The elites lost (de facto) control of lands by the inception of SF projects. For villagers these lands were grounds for fuel-fodder collection and pasturage. The allotment of plots to a specific group of local people (which is practically tantamount to privatization), has reduced villagers' share of the resource. In the face of the country's overwhelming landlessness and dire poverty (see Appendix 1), such privileged allocation of land to a tiny section of local people (that is, the SF farmers) is likely to provoke envy and discontent among other sections. During informal interviews with government officials, politicians, villagers, and academics, a strong opinion was that "collective ownership" or "long-term usufructuary rights" might be better alternatives, as compared to individualistic ownership of SF lands. I will elaborate these views in Chapter 8.

In sum, land tenure seems to be a complex issue for SF. Tenurial security is a major motivating force for farmers. But providing permanent title to land alone has little impact on farm productivity, or on the growth of collective effort among SF farmers. Farmers need institutional assurance and support (especially from the government) to fully utilize the potential rights and benefits associated with such ownership. Farmers who have received permanent land ownership (Betagi) and who are struggling for ownership (Chandra and Pomora), both are sceptical and uncertain about the future of their plots. They are suspicious of the government's good-will, and anticipate future conflict both with the government and local elites (see Table 6.3). Moreover, the rewards from SF are not encouraging enough to confine farmers only to SF activities. Therefore, land security needs to be backed up by technological and logistic support from the government and other concerned agencies, so as to make SF activities economically more viable than other off-farm alternatives. Vast majority of farmers in all the three projects have kept a portion of their land fallow, which is attributable to such reasons as uncertainty or insecurity of tenure, anticipation of future conflicts, economic hardship, and suspicion of the government's intentions (see Table 6.4).

Table 6.3 Apprehension about probable future problems over land tenurial status (as reported by farmers)

Nature of problem	Pomora				Betagi				Chandra			
	1		2		1		2		1		2	
	N	%	N	%	N	%	N	%	N	%	N	%
More conflicts and litigations	10	25	10	25	6	25	1	4	2	15	0	0
Conflict with government	18	45	2	5	0	0	2	8	7	54	2	15
Sabotage by elites	2	5	2	5	4	17	2	8	2	15	2	15
Other	0	0	2	5	0	0	2	8	0	0	0	0

*Notes: * 1 = very acute problem, 2 = acute problem. * In response to the question "Do you apprehend any problem regarding land-tenure ?", in Chandra 11 (84%) farmers replied in affirmative ("yes") and 2 (16%) farmers said "no"; in Betagi "yes" = 10 (42%) and "no" = 14 (58%); in Pomora "yes" = 30 (75%) and "no" = 10 (25%). * N = Number. * The Betagi farmers felt comparatively secured because they had been given permanent title to land in 1987. In Chandra and Pomora, farmers use the land on temporary lease.*

Women in general have not been able to reap the benefit of long-term tenure. The "social justification" (Nadkarni et al., 1989, p.173; see Chapter 3) of private or individualistic land allotment to SF farmers, at the cost of excluding other equally deserving sections of local people, is also questionable. Fortmann (1987a, p.9) has also argued that such a "shift of resources to [a particularly targeted] beneficiary population may occur at the expense of other poor people, thus resulting in a redistribution of poverty". The government's role in the study-areas is one of unfinished nature. It has allocated land to SF projects, but has not provided any clear land-use or tenure policies regarding the varied problems and complications associated with such allotments.

Immediately after farmers were allotted land, there was an upsurge of claim and opposition from local elites. The Betagi farmers reported that 26 law-suits were furnished against them by local elites, and they had to suffer great miseries in establishing their rights. In Pomora, besides the problem of elite opposition, I observed confusions and conflict even within the public offices (for example, between the forest and land revenue departments) as regards the

Table 6.4 Reasons for keeping the lands (plots) partially fallow or not optimizing the land-use (as reported by farmers)

Reasons	Betagi		Pomora		Chandra	
	No.	%	No.	%	No.	%
Economic hardship	5	21	6	15	3	23
Time constraint	2	8	2	5	3	23
Labour shortage	3	12	1	2	1	8
Tenurial insecurity	2	8	11	28	4	30
Not enthusiastic about farming activities	5	21	11	28	0	0
For pasturage and livestock-rearing	3	12	2	5	1	8
Disputed land	4	17	7	17	1	8

*Notes: * Though formally lands were allotted to farmers, many plots remain disputed properties by constant threat and interference from elites. * Some farmers do not find farm activities sufficiently rewarding and show more interest in off-farm jobs.*

demarcation and distribution of SF plots. This was manifested in a number of official correspondences.[1] Furthermore, in Bangladesh some colonial laws pose potential tenurial problems for SF, especially when a SF project is launched on public lands (Ali, 1987, p.16). The government has provided no decisive clarification on these potential dangers. In fact, the "unfinished role" of the state and the consequent impacts on SF have been observed in many parts of the Third World (see Riddell, 1987, p.10).[2]

Market Opportunities

As Chapter 5 has shown, forest use in my study-areas is dominated by market forces which include commercial loggers, urban traders and their rural partners, local elites, and a sizeable section of government and local government officials. The commercial network is strong, well-organized, and fortified by collaboration and connivance of the bureaucracy and politics. As anticipated in Chapter 3, the poor and weaker sections of local communities do not have any control over market forces, whom Parker et al. (1989, p.9) call the "distant elites" who variedly "manipulate [them] from a distance". SF Farmers and poor villagers, for example, work as low-waged labourers in logging operations, share-croppers, servants or workers in the elites' households, trades and businesses. The terms and conditions of these employment or engagement are exploitative, and heavily favour the employers (Chapter 5). The dominance of the market forces has also caused a proportional decline in local people's share in forests and in their bargaining power.

In some parts of the study-areas, market principle of "demand and supply" has started to dominate the transactions between elites and the poor, especially if the concerned person is not a long-term client or kin of the elite. With increased landlessness, poverty and unemployment, some of the local poor, especially the migrants or newly-settled villagers, are placed in an increasingly vulnerable position when the market price alone (without any patronage considerations) determines their services. Wood (1994, pp.84-86) has also observed that rural Bangladesh is now dominated by "capitalist market relations" which he calls the "new political order". Thus, in some parts of my study-areas, the "new order" has started to affect patronage links. Labourers are "hired at rates determined by demand and supply" (Wood, 1994, pp.85-86); and since the supply is in abundance, employers can manipulate the situation to their profits. On some occasions, I have noted that elites tend to withdraw themselves from the obligations entailed in a traditional parton-client relationship. This signals that the limited patronage support which the poor used to get is also becoming commercialized and market-oriented. In more general terms, Wood (1994, pp.85-86) has noted that in Bangladesh the state is the greatest patron of this "new order" which has emerged "as a substitute for social order of the village". In my study-areas, however, market relations have not yet completely substituted the patronage network, and therefore, the "new order" is not in full-bloom. But, there are signs of such a transformation in some parts of the localities. I have seen labourers who wish to comply with almost any condition set by the merchants to work in timber logging. In the face of dire poverty and unemployment, labourers are ready to shoulder all risks involved in such operation in return for meagre cash. These desperately needy sections of local people have largely submitted themselves to market forces for deciding their works, conditions and wages.

The facilities brought by market in the study-areas (for example, better transportation, modern logistics for timber extraction, more exposure to urban market centres) are benefiting elites and a section of the well-off farmers.

There is no cooperative or collective effort for the marketing of SF products. Farmers largely depend on middlemen for marketing. In Betagi-Pomora, for example, as Quddus et al. (1992, p.44) note:

> middlemen play the most important role in determining prices. Usually, middlemen survey the market and they themselves fix the prices by size and quality. Once the price is announced, middlemen stick to it very rigidly. Sellers [SF farmers] virtually have no choice.

Any attempt by farmers to market their products collectively, meet with opposition from the network of middlemen and traders. The Betagi farmers, for example, when trying to sell lemon directly to a city-market in Chittagong "were openly discouraged and harassed by the established *Aratdars* [organised group of traders] and subsequently had to retreat" (Rahman and Khan, 1994, p.71). Although lemon is the principal horticultural (cash) crop in Betagi-Pomora and the most dominant farming practice, farmers still have not been able to establish the minimal control over lemon market. During peak seasons, the price of lemon goes down to Taka 8-10 (per 100 lemons) against the usual price of Taka 90-130. Farmers are helpless against such market fluctuations. There is no storage facility (to preserve vegetable and horticultural products) in the area. Therefore, in many instances, "huge quantity of lemon [are] kept unharvested, ultimately to rot" by the utterly frustrated farmers (Rahman and Khan, 1994, p.71). The situation of the commercially valuable trees is even worse. There is "an overfull demand" (Maslekar, 1983, p.195) for forest trees. But in the absence of an organized marketing system and logistic support from the government, the timber market is controlled by the aforesaid commercial network. Between the producing farmer and the consumer, a chain of middlemen operates at different levels which includes village traders, agents of forest industries and urban merchants, contractors, and commission agents. In the process of passing through different stages, the ultimate market price of the product shoot up. The profit is usurped by intermediaries and the original producer (the farmer) remains as unjustly paid as ever. For example, I found that a Betagi farmer (of Plot No. 269/5) sold 40 cft. of timber and 60 maund (1 maund = 37.32 kg) of fuel wood to a local middleman at a lump-sum price of Taka 2,500. In response to my enquiry, the middleman reported that he would sell the fuel wood and the timber to the agents of a local brick-kiln and an urban furniture shop respectively. He hoped to fetch a price of Taka 5,000 for the lot; and suggested that the market price for the products ultimately would reach something to the tune of Taka 6,500.

The government's support for farmers in promoting cooperative marketing and in developing a market network is largely absent. There is a complete lack of logistic back-up and market information from the government. Rather the plethora of governmental regulations regarding marketing and transportation of forest products have deterring and demoralizing impacts on farmers. Farmers think that the rules are well-beyond their comprehension and serve as leverages for harassment and extortion by government offices (see Table 6.5).

Table 6.5 Farmer's perception regarding selling and transportation rules and procedural formalities (as commented by farmers)

Farmers' comments	Betagi		Pomora		Chandra	
	No.	%	No.	%	No.	%
"Beyond our understanding"	3	12	4	10	2	15
"The rules give opportunity for corruption or extortion to government offices"	16	67	28	70	5	38
"Time consuming", "Unnecessary delay and harassment"	4	17	6	15	2	15
"No comments"	1	4	2	5	4	31

Farmers in Betagi-Pomora do not dare to market their product themselves and, as I have mentioned before, rely on the middlemen to "deal with the bureaucracy" (Saxena, 1991, p.164). What happens if a farmer ventures to sell his products by himself and try to tap market potentials directly? In quest of an answer, I travelled to a private farm at Hathazari in Chittagong to witness a horrifying case of bureaucratic hassle and extortion inflicted upon a private farmer, Mr. Sikandar Ahmed. Ahmed planted about 100,000 trees in his farm and appealed to the forest department to cut 5,000 matured trees in October 1990. After one year of persuasion the forest office permitted him to fell only

2,500 trees on 14.12.1991. He was given an impractical time of 5 weeks (up to 25.1.1992) not only to find customers, to mark and fell 2500 trees, transport them over the jungle and bring them to market; but also to fulfil the lengthy bureaucratic formalities. The forest laws required that each sellable log had to be marked by the Forest Range Office (which was located 7 miles off his farm) and a Transit Pass be issued by the Divisional Forest Office in Chittagong (18 miles off). Expressing his agonies while trying to cope with the bureaucracy, Ahmed (Letter in The New Nation, 10.2.1990) wrote:

> Needless to say all these "over-worked" officers are exceedingly busy men weighed down by "important" matters of state reposed on them and have scant free time to spare for menials who wish to legally cut down their own trees.

But had he "stoop to dispose the trees illegally to the numerous well-connected touts who monopolise this trade, then there [would have been] no problem whatsoever" (Ahmed's letter to the Mayor of Chittagong on 26.4.1992, p.2). Ahmed could never finish his journey through the weary detour of red-tapism. While he was busy in coming to terms with the government, local *mastans* (whom he mentioned as "touts"), in alliance with local police and forest offices, started robbing his trees and encroaching on his land. By the early 1993, all trees were looted, the farm house was set on fire, and Ahmed had to retreat to Dhaka on the face of life-threat. His "personal appeals to the prime minister and minister of forest evoked no response, not to speak of *Amla* [public bureaucrats] who implemented these outrageous regulations" (Letter in The Friday, 20.2.1992). Ahmed is an well-educated and respectable gentlemen, who received the "International Rolex Spirit of Enterprise Award" for his farm forestry. But even his intellect and status had little appeal to the bureaucracy and its allies in village politics. From this case, the possible plight and chances of general farmers (in securing a just market for them) become easily understandable. As regards the laws of marketing of forest products, Ahmed (Letter in The Bangladesh Observer, 19.2.1991) noted that,

> our archaic laws are there only to discourage the law abiding while acting as an incentive to the professional poacher....Who would be foolish enough to plant trees, nourish and protect them...and then have to grease lowly palms to enjoy the fruits of his own labour?

Therefore, it is no surprise that the "law-abiding" farmers of my study-areas never dare to "violate" the ever-stretching domains of the state and its allied commercial forces, by trying to do something on their own. Recently, even the village timber traders (who normally act as allies to the government) have become vocal against the harassment, procedural complications and massive corruption by public offices (see the public-circular issued by the Village Timber Traders Welfare Association, Chittagong, n.d.). Besides, in the wake of the campaign for SF, there is a call for "social marketing" by foresters who are supposed to be "improving their acceptability and image" by "serving and providing necessary support" to farmers in marketing their products (Maslekar, 1983, p.197). The planners in the government also consider that private forestry has the "best prospect of success" (FMP, 1993, p.ix) and call for providing increased incentive towards private afforestation (GOB, 1990; FMP, 1992; 1993). But in context of the evidences and cases discussed above, such calls appear to be a cry in the wilderness, serving rhetorical purposes only.

In sum, market in the study-areas are well beyond the control of SF farmers. They work under highly unfavourable conditions set by market forces. The benefits of market are usurped by elites. Unjust price, bureaucratic harassment, lack of government assistance, restricted market facilities, lack of transportation and logistics are the major problems faced by farmers in selling SF products (see Table 6.6). Markets are playing "exploitative and retrogressive functions" (see Chapter 3) which are manifested by the monopolistic control of elites, unjust prices, lack of market information, and exploitation by intermediaries. In rarely successful cases of Asia, for example, in Cebu (Philippines), farmers have been able to establish direct contact with urban commercial markets and to secure a fair price (Bensel and Remedio, 1994; also see Malla, 1993 for the Nepalese experience). But in my study-areas, farmers have virtually no control over market and, therefore, are vulnerable to varied manipulation and exploitation by market forces. This is attributable to many reasons as have been discussed above, including the relatively poor condition of farmers (as compared to the Cebu farmers who are organized rural landowners) and the strength and pervasiveness of the commercial network in Bangladesh, patronized by the state itself.

Notwithstanding the increased theoretical emphasis on the importance of market for participatory forestry schemes (see Beldt et al., 1994, p.7; Bensel and Remedio, 1994, pp.8-9; Maslekar, 1983, pp.197-198; Magno and Ali, n.d., p.39), marketing and selling of SF products remain the most ignored phenomenon in SF in the study-areas (like in many parts of south Asia; see for example Saxena, 1992; Malla, 1993; Messerschmidt and Hammett, 1993; also see Chapter 3).

Table 6.6 The commonly-faced problems in selling forest products in the localities

Problems	Sites	Degrees of acuteness					
		1		2		3	
		No.	%	No.	%	No.	%
Low or unjust price	Betagi	16	67	2	8	0	0
	Pomora	28	70	8	20	2	5
	Chandra	7	54	3	23	0	0
Difficulty in transportation	Betagi	5	21	2	8	0	0
	Pomora	6	15	2	5	0	0
	Chandra	5	38	2	15	1	8
No daily market facility	Betagi	0	0	2	8	0	0
	Pomora	2	5	10	25	6	15
	Chandra	1	8	0	0	0	0
Lack of government assistance and market information	Betagi	3	12	8	33	5	21
	Pomora	4	10	8	20	6	15
	Chandra	0	0	5	54	2	15

*Note: * Degrees of acuteness (as assessed by respondents): 1 = most acute problem, 2 = very acute problem, 3 = acute problem.*

The Role of Public Policies

Since SF "locates the main thrust of forestry activities at village level, gives subsistence needs new weight and ushers a new relationship between foresters, forest users, and forest, it leads inevitably to a complete reformulation of forest policy" (Shepherd, 1992a, p.24). My experience, however, suggests that such a "reformulation" still remains a remote possibility in the field. There are major implications of government policies for the functioning of SF. In this brief discussion, I can only report on some practical instances from the study areas.

Although government policy papers attach great importance to SF, the performance of SF plantations are not in conformity with such enthusiasm. The following table shows the achievements of three major components of SF in Bangladesh (namely, "woodlot and agroforestry plantations", "strip plantations", and "seedings distribution") during the last three Five Year Plans (1973-90). It is evident that the achievements have been nominal and they hardly reflect the lofty targets.

Table 6.7 Achievements of the major public SF activities (1973-1990)

Type of activities	1973-80	1980-85	1985-90		
			Target	Achieve ment	%
Woodlot and Agroforestry plantation (ha)	-	2932	15628	5413	35
Strip Plantations (km)	-	3410	4980	4241	85
Seedling Distribution (mil)	26.7	21	76	42	55

Source: modified from GOB, 1990, p.ve.2 (Table 1). Note: 1973-80: First Five Year (1973-78) and Two Year (1978-80) Plans; 1980-85: Second, and 1985-90: Third Five Year Plans.

As already noted, governmental decisions or policies on land and tree tenure, sale, transport, and market issues of SF are inconspicuous and make room for bureaucratic harassment and elite oppression on farmers. Similarly, a clear policy on the "benefit sharing" arrangements for SF farmers is yet to be developed. For some projects, there are basic guidelines as regards benefit-distribution; but they are not properly implemented in most cases (see for example Anon., 1991a, p.83; Manandhar, 1987, p.38; FMP, 1992, p.22). In many instances, as the Forestry Master Plan notes, the agreements between the government and farmers "have not been fully honoured...which has made farmers suspicious about [the government's] intentions" (FMP, 1992, p.22). Farmers have been in "confusion as to whether they would really receive the benefits promised" (FMP, 1992, p.vii). In Pomora, no such benefit sharing contract is in existence, adding fuel to the long-standing mistrust between foresters and farmers. In Chandra, although such an agreement officially exists, farmers have not yet received any document (which has made them suspicious about the government's intentions). In general, lack of or vague benefit sharing mechanism has caused sufferings to SF farmers. In the Manikganj SF Project, for example, 165 participants planted trees under the directives of the forest department alongside a 10-mile-long public road. In 1987, after 5 years of hard labour and enthusiasm, when farmers were preparing to harvest their planted trees, a dispute arose between the forest and the "roads and highways" departments. The highways department claimed ownership of the land (and trees grown on it); and not only deprived farmers of the trees, but also ruthlessly evicted them. A minister of the government himself reported the incident and called it an "unacceptable" case of "outright deprival" (Tour report of the Land Minister, Ministry of Land, Dhaka, 1987 (in Bengali), cited in Anon., 1991, Annex-xi). Similarly, in the North Bengal project, Manandhar (1986, pp.2-3) observed that "in the absence of written commitment" of benefit sharing from the government, "some participating farmers left the farm". Besides the ambiguity over benefit-sharing arrangements, there is no strong effective policy directive to ensure coordination within the public bureaucracy. Inter-departmental coordination and communication within concerned public offices remain a major constraint on SF, causing frustration, confusion and harassment to farmers (see Anon., 1991, p.41; Task Force, 1987, pp.102-104; Anon., 1991a, p.84; FMP, 1992, p.vii; 1992a, p.39; Kader, 1989, p.21; Lai, 1990, p.96). In Betagi and Pomora, I have noted a "cold war" between the forest and the revenue departments over authority and ownership of SF lands. Some revenue officials expressed their annoyance with the forest department's activities. Bhuiyan (1994, p.30), among others noted above, also observed a constant *tana-hechra* (tussle) between these two departments in the Madhupur Garh area (of which Chandra is a part). Similarly, Khan (1994b, pp.138-145) noted a general lack of coordination between the forest and other departments (especially, the police), and also between the central and field offices of the forest department.

Some public policies are not in conformity with field realities. For example, the government has put a complete ban on felling of timber since 1989 (Choudhury, 1993, p.38). Besides academics and villagers, I have found that some senior foresters themselves consider this policy to be an act of folly. Because, first, the government is in no position to execute this policy with its meagre (and sometimes inefficient or unwilling) manpower and resources, Second, it is completely regardless of the existing demands of forest products and the field-realities. A national Task Force on Environment Policy (Task Force, 1991, p.125) notes:

> [the] moratorium on extraction from government forests without making any alternative arrangements for supply of forest produce for consumption...has resulted in fast escalation of price and general shortage of wood in the country. This has also brought additional pressure onsting forest resources as illegal removals have become more profitable under current circumstances.

With the official closure of legal sources of timber, this policy has consequently put renewed impetus to underground commercial logging. Field foresters are pushed into a more challenging situation by this policy, because this has made the loggers more desperate. One officer of the Chittagong Forest Division told me:

> now the loggers are reckless -- they clear off patches of forest overnight....Since there is no opportunity of official [legal] extraction, the whole timber-business has gone underground.

This policy has also put additional pressure on SF farmers. They are constantly approached by loggers for trees in their plots; and often loggers make sure that farmers can not reach the open market. Loggers purposely create a situation which compels farmers to sell their products to them (loggers) at unjust prices and terms. If farmers do not comply with their demands, loggers send armed *mastans* to plunder the trees. During my stay in the study-areas, I observed several of such incidents. From Nepal, Malla (1993, p.30) also reports that the current government policy, which bans felling of trees on public lands, has intensified "black market" activities and has put increased pressure on private tree growers. Similarly, the ban on using firewood in brick-kilns in Bangladesh (by the Act 8 of 1989) seems to be honoured only in its widespread violation. In the study-areas brick-kilns absorb the bulk of firewood, supplied by the illicit logging network (Chapter 5). The ban is of no fruit; except it has served as a lever to a section of the bureaucracy for greater corruption. I talked to some brick-kiln managers to know their view of the situation. I have been reported that firewood is the cheapest and commonly available source of fuel. Any brick-kiln can make an agreement with any merchant of the logging network to supply the required quantity of wood for the whole production-season. The supply is secured, efficient, economic, and flexible (for example, brick-kilns can get firewood on partial payments or instalments; they can even get some loan from merchants). On the contrary, if they have to bring charcoal (the alternative fuel-source) from the concerned government offices (based in Chittagong city), as the brick-kiln managers told me, the cost of "transportation is high", the "incidental expenses of satisfying the officers are exorbitant" and the "delay and harassment are intolerable". The policy failure on the issue of restricting fuel wood in brick-kilns has been widely documented in mass-media (see for example The Janakantha, 16.2.1994; 20.2.1994; 28.2.1994; 3.5.1994; The Purbakon, 22.4.1990; The Azadi, 9.5.1994). In fact, foresters are also alive to the widespread use of fuel wood in brick-kilns; but they report that the policy on control of firewood is not backed-up by required legal authority for the forest department to take action against offenders. For example, in a letter (No.684/28-13, dated 2.2.1994) to the Deputy Commissioner of Chittagong, the Divisional Forest Officer noted that his office did not have the authority or resources to execute the policy of prohibiting fuel wood in brick-kilns. A similar ban on extraction of trees in the central Sal forest region in 1972 (Das, 1982, p.11), neither was effectively enforced nor could stop the devastation caused by "illegal exploitation" (GOB, 1990, p.ve-2; also see Byron, 1984, p.1). Thus, policies are often impractical and devoid of the consideration of field-realities and the capacity of the implementing agencies. Moreover, the government seems to be somewhat naive in taking into account the past experiences while formulating future policies.

In other instances, the activities of local forest offices do not conform to the policies set by themselves. For example, in one site I saw a forest floor was completely burnt and people ("encroachers") were evicted by foresters in preparation for new plantations; although foresters, in principle, vehemently oppose "slash and burn" and many local people were chased for this. The evicted people came to complain to the local Union Council. While discussing with local government leaders and the evicted people, I was told that the reason for the deviation from policy was financial: burning saved the cost of labour for clearing the forest floor, and the saved money could then be (mis)appropriated. Furthermore, some fickle decisions of the central government cause great inconvenience in field. In Betagi, for example, after farmers started to nestle down on their newly allotted plots, the government revoked some plots on "state-purposes" (no formal information exists in the project office about this incident; some villagers and local government leaders told me that the plots were reallocated to a foreign government as a "token of goodwill"). The displaced farmers were pushed into inconvenient or remote plots and got into fresh conflicts with elites over demarcation and settlement in the new territories. On the other hand, the government (or the new allottee) never used the lands, which presently lay vacant and barren.

In sum, government policies on issues of major significance to SF -- such as land tenure, market, benefit-sharing arrangements, and inter-departmental coordination -- have largely been ineffective and ambiguous. Furthermore, some policies (for example the ban on timber felling), which are devoid of the consideration of field realities, make the tasks of SF farmers and foresters more difficult. SF's lofty goals are not supported or complemented by a proper policy back-up.

The Role of Forestry Organization

Chapter 3 has reviewed the major institutional and attitudinal shortcomings of forestry organizations in south Asia which are detrimental to the success of participatory development initiatives including SF. It has been unequivocally established that there needs to be a complete reorientation of the forest department as the principal facilitator of SF (FMP, 1992, pp.13-14; GOB, 1990, p.ve-5; Lai and Khan, 1992, p.38; Task Force, 1987, p.99; Ahmed, 1993, pp.5-6; Khan and Hasan, 1995, pp.57-58; Singh and Khare, 1993, pp.280-281). In essence, the "re-orientation" may be seen as a transformation of the "foresters-as-enemy to foresters-as-facilitator" (Shepherd, 1992a, pp.20-21). In view of the above, this section investigates two questions: first, is there a change emerging in the attitude and outlook of foresters in performing their role in SF? Second, what are the factors that inhibit the growth of good "forester-people" relations in field?

Regarding the first enquiry, I have asked forest officials a few simple questions to gain a general idea about their attitude and orientation towards SF and local people. The questions are: what are the fundamental reasons for practising SF? what should be the role of participating farmers in SF? what should be the nature of land-tenure in SF? what they (foresters) think about their principal responsibilities? and what are the causes of current deforestation? In response to the first question, 60% of the respondents (foresters) think that SF's main purpose is to reduce pressure on "reserve" forests, and it should serve as a "buffer zone" between state forests and the "invading encroachers". Only 33% respondents view SF as a means to meet local demands (Table 6.8). Similarly, 60% of them think that local people are not competent enough to take decisions on a "technical subject like forestry". Therefore, people should be happy with the benefits generated from SF, while the decisions and implementation of the SF projects should be shouldered by professional foresters (Table 6.9). As regards land tenure, 53% of them believe that land (used for SF) should remain under the government's control, but farmers should be given SF's benefits including the generated products. Another 40% think that farmers can only be given usufructuary rights, and not land-ownership (Table 6.10). 40% foresters believe that their main tasks are to "conserve government forests and to check all destructive activities and illegal encroachers therein". Another 40% think that they are to ensure greater revenue from forests in the "national and public interest" and to make "optimum utilization of forests through scientific forestry". A meagre 20% of them think their principal job is to provide opportunities for meeting local demands and to help building mass-awareness in favour of forestry (Table 6.11). 73% officers blame the pressure of local population (including "encroachment") to be the principal cause of current deforestation. Only 2 out of 30 foresters have opined that government policy failures might be a major cause of deforestation (Table 6.12).

Table 6.8 The fundamental reasons for practising SF (as reported by foresters)

Reasons	Number	%
To reduce pressure on public forests and to act as a "buffer" to state-reserves	18	60
To meet local demands and to foster better relations with people	10	33
Other reasons	2	7

*Notes: * For tables 6.8 through 6.13, n = 30 (30 forest officials from different levels of the administrative hierarchy were interviewed. Chapter 4 has the details of the interview methodology).*
** "other reasons" include "environmental conservation", "rehabilitation of denuded hills", and "increased forestry production".*

Table 6.9 Suggested role of farmers in SF projects (as suggested by foresters)

Suggested role	No.	%
Farmers will plan, implement and evaluate SF projects	2	7
Farmers will enjoy benefits, but the government will plan and implement projects	18	60
Farmers will plan and the government will provide technical consultancy and assistance	8	26
The government will plan and implement, but people through their political representative (local government leaders) will evaluate project performance	2	7

Table 6.10 Suggested nature of land ownership and tenure in SF (as suggested by foresters)

Suggested nature	No.	%
Government will own land and implement SF projects; and farmers will be the beneficiaries	16	53
Government will own land, but farmers may be given usufructuary rights	12	40
Permanent title to land should be given to the farmers	2	7

Table 6.11 Principal responsibilities of a forester (as conceived by foresters)

Responsibilities	No.	%
To conserve public forests, check all illegal activities (especially "encroachment")	12	40
To earn more revenue from forests in national interest and to ensure optimal utilization of forest through "scientific forestry"	12	40
To foster mass-awareness about forestry and to meet local demands	6	20

Table 6.12 Principal causes of current deforestation in Bangladesh (as conceived by foresters)

Causes	No.	%
Pressure of local population (including "encroachment")	22	73
Illegal commercial logging	6	20
Failure of public policies and administration	2	7

From the above responses, it becomes evident that the desired "bureaucratic reorientation" of the forest department remains a distant possibility. To foresters, SF is yet another technical programme, planned and run by technically qualified professionals, with the end of benefiting people. Participation in SF, in this sense, is an mere invitation to local communities to take part in a technically-oriented and bureaucratically-managed afforestation programme. Forest officials are, in fact, "bureaucratic elites" (see Chapter 3). They are lineal descendants of the Superior Forest Service of Pakistan and the Imperial Forest Service of British India. They are trained and indoctrinated in a techno-professional way to assume the role of "conservators of forests", not as "facilitator of forests" (Gronow and Shrestha, 1991, p.6; Shepherd, 1992a, p.21). Foresters, especially the senior officials, cherish their affiliation with the early predecessors and are largely imbued with elitist value-orientation. They represent the upper echelon of the Bangladesh society (especially land-owner, government official, and business elite classes). They hail from high-status families in terms of land-holding, income and education (Table 6.13).

Almost all of my respondents have two post-graduate degrees and many of them are trained in overseas universities. They are a highly educated group "to the extent of being very unrepresentative of their society and extremely exclusive" (Ahmed, 1980, p.51). They lead a fairly high life-style with government-supplied vehicles, furnished lavish *bungalows*, well-decorated offices, a host of subordinate staffs and other subsidized facilities. The life and working environment of a forest officer, like his partners in civil administration, police or judicial services, are marked by "a feudal type splendour and distinction" (Ahmed, 1980, p.60). Ahmed's (1980, p.181) observation on bureaucratic elites in the civil administration of Bangladesh largely applies to forest officials:

> They are deeply imbued with guardianship orientations; they favour a kind of limited popular participation within the framework of a paternal rule. They

are...elitist.

Table 6.13 Demographic characteristics of the respondents' (forest officials') parents

Characteristics	No.	%
Father's (or legal guardian's) profession		
Business	6	20
Public service	22	73
Professionals (including lawyers, teachers, medical-practitioners and engineers)	2	7
Father's (or legal guardian's) land-holding		
less than 3 acres	0	0
3 to 6	4	13
6 to 10	8	27
10 to 15	12	40
15 and above	6	20
Father's (or legal guardian's) monthly income (at current value)		
4000 to 6000 Taka	2	7
6000 to 8000	14	46
8000 to 12000	2	7
12000 to 20000	6	20
20000 and more	6	20

Furthermore, foresters are "exclusive in societal term" (Ahmed, 1980, p.47) because they are separated from the mass by their education, their origin in upper classes and their professional training and life-style. In line with the general trend of the country's public administration, foresters lay emphasis on "administration" rather than on "public" on the plea that, "administration is a technical matter and only those who are specialists...can have a say in it" (Anisuzzaman, 1979, pp.90-91). Local people are "subjects" to them, not "citizens" (see Chapter 3). Thus, with their "elitist-orientation" and "exclusiveness", forest officers are at odds with the idea of fostering a "participatory environment" (Gronow and Shrestha, 1991, p.5) for SF. However, as I have often mentioned, a gradual realization about the need for involving local communities in afforestation schemes has emerged among senior officials. But this "realization" stems from the fact that the government can no longer manage forests singlehandedly. Public officials, especially foresters, need to make some room for compromise with people. This does not necessarily speak of their changed attitude or increased trust in people's competence. Rather the realization (which contributed to the rise of SF) may be seen as a strategy, by which foresters are trying to save their endangered domains (land and authority) by coming into some sort of truce with local people. Therefore, as my observations suggest, the government's emphasis on SF is more of a sign of, what may be called, a "strategic compromise" than that of "bureaucratic reorientation" (Korten and Uphoff, 1981) or "participatory styles" (Chambers, 1983, p.212).

The "forester-people" relationship is of paramount importance to SF. At the micro level, some factors thwart the growth of participatory environment and harmonious forester-people relations. First, the working environment in forest offices is rigidly "hierarchical, directive and punitive" (Gronow and Shrestha, 1991, p.19). Innovation and new ideas about working with people, developed by field staff, are not entertained by the superiors. "In meetings subordinates are upbraided, cajoled and given orders" (Chambers, 1983, p.211). The "rigid paramilitary orientation of forest service, with its restrictive and primitive approach" (Lai and Khan, 1992, p.38) blocks upward mobilisation of innovation and planning on the one hand, and inhibits public dealings and access on the other. In such a rigidly hierarchical arrangement, "senior officers do not learn from subordinates and subordinates do not learn from their rural clients" (Chambers, 1983, p.211). The planning practice, in this context, remains essentially top-down and centrally "conceived-administered-evaluated" (Anisuzzaman, 1979, p.109), restricting one basic requirement of participatory programmes such as SF (Shepherd, 1992a, p.22; 1993, p.320; Gronow and Shrestha, 1991, pp.4-10). In planning, the emphasis is on visible quantifiable results and statistics, rather than on such qualitative issues as social welfare, equity, and participation (the "statistics-consciousness" (Tepper, 1976, p.44) of public offices is further intensified by international donors' pressure, which I will subsequently discuss; also see Chapter 3). Second, though training is repeatedly argued to be the precursor of participatory environment (Shepherd, 1992a, p.22; Task Force, 1987, p.104; Magno and Ali, n.d., p.11; Anon., 1991, p.41; Ahmed and Bhuiyan, 1993, p.5; FMP, 1992, p.viii; Abedin et al., 1990, p.148; ADB, 1989, p.11), majority of field staff concerned with the SF projects have no or very little training on SF and related subjects. In the Pomora Forestry Extension Centre (the office responsible for supervising the Betagi and Pomora SF projects), for example, out of 5 personnel, only 1 had a 21-day-long short training on SF. Moreover field forest offices constantly suffer from staff shortage. The limited staff, busy otherwise in their routine duties, virtually play no effective role in SF projects. Fourth, SF assignments, as compared to the so-called "territorial postings" (which entail more power, authority and facilities), are not generally attractive to foresters. Most of them consider involvement in SF as a "punishment placement". Besides senior officials, similar attitude is also prevalent among field-staff. During the field-work, I found at least 3 Forest Guards who had been transferred to the Pomora Extension Centre and were facing temporary suspension for alleged offenses in their previous postings. In this case, SF indeed served as a "punishment" assignment. These demoralized and punished staff, needless to say, are hardly effective in meeting the challenging tasks of SF. Fourth, the frequent change of staff and officers, which "prevents officers from acquiring intimate knowledge" of the locality (Kaufman, 1960, p.238), remains a major problem for SF in Bangladesh (Manandhar, 1987, p.36). The knowledge gathered by a "local socio-economic analysis", which is sine-qua-non "for working with the poor", is lost when the concerned officer is transferred (Wood, 1994, p.468). The Officer in Charge of the Pomora Extension Centre (who was also in-charge of the SF projects), was changed 5 times in just over two years time (1992-94). Furthermore, generally aged officers (50+) were assigned to head the Centre, some of whom went to retirement from this post. Therefore, no wonder that farmers witnessed a gradual decline in the official enthusiasm for SF in recent years and were rarely visited by the Officer-in-Charge. Fifth, the discrepancy between forest officers (the "first class" officials) and subordinate staff (in terms of power, status and privilege) is wide. Staff have to live in remote territories with acute shortage of logistic facilities. They are so poorly paid that even the basic amenities of life become luxuries (which also partially explains the widespread corruption in local public offices including forest offices). Virtually no reward or motivation system operates, though it has been argued that "dynamic, innovative and risk-taking performance cannot be sustained indefinitely without some prospect of recognition" (Wood, 1994, p.469; also see Task Force, 1987, p.109; Chowdhury and Hossain, 1990, p.104). Sixth, the essentially punitive and orthodox laws and policies "turn foresters into policemen and villagers into notional thieves" (Shepherd, 1992a, p.19), and add fuel to the long-standing antagonism between foresters and villagers (Ahmed and Bhuiyan, 1993; Lai and Khan, 1992; Khan and Hasan,

1995; Poffenberger and Singh, 1992). Seventh, although there has been increased emphasis on the centrality of understanding the social factors and forces for the success of SF (Cernea, 1989; Burch, 1988; Clausi and Bopp, 1989; Lee et al., 1990; also see Chapter 3), foresters in my study-areas still continue to treat afforestation (even when the plantation is on a village-land or on a SF plot) as a "technical matter". The egoistic views, such as "the forest department is the *only* public agency which has *all* the technical knowledge on forestry" (CFD, 1994, p.1 (introduction), emphasis added), undermine people's competence and access. These views also embolden traditional bureaucratic convictions that "people are not experts" (Anisuzzaman, 1979, p.91); and that, without central control, their involvement in development projects may lead to a "chaotic situation" (Ahmed, 1980, p.171). Majority of SF farmers also consider that the forest department's attitude towards them is "authoritarian" and "formal" (Table 6.14). Eight, as I have discussed earlier, in the absence of vigilance from higher authorities, field forest offices' service to farmers erodes and they tend to return to their usual naivety in operation. The vivid loss of enthusiasm has negative effect on farmers. It makes them suspicious of the credibility, commitment and continuation of the government's support. In the Betagi and Pomora projects, the forest department's support has gradually declined, and in the last two years it has been minimal. But in Chandra the situation is better, mainly because Chandra is still in the lime-light of attention from the higher authorities (Table 6.15).

Table 6.14 Farmers' evaluation about the forest department's attitude towards them

Sites	Attitudinal manifestations							
	Authoritarian		Formal		Friendly		No comments	
	No.	%	No.	%	No.	%	No.	%
Betagi	6	25	8	33	4	17	6	25
Pomora	14	35	11	28	5	12	10	25
Chandra	2	15	4	31	5	39	2	15

*Notes: * The terms (such as "authoritarian") are deduced by me, following farmers' comments and views expressed in the course of interviews and discussions. These comments were originally in local dialects, which were* *subsequently translated into the above terms reflecting nearest meaning of the farmers views.*

Table 6.15 Farmers' evaluation about logistic supports and services from the forest department during the last two years

Sites	Frequency of support and services									
	Regularly		Occasionally		Rarely		Never		No comments	
	No.	%	No.	%	No.	%	No.	%	No.	%
Betagi	1	4	4	17	11	46	6	25	2	8
Pomora	0	0	9	22	13	32	12	30	6	15
Chandra	2	15	7	54	3	23	1	8	0	0

*Notes: * The last two years prior to the time of data collection were 1992 and 1993. * "logistic supports and services" include financial input, technology, material input, advice and counsel, as appeared necessary to the farmers.*

Ninth, as Chapter 5 suggested, the enmity between foresters and local people has a historic origin. The very establishment of forestry organizations in the colonial days announced the gradual reduction in local rights and access. Local foresters have been the representative and bearer of these unwelcome actions and news to the locality.

Similar to many south Asian instances (Chapter 3), the forest department in Bangladesh suffers from major institutional maladies (for a list of major problems, see FMP, 1992a, pp.43-44). Its "bureaucratic structures are grossly inadequate" and "on the whole do not know how to get people's participation" (Chowdhry, n.d., p.66). The "attitudinal change in the forestry service does not appear to be deep rooted" (Sarma, n.d., p.145); and the forester-people relation is also in a fragile state. I will close this section by quoting a remark (made during an interview with me) of a mid-level forester, which amply brings home the forest department's attitude and orientation towards SF (emphasis added):

> We do like to *invite* the public to join *our* programmes...; but *they do not know what is good for them*....In fact, are they supposed to know?....*Forestry is a purely technical matter, it is Science -- no layman's business.* We left the choice of species

> to people in some instances; they spoiled the whole site by unscientific planting and poor cropping....So I think, people should be given maximum benefits and share of products, *but they do not have the skill to make decisions on plantation programmes*....We work in the field, we can't be romantic like others.[3]

The Role of International Donor Agencies

The SF projects are predominantly supported and funded by international donors. The following table depicts the major on-going SF projects in Bangladesh and their fund-sources.

Table 6.16 Major public SF projects in Bangladesh and their fund-sources

Project title	Project costs in Lakh Taka					
	Local (GOB)		External finance		Total	%
	Amount	%	Amount	%		
TANDP	2350.00	12	ADB: 16292.18 UNDP: 12055.82	83 5	19698.00	100
SALT	0	0	ICIMOD: 3.95	100	3.95	100
FRMP	3486.76	16	WFP: 194.40 IDA: 18007.70	1 83	21688.86	100
ARDP	5.00	6	Ford: 78.39	94	83.39	100

*Notes and Data Sources: * 1 Lakh = 100,000. * GOB = Government of Bangladesh, TANDP = Thana Afforestation and Nursery Development Project (TANDP, 1994, p.5), SALT = Sloping Agricultural Land Technology Project (SALT, 1993, pp.2-3), FRMP = Forest Resources Management Project (FRMP, 1992, p.7), ARDP = Agroforestry Research and Demonstration (ARDP, 1989, pp.1-2). ADB = Asian Development Bank, UNDP = United Nations Development Programme, ICIMOD = International Centre for Integrated Mountain Research and Development, IDA = International Development Agency, WFP = World Food Programme, Ford = Ford Foundation.*

The above table manifests the magnitude of external dependence of Bangladesh SF. Therefore, understandably donors have a profound influence on governmental activities and policies.

Mirroring the observations of Chapter 3, I have found, for example, in Chandra, that foresters are chasing physical targets of seedling distribution and raising plantations to appease critical donors especially the expatriate officials, who, according to Wood (1994, p.429), have the "unfortunate...attitude of policing their aid to Bangladesh". During informal interviews, many senior officials have reported that, like in most Third World countries (for example in Madagascar), donors "deal in a hasty, high-handed and unsympathetic way with national-level institutions" (Shepherd, 1993, pp.319-320). Consequently, extra care is taken in keeping records of the foreign-funded projects. The concerned officers in Chandra are very particular in filing information on the achievement of physical targets, time schedules and quantitative statistics. In this way, officers also reduce "risk" and secure their position in the face of central monitoring (see Chapter 3). In fact, as Tepper (1976, pp.44-45) notes, this has been the general pattern of bureaucratic behaviour in the country:

> The Weberian-rational bureaucratic model, in its south Asian incarnation, has shown itself in [Bangladesh] to be statistics conscious; the tension repeatedly, perhaps inevitably, has been resolved in favour of a strong paper superstructure and a weak organisational base....Pressure to produce visible results, combined with relatively limited governmental funds and a capital-starved economy, leads to subversion of the social welfare motives.

Therefore, it is no surprise that, although very particular about statistics, public officials (including foresters) are often unsure about the demands, likings and living of local people (especially, of SF farmers). Obsession with physical targets and donor's satisfaction, not only over-shadow people's choice and local conditions, but also give "little thought for long-term sustainability of new arrangements" (Shepherd, 1993, p.320). Lai (1990, p.91), an expatriate consultant himself, observed that in Bangladesh the "external models" of SF often designed by expatriate experts emphasize "maximization of production", but "seldom do fully take into account local conditions, practices and needs". In further worse cases, for example in some parts of Madhupur Garh, foreign funds have even contributed to accelerated deforestation (The Courier, May 31-June 6, 1991, p.26):

> ADB [Asian Development Bank] funds have worked as an incentive to hasten degradation...; the funds were in the pipeline for fresh plantation which encouraged the illegal loggers in connivance with corrupt FD [forest department] officials to completely denude the last standing trees.

Conclusion

SF has to be seen in context of surrounding political-economic structures and relations (which I refer to as the frame of SF). This chapter discussed the implications of some of these variables which strain and filter SF's stated goals. It reaffirmed the argument of Chapter 3 that SF's goals are shaped and conditioned by surrounding frame, and ultimately may produce results which could look very different from the initial formulations.

The frame of SF in the study areas, as this chapter has revealed, is not generally conducive to the achievement of SF's stated goals. The broader social milieu, in which SF schemes operate, shows highly unequal social relationships, structures of exploitation and stratification. Rural elites are allied to central and regional politics and bureaucracy, and have successfully utilized these structures to their vested interests. Consequently, they sometime coopt the benefits of SF, hinder farmers' access to input and services, and continue to enlarge their network domination and extortion. Furthermore, other variables (such as market, government policies, land and tree tenurial arrangements, foresters' and donors' roles) have also not been conducive to SF. Farmers do not have control over markets and they work under highly unfavourable conditions set by market forces. Public policies on land and tree tenurial issues are ambiguous and make room for oppression of farmers by unscrupulous quarters. The government's institutional and catalytic support and services have gradually eroded, and today, these are almost absent. Foresters with their "elitist exclusiveness" are at odds with creating a true participatory environment for SF. They have a different explanation of SF, where they invite people to act within the strict limits of a bureaucratically designed-executed-monitored programme. Donors' high-handed policing attitude has resulted in the government's obsession with physical targets at the cost of over-shadowing local demands and choices. Thus, these variables have rather supported the dominant-elitist forces in fortifying their net-work of manipulation and extortion in the study-areas.

In such a competitive and unfavourable working environment, the inputs and goals of SF are most likely to falter and deviate, and are likely to produce only marginal results. Keeping in mind the above context, in the next chapter I will examine the level and significance of SF's achievements against its stated objectives.

From the preceding discussion on the generally unfavourable frame, the stated goals of SF seem largely unrealistic. The next chapter, therefore, also raises the subsequent question: why then the government and donors are pursuing such unpromising goals? Are there any tacit purposes behind the continued pursuit of SF? It sheds light on these purposes which also help to partially explain why SF is being patronized despite its unfavourable working-context and limited achievements.

Notes

1. See for example Divisional Forest Officer (Chittagong)'s letters 1308/13-14 dated 16.3.1992; 2159/13-14 dated 21.4.1992; 2159/13-14 dated 21.4.1992 and Officer in Charge (Pomora Extension Centre)'s letters 32/pom-29 dated 23.4.1984; 10/pom-13 dated 16.3.1992; 16/pom-13 dated 12.4.1992.
2. Khan (fc.) partially draws on this section.
3. Selected findings of this section were included in the *Proceedings of the Symposium on Resource Management: Issues, Visions, Practice* (5 - 8 July 1996), held in the Centre for Resource Management, Lincoln University, Canterbury, New Zealand. Khan (fc.[a]) is an expanded and thoroughly revised version of the original Symposium presentation.

7 Social Forestry at Work: The Achievements

Introduction

This chapter analyzes the significance and level of SF's achievement in the light of its selected stated goals. Chapter 6 concluded that the surrounding frame of SF is not conducive to achieving its lofty goals; and predicted that in such an unfavourable surrounding, SF's goals are likely to falter and produce only marginal results. The following discussion substantiates the prediction and shows that SF's achievements are marginal, limited, and far from the initial formulations. Then it proceeds to probe the subsequent question: why do the protagonists of SF still pursue such unpromising goals? In response, it is argued that SF is effectively serving certain strategic but tacit interests of the government and international donors. For the government, SF serves as a fertile ground for attracting foreign fund. At the same time, it provides the opportunity to manage and reassert governmental control over encroached forest lands. It also has a populist image allowing state officials to serve as "official providers" in the forest sector. The image is equally important overseas, and allows donors to disburse funds on behalf of apparently popular and equitable programmes. Furthermore, for donors SF projects (being heavily dependent on external fund and advice) offer prospective avenues for exerting control over the government, for influencing departmental behaviour (the weaker and unequal partner), and for promoting their ideas and innovations. It seems that as long as SF fosters these "subterranean" purposes, albeit its limited achievement of the stated goals, it will continue to prosper.

As noted in Chapter 6, my case-study projects (Betagi, Pomora and Chandra) have borne quite lofty goals in line with the country's general trend. These projects are considered part of the overall rural development strategy of the government; and therefore, their goals reflect a wide range of rather generalized targets in community and rural development. The goals of participation, empowerment and autarky, meeting local demands, promotion of cooperation and equity, raising farmers' income, living standard and women's status, rehabilitation of degraded public lands -- are frequently found in varied project papers and documents in differential languages and styles.[1] It is impossible to evaluate the achievements of all of these wide-ranging goals. The specific goals and corresponding achievements analyzed in this chapter are: improvement of women's status; growth and development of cooperative organization; increase in income, productivity and standard of living for the participating farmers; higher level of literacy, adoption of family-planning and general awareness; promotion of social equity; fulfilment of local forestry demands (especially, demands for fuel wood and timber); and farmers' participation in project cycle. The chapter explores the level and significance of achievement of each of these goals. It shows that, among these goals, a marginal rise in farmers' income and some degree of social recognition for them are the only notable achievements. Majority of the goals have remained unrealized. Then, by way of conclusion, the subterranean purposes of SF are revealed.

Women's Role

Mirroring the general findings of other studies (for example Jansen, 1986; Ahmed, 1992; Wallace et al., 1987; FAO, 1989; Khandakar, 1994; Mies, 1986; Musgrave, 1990; Quddus et al., 1985), women in my study-areas are "invisible" workers, toiling from dawn until dark without recognition of the worth and economic value of their labour. Their work (labour-inputs) are basically of two broad categories. The "household chores" which include cooking for the whole family, child care, cleaning house and utensils, washing clothes, fetching water, livestock and poultry rearing, firewood collection, and handicrafts. The "agroforestry activities" include managing home and horticultural gardens, and processing harvested crops (mainly paddy). They also help male (family) members in various forest (farm) plantation work which include preparing land for planting; digging narrow ditches for watering seedlings; preparing seed beds; collecting cow-dung to use as fertilizer; making protective gabion; and grafting particular plants (also see Quddus et al., 1992, pp.35-36 for details of women's activities in Betagi and Pomora projects). The magnitude of women's work in my study-areas, except slight variation in specificity and nature, are more or less in conformity with the findings of other detailed studies on Bangladeshi rural women's working patterns (for example Ahmed, 1992; Abdullah and Zeidenstein, 1982; Wallace et al., 1987; Hartmann and Boyce, 1983; also see Hannan and Islam, 1986 for an annotated bibliography on women in agriculture). Women on an average work for 15-16 hours. The following case from Pomora depicts the daily life of a typical housewife in SF project (see Table 7.1). Banu Bibi (aged 36) left bed at the first hint of dawn, and then her routine (daily) work continued as shown in Table 7.1.

In line with the general experience of rural Bangladesh, women in my study-areas, thus, work hard and are confined mostly around homesteads. Women's activities occur in a male-dominated unequal surrounding where their works are often unseen and undermined (Ahmad, 1991, p.1; Chaudhury and Ahmed, 1980, Chapter 2; Hartmann and Boyce, 1983, Chapter 7; Mies, 1986, p.54; Quddus et al., 1985, pp.32-33). In general terms, cultural norms, social tradition, illiteracy, lack of training, and unequal access to assets have forced them into a precarious state of dependency and subordination (GOB, 1979, p.364; 1990, pp.xiii-4; White, 1992, p.88; Chaudhury and Ahmed, 1980, p.5).

In particular, the major constraints faced by women in my study-areas are: first, a non-recognition of their work and its economic worth. My male respondents, although quite aware of the magnitude of women's work, often belittled them on the plea that these activities are *Ghoror Kam* (household work), implying that they do not generate direct income. Second, women's access to resources (such as land, livestock, material properties) are unequal and limited. Even in Betagi where the (farmers') wives are officially the co-owners of the allotted lands, most women are unaware of their share, rights and privileges (see Chapter 6). In Pomora and Chandra women are completely left out of land contractual arrangements. Tree and land tenurial rights for women are obscure, ambiguous, and far beyond their comprehension or control. Even their very personal assets (such as jewellery or ornaments) are sold by their husbands in face of family crises. Third, women are generally illiterate and have virtually no access to technology and inputs (supplied by the government or other external agencies). Women do not attend the meeting of the farmers' cooperative (*Samity*) in Betagi-Pomora because of *purdah* restrictions.[2] Government agriculture or forest extension staff deal with male members only. Female facilitators, appointed by government offices, especially by the family planning department, are largely absent in SF areas. The child and medi-care facilities, although urgently needed, remain beyond the reach of women. Hence, technological innovations and logistic supports do not reach women directly. Fourth, women's involvement in outside employment are not encouraged in the country-side, primarily due to customary gender division of labour and *purdah* values (Ahmed, 1992; Chaudhury and Ahmed, 1980; Gardner, 1994; Hartmann and Boyce, 1983; Hocking, 1991; M.A. Khan, 1985). In the Betagi-Pomora projects, women do

Table 7.1 The typical daily activities of a woman in a SF project (as reported by the respondent)

04.55-05.10	Saying morning prayers;
05.10-05.30	Cleaning rooms and furniture;
05.30-05.45	Mopping up front and back-yards of the house;
05.45-06.05	Cleaning *Goal* (cow-shed), poultry-shed, and feeding livestock;
06.05-06.45	Washing utensils, cleaning kitchen, and preparing food for the family;
06.45-06.55	Calling children from bed, cleaning them and getting the school-going children ready;
06.55-07.15	Feeding the family (breakfast);
07.15-09.30	Working on home-gardens (vegetable and horticulture);
09.30-11.00	Helping male-members in various forest plantations;
11.00-12.45	Preparing lunch for the family;
12.45-01.10	Feeding, washing and cleaning livestock;
01.10-01.40	Feeding the family (lunch);
01.40-02.10	Cleaning utensils, sweeping kitchen, and leisurely discussion with family members;
02.10-02.40	Child-care, bath, cloth-washing, and saying noon prayers;
02.40-04.00	Firewood and fodder collection;
04.00-05.30	Working on home gardens and nurseries;
05.30-06.00	Securing livestock, feeding and cleaning them;
06.00-07.00	Preparing raw-tea for male-members, and light food for children, checking all the doors, lighting up lamps, and saying evening prayers;
07.00-08.30	Preparing diner for the family;
08.30-09.05	Feeding the family (diner);
09.05-10.00	Cleaning and sweeping utensils and kitchen, and preparing children for bed;
10.00-11.15	Handicrafts;
11.15-11.40	Final check of the house and homestead, night-time prayers, and bed-time preparation.

Note: I was not allowed to directly face the woman (Banu Bibi) and was accompanied by her elder brother. Banu sat behind a Dhal (purpose-built screen) during the conversation. The brother suggested the work-timings in consultation with her.

not participate in external (beyond homestead and farm boundaries) paid-works (for example, working in road construction, repairing culverts, public road-side plantations) sponsored by the government or other agencies. In Chandra, however, I have seen women (mostly from the poorer class) working in road construction schemes. This is mainly because these women are mostly *Barmans*, successors to the hard-working *Kooch* tribe, and are less exposed to socio-religious taboos on female work. Besides, Chandra, unlike Betagi and Pomora, enjoys wider influence of and exposure to the activities of a number of non-government voluntary organisations (mainly, Proshika and Grameen Bank). The exposure has somewhat diluted the immediate social restrictions on female work outside homesteads and has offered some employment opportunities to women. However, it should be noted that external connections are not an unmixed blessing. Even where women are working outside, they are poorly paid as compared to men and get into the contracts in unequal terms. In other words, even when women are engaged in some (limited) market contracts (mainly in the forms of paid employment, and selling livestock and SF products through middlemen), they do not have any substantial control over such transactions. White (1992, p.84) has also found that women's limited market "transaction is deeply unequal and serve to reproduce inequalities between two parties". Fifth, women do not have their own organization or cooperative forum to foster and articulate their views, demands and interests which is crucial for empowering them (FAO, 1989, pp.20-21).

Although women's activities are mostly confined around homestead, they have some informal market and social connections. Women meet each other in course of fuel-fodder collection or visiting neighbours. They sell livestock either to local (known) vendors or to weekly *haats* (village open-markets) through a male member of family. Village beggars (mostly destitute women) who travel from door to door are also a source of information to them. Besides, women share-tend livestock as additional income-source. Typically, villagers give goats, cows or paultry to (SF) project women to tend (In project areas forage is relatively abundant as compared to neighbouring villages). Women value these social relations and pursue their contacts as much as possible. The various links (such as the links developed through informal market transactions, share-tending and kinship) are their "social capital" and "represent crucial locus of female power" (White, 1992, p.93).

As far as their homestead plantations and routine work are concerned, *purdah* does not seem to be a major impediment. Women seem to be quite comfortable in working around their farms within *purdah* rules. They themselves consider *purdah* as a symbol of social status, and therefore, "typically choose to work within their own homesteads" (White, 1992, p.25). The poorer women who work publicly are looked down upon by the richer ones. Therefore, the popular academic contentions that "women must be brought outside their home" (Chaudhury and Ahmed, 1980, p.185), or *purdah* "is the main obstacle to their progress" (Quddus et al., 1985, p.2) are in contrast with the practical thoughts and perceptions of women. Besides, *purdah* rules are malleable in SF areas. In line with the experience reported by a number of studies (Gardner, 1994; White, 1992, pp.22-25; Hartmann and Boyce, 1983, pp.86-87; Begum, 1988, p.12), I have found that adherence to *purdah* is contingent on such factors as the age and social status of particular woman, and the economic status of her household. Young wives tend to be very fastidious, while aged women with superior household position (for example, grandmothers or mothers-in-law) enjoy flexibility in *purdah* boundaries. Poorer women are forced to seek outside employment because "who can bother with *purdah* when the belly's empty?" (Gardner, 1994, p.65), although thereby their social status withers. In sum, though *purdah* restricts women's mobility to formal market sectors, they tend to follow it for a number of reasons. First, it reflects increased status for the project women in neighbouring villages and emboldens the female image of modesty (Gardner, 1994, p.65 and Hartmann and Boyce, 1983, p.87 noted similar experience in other parts of Bangladesh). Second, it does not pose any serious problem in SF activities around the homestead. Third, from family welfare point of view, women's homestead activities, especially child care, are socially more valued than outside employment. Hence, echoing women's own beliefs, Ahmad (1985, p.38) notes that, "more outside work may not necessarily be a good thing for family welfare".

Women consider tree planting as an auxiliary source of income and social status. They choose trees of economic value (for cash income) and fruit trees (for own consumption and gift to neighbours). Hocking (1991, pp.17 and 22) also notes that women prefer trees as "income supplement" and "symbolic prestige". Interestingly, fuel and fodder trees are not their "urgent priorities" (Hocking, 1991, p.15). In the study-areas, women do not use fuel wood for domestic consumption, rather sell them to brick-kilns for cash. Besides, they have many substitutes for fuel-fodder (also see Hocking, 1991, p.15). Although they have some voice in decisions about planting of horticultural and home gardens, majority of women leave the decision of selling the products (and control of sale-proceeds) to the male heads of their family. Quddus et al. (1992, pp.54-55) also note that in Betagi-Pomora two-thirds of women do not exercise any right over their own products. Besides, whatever meagre cash income women may generate through their informal transaction and production, it adds to family's income and is spent for family's welfare. Women's income is utilized for the benefit of their family and not for their personal consumption or benefit (also see Hocking, 1991, p.12; White, 1992, p.79).

Women have reasonably elaborate knowledge on home-gardens, horticultural species and their uses.[3] They also show spectacular knowledge of plantation lay-out, for example, planting bushy shrubs to shield women's bath and toilet; aesthetic and optimum use of back and front yards by home gardens, building "living-fence" with bushy plants, and building animal corral by regenerating shrubs. Indeed, the fact that women are fairly knowledgable on some aspects of forestry and plantation is being increasingly recognised (see Hocking, 1991, p.15; FAO, 1989, pp.15-16). But foresters are not aware (or neglectful) of women's indigenous knowledge. When asked whether he contacted any women while choosing the species for SF plantation, one Beat Officer echoed the conventional answer:

> What can they possibly know about forestry? Forestry is pure science; one has to spend years in college and libraries just to get a basic idea -- a "feel" -- of the subject....If women or farmers know so much about forest science, they should come and join the [forest] service!

SF intervention has brought little qualitative change in the general pattern of women's life and status. No substantial reshuffling has been observed in their subordination, dependence and unequal status. However, at a more minor level, SF has brought some subtle changes. First, it has eased the task of fuel-fodder collection. Majority of women can, at least partially, acquire fuel and fodder from their own (SF) plots, and thereby, can (partially) avoid travelling to public forests and village (common) lands. Second, it has also facilitated the task of livestock rearing. Women can tend animals in their own lands. Third, SF has provided some opportunities for cash income for women (for example, share-tending livestock), and thereby, has contributed to increased economic status of their families (This has also, to a limited degree, ameliorated women's status in respective families as "earning members").

However, rehabilitation on hilly plots (through SF) have brought about some disadvantages as well. First, life in a hill is much harder and the additional manual work needed in a hill-life fall mainly on women. Second, living in hills is generally looked down upon and often ridiculed by (plain-land) villagers. Women are particularly sensitive to such comments and attitude of neighbouring villagers. These sentiments are conveyed to them, when (SF) project women meet village womenfolk on some occasions; for example, when a village woman visits a project woman to make contracts for share-tending of livestock, or when beggars with varied information move from door to door.

In summary, though women contribute significantly to homestead and forestry activities, their role is hardly recognized. They lack access to resources and technology, and are largely left out of formal economic sectors. They work within the broader frame of deep social inequality, subordination and dependence. However, they have developed some informal socio-economic links and wield some power through such networks. But such links "do not indicate women's independence....Rather, women's social and economic activities are in fulfilment of their family roles; providing for the reproduction of the household unit" (White, 1992, p.92). Whatever resource and power they possess, women sincerely devote them to family welfare. SF has brought no qualitative change in the basic working environment of women. Economic decisions in the household are largely confined to male members. Women lack organizations to articulate their interests and voices in SF projects. However, SF has eased some of their traditional work and marginally improved their economic position.[4]

The Growth of Cooperative Organizations

As Chapter 4 has noted, there are farmers' cooperative associations (*Bhumiheen Samities*) for running the day-to-day affairs of the Betagi and Pomora projects. These *Samities* deal with farmers' (field) problems, liaise with government offices, and attempt to resolve internal conflicts among farmers. In Chandra, however, such a project-wide association, covering all farmers, does not exist. Instead, there are some small informal groups (each group consisting of 5-8 farmers), facilitated by a voluntary organisation (Proshika). These externally induced groups are mainly formed as prerequisites for credit scheme operation; and they are not particularly meant for SF. SF activities are designed and managed by local forest office in Chandra. Therefore, my discussion on the growth and performance of farmers' cooperative institutions primarily concerns Betagi-Pomora *Samities*.

During the initial years of the projects, Betagi-Pomora *Samities* were very effective, and cooperative management had the following clear manifestations. First, farmers acted collectively to ensure logistic support and services from the government. Second, their collective effort helped them in tackling elite opposition. From the very inception of the projects, farmers got into relentless conflicts with local elites (who used to exercise de facto control over the lands prior to the initiation of SF projects). Law-suits were filed against farmers. In many instances their farm-houses were burnt, crops damaged, and trees chopped down. Farmers united (through the *Samity*) against such oppression. They collectively patrolled their farms round the clock, raised common funds to meet the costs of litigations, and acted together on the face of armed attack on any of the fellow farmer. Third, farmers also performed collectively in facing natural hazards and problems in the area. They jointly guarded their farms from the attack of wild animals (especially of porcupines and boars). During periodic calamities (like cyclonic storms and drought), farmers cooperated in post-damage rehabilitation of the farms, and in approaching government sources of relief and services. Besides, some infra-structural developments were also made through the *Samity*. For example, farmers pooled together fund and labour (and also ensured assistance from government) to establish a primary school at Betagi for their children. Similarly, the Pomora farmers successfully negotiated with a voluntary agency in establishing a school.

The cooperative efforts during the initial years, which sprang out of particular contextual crises (such as elite opposition, natural calamities and various scarcities), were vital for the very survival of the projects. Cooperation had some dramatic effects on farmers' progress. First, with the launch of the projects, there was a qualitative change in the economic status of farmers. Before the induction of SF projects (in the late 1970s), the average annual income of farmers ranged between Taka 13,000 to 15,000. By 1991, it rose to Taka 47,500 in Betagi and 27,500 in Pomora (Rahman, n.d.). With increased economic solvency, farmers' living standard was also elevated. Second, acting collectively, farmers were able to increase their socio-political status and recognition in the locality. They were seen as a prospective power-base by local government leaders and other elites (see Chapter 6). In political activities, especially in Union Council elections, the farmers' *Samity* emerged as a significant "pressure group" in the local power-struggle.

Today there has been a drastic decline of collectivism and cooperative spirit among farmers. The following reasons can largely explain the erosion of cooperation. First, the *Thana* Selection Committee of Betagi and Pomora projects, for all practical purposes, has ceased from operation. After 1989, meetings became erratic and irregular; and since 12.12.1992 no meeting was held at the *Thana* level. The government's institutional support for the projects are nearly absent now. As Chapter 6 has shown, the catalytic support of government agencies has waned with the passage of time, primarily due to the absence of central vigilance (from higher authorities). Many researchers (Quddus et al., 1992, pp.112 and 114; Rahman, 1991, p.23; Task Force, 1987, pp.48-49; Magno et al., 1992, Chapter 8; Alim, 1988, p.150; Ahmed and Azad, 1987, p.42) also maintain that the government's facilitative support is needed to sustain and institutionalize local cooperative initiatives. Second, with the amelioration of economic position and living standard, farmers are tending to split up and move towards individualistic pursuits. Economic solvency has revived the sense of social and status differentiations among farmers, which was dormant in the initial years of crises (see Chapter 6). Farmers have overcome the initial crises by cooperation. The problems are not so acute now. As the crises have subsided, so has the community feeling which hold the farmers together in the face of emergencies and scarcities. Third, as Chapter 6 shows, the factional politics and conflicts among farmers have been intensified in recent years. Under the spell of the broad pattern of factional politics of the village, corresponding factions have emerged among farmers. Patrons maintain their own client-groups in the projects. Such divisions have caused rife in solidarity. During *Samity* election in January 1993, factional conflicts even burst into, what Zaman (1984, p.23) called, "armed gansterism". Indeed, mirroring the observations of major village studies, I have found that factional conflicts and patronage network have diffused any strong manifestation of cooperation; and caused, what Bertocci (1976, pp.46-47) called, "structural fragmentation" in the study area (also see Zaman, 1984, pp.21-24; Hartmann and Boyce, 1983, pp.203-206; Jansen, 1986, pp.297-304). A World Bank report (World Bank, 1979, p.6, cited in Hartmann and Boyce, 1983, p.203) clearly notes that "inter-group rivalries and competition often make it difficult to organise several competing patron-client groups into a single ...cooperative".

I talked to farmers about the fall of cooperative endeavour among themselves. They themselves have reported that with the rise of economic standard, the better-off farmers are neglecting *Samity* affairs. They do not attend *Samity* meetings. The leadership of *Samity* has also become controversial. The leaders have their own client groups among farmers and are engaged in factional politics. Majority of farmers reported that rarely the leaders inform or discuss with them about the *Samity's* tasks or performance (Table 7.2). Similarly, majority of farmers are not convinced about the leaders' commitments towards their welfare (Table 7.3).

Table 7.2 ***Samity* leaders' frequency of consultation with farmers (Q. Do the leaders inform or discuss with you about the task or performance of the *Samity*?)**

Projects	Frequency of Discussion									
	1		2		3		4		5	
	No.	%	No.	%	No.	%	No.	%	No.	%
Betagi	0	0	4	17	14	58	5	21	1	4
Pomora	2	5	15	37	14	35	8	20	1	3

*Notes: * 1 = regularly, 2 = occasionally, 3 = rarely, 4 = never, 5 = no comments. * General notes on tables: Tables in this chapter, unless otherwise stated, are based on the field-work. * Betagi: n = 24, Pomora: n = 40, Chandra: n = 13. * The percentage figures are rounded up to nearest whole number.*

Table 7.3 **Farmers' evaluation about commitment of *Samity* leaders towards their welfare**

Degrees of Commitment										
Projects	1		2		3		4		5	
	No.	%	No.	%	No.	%	No.	%	No.	%
Betagi	0	0	8	33	13	54	2	8	1	4
Pomora	1	2	10	25	21	53	5	13	3	7

*Notes: * 1 = very committed, 2 = committed, 3 = not so committed, 4 = indifferent, 5 = no comments.*

Furthermore, farmers' also do not seem to be very enthusiastic in attending the voting sessions for selecting leaders (Table 7.4). Voting for leadership has become irregular. Leaders are clinging to their positions beyond the scheduled period. *Samity* meetings have become very irregular (Table 7.5). Farmers report that these meetings are largely dominated by the leaders and rich farmers (Table 7.6).

Table 7.4 Farmers' attendance in the last three *Samity* elections

Projects	Election 1		Election 2		Election 3		None	
	No.	%	No.	%	No.	%	No.	%
Betagi	7	29	2	8	5	21	10	42
Pomora	5	13	15	37	12	30	8	20

Table 7.5 Farmers' attendance in *Samity* meetings (in the last 12 meetings)

Number of meetings	Pomora		Betagi	
	No.	%	No.	%
1 - 3	4	10	3	13
4 - 6	8	20	2	8
7 - 9	10	25	5	21
10 - 12	11	28	0	0
0 (no attendance)	7	17	14	58

Table 7.6 Dominating voices in *Samity* meetings (as perceived by farmers)

Dominating agents (voices)	Pomora		Betagi	
	No.	%	No.	%
Government representatives	3	8	3	13
Local elites (or patrons)	3	8	3	13
Rich farmers and *Samity* leaders	24	60	12	50
General farmers	7	17	2	8
Special guests or visitors	3	8	4	16

The initial "community culture" which was manifested by regular voting, attendance in the meetings, accountability of the leaders, consultation between leaders and farmers -- is gradually fading away. When confidentially asked if they were satisfied with their roles in the *Samity*, 63% of Betagi and 70% of Pomora farmers expressed their dissatisfaction for such reasons as domination by the powerful members, irregular meetings, little chance of discussing their own problems, and negligence towards them (Table 7.7).

Table 7.7 Farmers' satisfaction or dissatisfaction of their roles in *Samity* (Q. Are you satisfied with your role in *Samity*?)

Response: "yes" (satisfied farmer): Betagi = 9 (37%); Pomora = 12 (30%).

Classification of the "satisfied" members	Betagi		Pomora	
	No.	%	No.	%
Relatively rich farmers	1	4	2	5
Poor and vulnerable farmers	2	8	2	5
Beneficiaries of current leadership; members of faction (in group politics); kin of the leaders	6	25	8	20
sub-total	9	37	12	30

Response: "no" (dissatisfied farmer): Betagi = 15 (63%); Pomora=28 (70%)

Causes of dissatisfaction	Betagi		Pomora	
	No.	%	No.	%
Dominated by rich and powerful members	5	21	8	20
Irregular meetings	3	13	7	18
Little or no chance of discussing problems	4	17	6	15
Feels neglected	2	8	4	10
No comments	1	4	3	7
sub-total	15	63	28	70

*Notes: * The classification of "satisfied" members is made by the author by an in-depth observation of their interest, links and motives. The causes of dissatisfaction are reported by respondents.*

On further investigations, the minority, who said they were satisfied, mostly turned out to be the rich farmers and the beneficiaries of current leadership. A few poor farmers also reported their "satisfaction". As my observation and experience suggest, they did so in an attempt to show their allegiance to the leaders; simply because they were too vulnerable to offend the power-holders.

In Chapter 3, from a review of literature, I have identified some conducive variables or conditional factors which facilitate cooperation and collective efforts. In light of that discussion, the following table briefly examines the availability, nature and role of such facilitative variables in my study-areas (Table 7.8).

Table 7.8 The nature and role of selected facilitative factors for cooperation in the study-area

Factors	Their role and nature in the study-area
State's role	Initially the government was supportive of the growth of *Samity* (farmers' cooperatives organisation) and provided necessary logistics and services. However, at present, governmental support is nearly absent.
Nature of resource	There is conflict about the demarcation of land in the SF projects. Customary laws on land and tree-usage are also ambiguous. Therefore, the resources and their use are not conspicuously defined.
Nature of production process	SF as a production process requires investment and restriction on use at present in return for benefit in future. The gestation period is long and collective benefit is not immediate. In such a situation, cooperation becomes less likely.

Pattern of Accountability	*Samity* leaders are deeply engaged in factional politics. Manifestations of an accountable leadership (for example, consultation with general farmers, attendance in meetings, reporting back to farmers about the performance of *Samity*, regular voting) are largely absent.
Relation with local government institutions	*Samity*'s relation with local government is one of competition and cold-conflict. Local government leaders maintain factional/client-groups among farmers; and the services from local government offices are confined to these particular groups.
Control Structure	*Samity* is not strong enough to enforce a reward or punishment system on its members. For example, the rich farmers mostly do not attend meetings and hardly perform any task assigned to them by *Samity*. Still the *Samity* has no practical means to punish them. Rather it is being dominated by the rich farmers. Furthermore, *Samity* has no organized "control and detection" system of offence and crimes (for example, trespassing, illicit felling of trees, violation of land-boundaries, violation of *Samity* rules). Such crimes are mostly detected by the victim himself and the *Samity*'s role in systematic detection and resistance is insignificant (see Table 7.9).

The above table suggests that the facilitative factors, necessary for promoting cooperation, are largely absent or ineffective in the locality. *Samity*'s role has been nominal in institutionalizing a cooperative culture.

However, rather contrary to the findings of some studies on the projects (for example Quddus et al., 1992, pp.55-56), I have found that Betagi farmers, who are comparatively well-off than Pomora farmers, have become more individualistic; while Pomora farmers are showing some degrees of cooperation for specific purposes. Two crises are instrumental behind the collective activities of the Pomora farmers. First, they have not yet received permanent title to land and the crisis of insecurity has inspired them to continue their collective struggle for permanent ownership. Second, they are still under continuous threat of conflict with local elites; and farmers know that they cannot tackle the elite opposition individually. The conflicting factions in Pomora (for example, the Azam group versus Rahman group) have strategically joined under a temporary cooperative platform to press ahead their common demands.

Table 7.9 Detection and control of offenses (crimes) around the farm (as reported by farmers)

Detected by	Betagi		Pomora	
	No.	%	No.	%
Local elites	2	8	4	10
The *Samity*	6	25	5	13
Government and local government offices	1	4	2	5
The victim himself	15	63	29	72

Therefore, one can see that, the widely applauded "spectacular cooperative spirit" (see Ahmed and Azad, 1987; Miah, 1989; Alim, 1988) of the early days of the projects is, in fact, to use Attwood's (1988, p.77) words, "selective and negotiated alliance". Scarcities, crises and risks, as in other instances of south Asia (see Chapter 3), have contributed to collective action in the projects during the initial years. The government also provided the necessary "selective incentives" (Olson, 1971, p.167), in the forms of logistic support, reward for good performance, and assurance of tenurial security -- to boost farmers' cooperation. But today such inducement is nearly absent. Crises and problems bind farmers together for a particular period; and with the normalcy of the situation they tend to abandon communalistic behaviour. However, though (farmers') inter-factional relations are often rife with conflicts, on the face of particular problems (the problems which commonly affect all, and which can only be managed by acting collectively), the factions may align together leading to some degree of cooperation. Such strategic alignments have been observed in Pomora, who are struggling through more crises than Betagi farmers. But as Zaman (1984, p.21) has noted, "such alignments are for specific purposes, and their scope of joint action are limited and short-lived" (also see Jansen, 1986, pp.297-304; Chapter 3 of this book). The initial collective spirit among farmers which was sparked by crises had not been institutionalized through a proper organizational structure. Factional conflicts, patronage influences and lack of external catalytic support have primarily hindered the perpetuation of collectivism in the study-area.

Income, Productivity and Standard of Living

Probably, the most visible progress has been made in farmers' income and living standard. Prior to their involvement in SF, the Betagi and Pomora farmers were "nothing but paupers" (Ahmed and Azad, 1987, p.24). Before joining the projects, the annual average income of farmers in Betagi and Pomora were Taka 13,800 and 15,210, and the land-holding were 11.8 decimals and 35.5 respectively (Table 7.10). Share-cropping and daily-labour were their major occupations (Table 7.10). In Chandra too, farmers' status (prior to the project) was similar: the average annual income was Taka 12,800, land-holding size was 5.53 dec., and daily-labour was the major occupation (Table 7.10). Shortly after the induction of SF projects, substantial economic improvement was noted. In 1987, farmers' annual income rose to around Taka 18,000 in Betagi and Pomora (Ahmed and Azad, 1987, p.24). By 1991, annual income at Betagi ranged between Taka 30,000 and 65,000, and in Pomora, it was between 20,000 and 35,000 (see Figure 7.1; Data source: Rahman, 1991, p.19; also see Rahman, 1992, p.17; Miah, 1989, p.20). Farmers' living standard showed corresponding improvement. Their ability in spending on basic amenities (especially on medication, housing and clothing) increased considerably. Prior to their involvement in SF, farmers' expenditure on medication and housing were very negligible due to their abject poverty.

However, at present some degree of stagnancy has crept into the project, and the initial enthusiasm has been gradually eroding (see for details Khan et al., 1996). The income trend virtually shows no major increase in recent years. Currently, farmers' average annual income at Betagi and Pomora are Taka 49,990 and 32,898 respectively (Tables 7.11 and 7.12).

Table 7.10 Farmers' status prior to the initiation of SF projects

Projects	Annual Income (in Taka)	landhold -ing (in decimal)	Occupation (No. of respondents)						Liter -acy rate (%)
			a	b	c	d	e	f	
Betagi	13800	11.8	7	2	8	5	0	2	8
Pomora	15210	35.5	8	2	12	8	4	6	20
Chandra	10976	5.53	0	1	8	2	2	0	8

*Notes: * Occupation categories: a = Landless share-cropping farmer, b = Land-owners, c = Daily-labourers, d = Petty-business, e = Technical skill, f = Job/service. * Land-holding includes homestead land.*

Figure 7.1 Farmers' annual average income trend in Betagi and Pomora

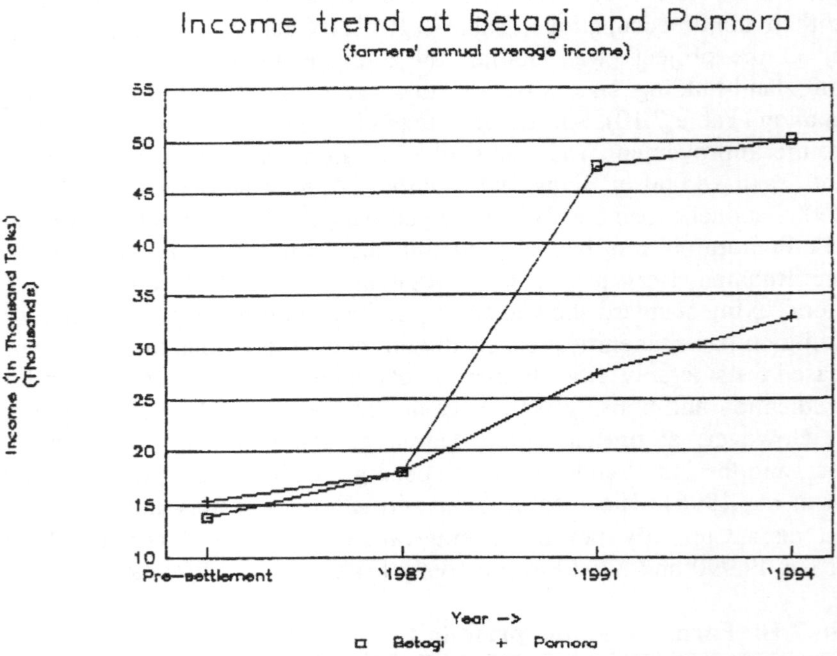

Data sources and notes: This figure is mainly based on my field-work data. Additional information sources are Ahmed and Azad, 1987; Rahman, 1991; and Miah, 1989. The income figures are at current value.

Table 7.11 Farmers' average annual income at Betagi (in Taka)

Farm Sources	Average	%	Total	Stand Error
Fuel wood and small logs	10808	22	259392	227
Lemon	13680	27	328320	294
Fruits and horticulture	6840	14	164160	193
Vegetable	3632	7	87168	128
"Sungrass", bamboo and cane	1091	2	26184	74
Livestock and poultry	1949	4	46776	120
Non-farm sources	11990	24	287760	

*Note: * The "non-farm sources" mentioned in this table and in Tables 7.12 and 7.13 include earning from waged-labour, share-cropping and other off-farm employments, and also the contribution from family-members working overseas.*

Table 7.12 Farmers' average annual income at Pomora (in Taka)

Farm Sources	Average	%	Total	Stand Error
Fuel wood and small logs	5775	18	231000	603
Lemon	6110	19	244400	626
Fruits and horticulture	8310	25	332396	675
Vegetable	725	2	29000	70
"Sungrass", bamboo and cane	1715	5	68600	146
Livestock and poultry	2238	7	89540	241
Non-farm sources	8025	24	321000	

Planting of horticultural and long-rotation forest species has also declined. In Pomora, for example, farmers reported that in 1982 each household (on an average) planted 105 horticultural and 45 forest trees (saplings); whereas in 1993 the numbers of planting were reduced to only 24 and 7 respectively. Similarly, in Betagi 220 horticultural and 95 forest saplings were planted (per household/average) in 1982. In 1993, only 45 horticultural and 50 forest trees were planted. Presently, non-farm sources constitute a major source of income, though (as per the terms of allotment) farmers are not supposed to work outside their farms and are expected to solely devote themselves to SF (farm) activities (Rahman, 1987, p.2; 1992, p.10). My survey suggests that farmers' engagement in farm and plantation activities has gone down to 9 labour/hours per day (per family), from that of 17.35 labour/hours in 1988, as reported by Miah (1989, p.19). Furthermore, farmers are sending their sons abroad (mainly to the Middle East). The foreign connections upgrade their social and economic status; but reduce the supply of valuable family labour for farm development. Besides, quick cash benefits from the foreign employments further strengthen the (farmers') inclination to off-farm jobs. About 23 (15%) and 11 (13%) families respectively in Pomora and Betagi have at least one member working abroad.

In Chandra too, 65% of total annual income come from non-farm sources (mainly, from waged-labour). Agriculture and vegetable gardens are the main farm sources (Table 7.13).

Table 7.13 Farmers' average annual income at Chandra (in Taka)

Farm Sources	Average	%	Total	Stand Error
Fuel wood and small logs	773	4	10050	343
Agriculture crops (mainly paddy)	1454	8	18900	254
Nuts, beans and pulses	1038	6	13500	310
Fruits and horticulture	627	4	8150	330
Vegetable	1350	8	17550	269
Livestock and poultry	792	5	10300	106
Non-farm sources	11454	65	148900	2134

However, it may be remembered that Chandra project is still in its nascent stage. It will take a long time to tap income from other farm sources (such as forest plantations). Besides, Chandra farmers have a particularly strong tradition of working as waged or contractual labourers in agriculture and "house-building" (the skill of preparing the popular rural mud-houses). Therefore, it may be difficult to induce them solely to farm activities.

From the above, two broad observations can be made. First, contrary to the stated goals and terms, farmers are gradually turning to off-farm activities for quick cash returns. Second, long term "forest cropping" (of quality timber species) has been negligible. Again, a number of reasons can explain these trends. First, farmers do not find farm plantation and forestry activities economically rewarding enough to confine themselves solely to these activities. Second, farmers feel insecure to plant long-rotation forest trees because these are particularly targeted by the loggers for plundering (Chapter 5). From his survey of *Thana* Afforestation and Nursery Development Programme (the country-wide SF scheme), Chowdhury (1993, p.84) also reported that only less than 2% of his respondents (participating farmers) favoured long-rotation species. Third, in most cases, land tenurial insecurity and ambiguous customary rights over trees have reduced farmers' interest in farm activities. Fourth, with the declining logistic and institutional support from the government, the farm production technology has deteriorated, yielding low economic return. Recent studies have noted that the farms in Betagi and Pomora projects are silviculturally not sound; plantations are haphazard and plots have not been developed according to "optimum cropping pattern" (Rahman, 1991, p.21; Rahman, 1987, p.6; Zabala, 1990, p.177; Rahman and Khan, 1994; M. Mohiuddin and M.K. Hossain, Institute of Forestry, personal communication). Fifth, with improved living standard, farmers' expenditure and demands have also correspondingly increased. The increased consumption needs have caused greater demand for cash. Consequently, farmers are pushed to work beyond the farm to respond to their ever-growing demands of live. Sixth, farmers are beset with some chronic problems in their farms, including water scarcity; damage by pests and wild-animals; tenurial insecurity; danger of plundering of trees; lack of governmental support, and disputed land (Tables 7.14, 7.15 and 7.16).

Most of these problems are beyond farmers' control; and the solutions need substantial institutional support from the government and other external agencies. As has been already discussed, currently such catalytic supports are scarce or absent.

There has been a clear amelioration of farmers' quality and standard of living. From a state of dire poverty, they have reached a point where they can spend on nutritious food, clothing and medicine, which were well beyond their means prior to the projects. Even farmers occasionally can manage to make ostensible savings. For example, in 1987, Ahmed and Azad (1987, p.30) reported that Betagi and Pomora farmers annually saved Taka 3,000 and 1,000 respectively. However, with the passage of time it seems that the cost of better living standard has increased disproportionately to income. Improved living standard has brought the vices of renewed desire for consumption and luxury-expenditure.

Table 7.14 Major problems in the farm: Betagi

Problems	Degrees of acuteness					
	1		2		3	
	No.	%	No.	%	No.	%
Protection of trees	2	8	4	17	2	8
Land tenurial insecurity	0	0	0	0	2	8
Economic hardship	7	29	6	25	2	8
Pests, animals and diseases	4	17	5	21	4	17
Water scarcity	6	25	4	17	3	13
Lack of governmental assistance and logistic support	3	13	4	17	4	17
Land conflicts and disputed land	2	8	3	13	3	13

*Notes: * Degrees of acuteness: 1 = most acute problem, 2 = very acute problem, 3 = acute problem. * The problems described in this table and in Tables 7.15 and 7.16 were identified by the respondent farmers, and they attached value-judgements to the problems in terms of relative degrees.*

Thus, the surplus income, if any, is absorbed by the increased lust for consumption. I have observed that the rise in living standard has linked farmers to market places and urban centres. The dormant kinship ties with urban relatives have been rejuvenated. Proximity to urban and market places has widen their range of demand and the resultant expenditure. Consequently, as I noted, majority of farmers in Betagi, Pomora and Chandra were in deficit in 1993 (Table 7.17). The deficit-farmers are forced to borrow from local elites and *Mohazans* (private money-lenders). This, in turn, has reinforced the patronage and other dependency relations. A good deal of bank and other institutional loans, which farmers borrow for farm development purposes, are diverted and (mis)used to meet household deficits (Table 7.17).

Table 7.15 Major problems in the farm: Pomora

Problems	Degrees of acuteness					
	1		2		3	
	No.	%	No.	%	No.	%
Protection of trees	3	7	5	13	6	15
Land tenurial insecurity	12	30	1	3	3	7
Economic hardship	8	20	10	25	4	10
Pests, animals and diseases	4	10	9	23	5	13
Water scarcity	4	10	4	10	12	30
Lack of governmental assistance and logistic support	4	10	8	20	7	17
Land conflicts and disputed land	5	13	3	7	3	* 8

*Note: * Degrees of acuteness: 1 = most acute problem, 2 = very acute problem, 3 = acute problem.*

Table 7.16 Major problems in the farm: Chandra

Problems	Degrees of acuteness					
	1		2		3	
	No.	%	No.	%	No.	%
Protection of trees	1	8	2	15	0	0
Land tenurial insecurity	3	23	3	23	0	0
Economic hardship	3	23	2	15	2	15
Pests, animals and diseases	1	8	0	0	2	15
Water scarcity	3	23	1	8	1	8
Lack of governmental assistance and logistic support	1	8	0	0	2	15
Land conflicts and disputed land	1	8	0	0	1	8

Note: * *Degrees of acuteness: 1 = most acute problem, 2 = very acute problem, 3 = acute problem.*

Table 7.17 Number of deficit farmers and the sources of meeting deficits in household expenditure

Q. were you in deficit in meeting household expenditure during last year?

Projects:	Yes:	No:
Betagi	13 (54%)	11 (46%)
Pomora	27 (67%)	13 (33%)
Chandra	8 (62%)	5 (38%)

Sources of meeting deficits	Betagi		Pomora		Chandra	
	No.	%	No.	%	No.	%
Loan or grant from patrons	6	25	11	27	3	23
Loan or grant from kin	1	4	2	5	1	8
Loan from *Samity* or fellow farmers	1	4	3	8	0	0
Bank or institutional borrowing	2	8	4	10	3	23
Loan from *Mohazans*	3	13	7	17	1	8

Note: Year of data collection: 1994.

The primary heads of expenditure for the three projects are depicted in Tables 7.18, 7.19 and 7.20.

Table 7.18 Annual expenditure pattern at Betagi (in Taka)

Heads of expenditure	Total	Average	%	Standard Error
Non-farm expenses				
Housing	38400	1600	6	204
Health-care	91200	3800	13	159
Child-education	6960	290	1	25
Day-to-day expenses including food-stuffs	473280	19720	69	1349
Farm expenses	78000	3250	11	337
Total	687840	28660	100	

Table 7.19 Annual expenditure pattern at Pomora (in Taka)

Heads of expenditure	Total	Average	%	Standard Error
Non-farm expenses				
Housing	102600	2565	8	663
Health-care	176000	4400	14	624
Child-education	18080	452	1	107
Day-to-day expenses including food-stuffs	864000	21600	69	1301
Farm expenses	102040	2551	8	345
Total	1262720	31568	100	

Table 7.20 Annual Expenditure pattern at Chandra (in Taka)

Heads of expenditure	Total	Average	%	Standard Error
Non-farm expenses				
Housing	13350	1027	8	212
Health-care	9740	749	5	87
Child-education	4790	368	3	67
Day-to-day expenses including food-stuffs	126000	9692	72	763
Farm expenses	20300	1562	12	95
Total	174180	13398	100	

The expenditure pattern shows that investment on farm activities is very insignificant. Bulk of the income is absorbed by food-stuff and day-to-day expenses. The demand in these heads has increased sharply; because of the increased living-standard, inflation, and the ever-increasing prices of daily necessities. It may be noted that in the year of data collection (1994) the country's cost of living increased by 3.3%, and the rate of inflation was 2% (The daily Janamat, May 26 - June 1, 1995, p.15). Re-investment in farm production and expenses on children education are negligible. Farmers have reported a number of major problems in the locality which constrain their living standard and economic progress, including inadequate housing; water scarcity; lack of marketing and transportation facilities; poor health-care facilities; and unconducive law and order situation (Tables 7.21, 7.22 and 7.23).

Table 7.21 Nature of major problems in the locality: Betagi

Problems	Degree of Acuteness					
	1		2		3	
	No.	%	No.	%	No.	%
Inadequate housing	2	8	2	8	6	25
Scarce water and poor sanitation	8	33	6	25	4	17
Inadequate schooling for children	0	0	2	8	4	17
Inadequate market facilities and poor transportation	3	13	4	17	5	21
Inadequate health and medi-care	8	33	4	17	2	8
Deteriorating law and order situation	1	4	4	17	2	8
Lack of additional employment opportunities	2	8	2	8	1	4

*Notes: * Degrees of acuteness: 1 = most acute problem, 2 = very acute, 3 = acute. * The problems described in this table and in Tables 7.22 and 7.23 were identified by the respondent farmers, and they attached value-judgements to the problems in terms of relative degrees.*

Table 7.22 Nature of major problems in the locality: Pomora

Problems	Degree of Acuteness					
	1		2		3	
	No.	%	No.	%	No.	%
Inadequate housing	4	10	3	8	7	18
Scarce water and poor sanitation	10	25	14	35	8	20
Inadequate schooling for children	2	5	2	5	6	15
Inadequate market facilities and poor transportation	6	15	8	20	7	18
Inadequate health and medi-care	11	27	8	20	7	18
Deteriorating law and order situation	4	10	3	8	4	10
Lack of additional employment opportunities	3	8	2	5	1	2

*Note: * Degrees of acuteness: 1 = most acute problem, 2 = very acute, 3 = acute.*

In summary, with the settlement in SF projects, farmers have significantly improved their economic status and corresponding living standard. However, with the rise in socio-economic position, their demand and expectations of life have also widened, probably disproportionately to the income. The sharp rise in living standard, coupled with the country's rapidly increasing inflation and cost of living, has thwarted the pace of progress. In real terms (for example, net savings), farmers' economic upliftment has not borne much fruit. The earnings have been absorbed by increased consumption; and investment in productive sectors (such as in farm production) is nominal. Consequently, farmers are pushed to more off-farm involvement (mainly, share-cropping and waged-labour) under unfavourable terms and conditions.

Table 7.23 Nature of major problems in the locality: Chandra

Problems	Degree of Acuteness					
	1		2		3	
	No.	%	No.	%	No.	%
Inadequate housing	2	15	3	23	1	8
Scarce water and poor sanitation	4	31	2	15	3	23
Inadequate schooling for children	2	15	1	8	2	15
Inadequate market facilities and poor transportation	3	23	2	15	1	8
Inadequate health and medi-care	1	8	2	15	2	15
Deteriorating law and order situation	1	8	0	0	2	15
Lack of additional employment opportunities	0	0	3	23	2	15

Note: * *Degrees of acuteness: 1 = most acute problem, 2 = very acute, 3 = acute.*

Moreover, despite the substantial improvement in economic status (as compared to their previous situation), most farmers are still in deficit in meeting the increased demands of living. This has, in turn, intensified farmers' dependence on local elites and cemented patronage extortions. With the reduction of catalytic assistance from the government and the sharp rise in consumption demands, the pace of farmers' economic upliftment has faltered and the initial success could not be institutionalized. Consequently, farmers are finding it difficult to uphold the enhanced living standard which they initially achieved. Moreover, their living standard are currently threatened by a host of problems.

Literacy, Family Planning and General Level of Awareness

Initially, farmers in Betagi and Pomora were trained by the forest department in basic reading-writing skills and agroforestry activities. One third of the farmers received training on agroforestry in the Forestry Training and Development Centre, located in Kaptai (Quddus et al., 1992). But in the last five years, no follow-up or refresher training or extension support has been arranged. In the absence of periodic back-up and monitoring, the impact of training has been negligible. Many of my respondents, with a smile of apology, told that they virtually remembered nothing of the techniques they had learnt in the training sessions. Quddus et al. (1992, p.107) also note that "the purpose of training does not seem to have had any impact on adoption of agroforestry because no relationship between agroforestry income and training is observed". However, farmers' literacy rate has increased, because initially the project pioneers at Betagi-Pomora were particularly concerned about literacy, and introduced compulsory education in *Samity* meetings. This system of forced education is no longer practised today; and farmers have not improved their education beyond basic literacy. Education among the next generation (children of farmers) is comparatively better. There are two primary schools in Betagi and Pomora. Although there are two adult schools in the area for educating farmers, for all practical purposes these are non-operational. The illiteracy rate is still acute; 55%, 50% and 62% of my respondents (including their family members) in Pomora, Betagi and Chandra respectively are illiterate.

Family planning activities are negligible in the study-areas. In Pomora and Betagi respectively, according to *Samity* chairmen, approximately 25 farmers (out of 151; that is, 17%), and 4 farmers (out of 24; 17%) use contraceptives (mainly condoms). Virtually none of my respondents in Chandra have adopted any family planning measure. Some voluntary organizations are trying hard to raise the general level of awareness among farmers and their families. Proshika, for example, has set up two adult "consciousness building centres" (family planning and general training schools) at Kalampur (Chandra). The average family size at Betagi and Pomora are 9 and 7 respectively. In Chandra, the family size is 5. The low figure, however, is due to the fact that most farmers have migrated to the project from their original homes and part of their family remained there. "As they get permanently settled in the [project], those members would...join them and the size of the families would get larger" (Ahmed, 1994, p.10). The reasons for larger families and non-adoption of family planning measures are: first, the wide belief amongst farmers that women who have adopted contraception are "less attractive" (sexually less appealing); second, the desire to have sons (vis-à-vis daughters) for security in the old age; and third, increasing population means increased number of family labourers (vis-à-vis hired labourers) for farm and non-farm work.

Farmers' exposure to public media (such as radio, newspaper) is nominal. Only 23% farmers in Chandra and 16% in Betagi-Pomora (jointly) attend to public media (Table 7.24).

Table 7.24 Farmers' exposure to public communication media (as reported by farmers)

Media	Betagi and Pomora						Chandra					
	1		2		3		1		2		3	
	N	%	N	%	N	%	N	%	N	%	N	%
Television	2	3	2	3	3	5	0	0	1	8	0	0
Radio	7	11	3	5	6	9	2	15	1	8	0	0
Newspaper	1	2	0	0	1	2	1	8	1	8	1	8

*Notes: * Degrees of exposure: 1 = regularly, 2 = occasionally, 3 = rarely. * As because Betagi and Pomora are situated in the same locality and opportunities of access to public media are broadly similar, the respondents of the two projects have been jointly considered (n = 64). * 10 (16%) respondents in Betagi-Pomora and 3 (23%) in Chandra reported to attend to public media. * Televisions and Newspapers are available at larger tea-stalls in markets which are generally enjoyed by farmers (free of charge). Some farmers possess radios.*

Their conception of SF, to say the least, is inconspicuous. When asked what they meant by SF, I received varied (and sometimes weird) responses from farmers. Some examples are as follows:

> SF is to plant trees by forming "committees" under the auspices of the forest office (Nurul Huq, a farmer in the Betagi project);
>
> The forest department will give land and we will plant trees (Nazim Mustan, Pomora);
>
> SF means to plant the government's *malaria gach* [Eucalyptus] around our

homestead and also to grow vegetable (Dhirendra Barman, Chandra).

To most farmers, SF denotes a governmental afforestation scheme, where they can participate in planting and benefit-sharing under strict supervision and control of the forest department. In fact, farmers have been trained and indoctrinated into such an incomplete (and perhaps misdirected) understanding about SF. I have already analyzed foresters' attitude and conception of SF in Chapter 6; and here, the above discussion suggests that, farmers' understanding of SF mirrors the bureaucratic explanation.

Meeting Local Demands

One explicit goal of SF is to respond to local demands. Is SF meeting this goal in the study-areas?

No doubt that farmers have marginally increased their cash income by SF activities. But the principal products of SF, especially fuel wood and timber, have gone to the hands of local elites and merchants. Fuel wood, small logs and timbers grown in the projects are under almost monopolistic manipulation by the commercial network (see Chapter 5), and are absorbed principally by brick-kilns, tea-stalls, saw-mills and furniture shops. The quality-timbers are transported to city markets. Howes and Jabbar (1986, p.10), from their extensive enquiry into rural fuel shortages in Bangladesh, also conclude that the availability of wood for domestic consumption "is restricted by competing external demands of brick-kilns, urban users and local restaurants", who "make substantial inroads into the amount available for domestic consumption".

SF's role in meeting rural fuel wood demand is also insignificant. My experience is that, using wood as cooking fuel is a rare luxury which farmers cannot afford. I have rarely come across a situation where farmers used wood as fuel. All sellable logs are absorbed by commercial markets. Farmers mainly use leaves, barks, twigs, and (dried) shrubs as cooking fuel. Small quantity of animal dung and agriculture residues are also used (Table 7.25).

General villagers in the locality (out side the project) have little share in SF products. Timber and fuel wood are well beyond their reach. However, some villagers can buy vegetable and fruits from SF farmers (at comparatively cheaper prices than in open markets). Some others have agreements of share-tending livestock with farmers.

Table 7.25 Farmers' use and sources of cooking fuel

Fuel-sources	Betagi and Pomora		Chandra	
	No.	%	No.	%
Leaves, twigs, branches, shrubs etc.	44	69	7	54
Fuel wood and small logs	5	8	2	15
Agriculture residue	9	14	1	8
Animal dung	4	6	2	15
Other	2	3	1	8

*Notes: * As because Betagi and Pomora are situated in the same locality and opportunities of access to fuel sources are broadly similar, respondents of the two villages have been jointly considered (n = 64). * Though a respondent may use different sources of fuel at a given time, only the major source have been considered here. "Other" sources include kerosene and charcoal.*

Promotion of Equity

SF also planned to foster equity in asset creation and benefit distribution. It has been repeatedly argued that SF "is social for its equity orientation" (Rahman, 1991, p.6). The equity issues in SF are varied and wide. In what follows, I shall evaluate the nature of equity in the study-areas by recapitulating some practical examples which I have raised in the previous chapters.

In many instances, the projects have not been able to reach their target group ("landless poor") equitably. In Pomora, there are as many as 37% fake "landless"; some of whom are in fact rich landowners (*Jotdars*). Besides, as Chapter 6 has shown, local elites have co-opted some of the benefits specially designed for SF farmers. Highly unequal social relations and power structures allow very little room for equitable distribution of benefits. In some instances (for example, in Chandra), although the poor farmers have officially been allotted the plots, still they are under constant threat from elites, and have to bribe the elites or their *mastans* to ensure their existence and continuation in the plots. Ahmed (1994, p.7) has also witnessed this situation:

> some rich and influential persons [have] assumed...a role of patrons of these settlers and make promises to protect them in return for some fees for this patronage to be paid to them on regular basis.

Farmers, as I have discussed before, are still exposed to manipulation and extortion by unequal social structures (such as patronage, status differentiation, factionalism), which constrain the prospect of equity in asset creation and benefit distribution from SF.

SF has also caused dissatisfaction among general villagers. With individual allotment of land to a particular group of farmers, SF has practically "privatized" the lands (previously under de jure control of the government), which villagers used as common grounds for pasturage, fuel-fodder collection and other purposes. Although SF has benefitted some farmers, it has effectively reduced the rights and shares of a large number of equally deserving population (see Chapter 6).

However, with economic amelioration, farmers "as a group" have achieved considerable social recognition. They are seen as prospective support and power base for local politics and election systems. Thereby, farmers are, to a limited extent, participating in local power politics. These marginal improvements in farmers' economic and political status, may have some positive (although very limited) impact on social equity.

Participation in the Project-Cycle

This section briefly focuses on the involvement of the farmers in the project cycle: in planning, decision making, implementation and benefit sharing of SF projects. The Chandra project is planned and administered, to a large extent, by the forest department. All major decisions including selection of beneficiaries, choices of species, selection of sites and the silvicultural technology are prescribed by foresters and technical experts from the funding agency (Ford Foundation). There is, for example, a Plantation Journal for the project which prescribes detailed plantation schedules and methods to farmers "as per the directives and models suggested by higher authorities [of the forest department and the Ford Foundation]" (Plantation Journal, p.2, Kaliakair Range Office). Similarly, in Betagi-Pomora, the SF farms were developed through an Integrated Agroforestry Systems (see for details Alim, 1984) designed by professional foresters (Zabala, 1990, p.177; Rahman, 1987, p.2; Rahman and Khan, 1994, p.68). However, Alim (1988, p.152) reported that, during the initial design farmers were consulted.

The culture of collective planning and implementation has not developed. Initially, although under strict supervision of the forest department, farmers (to some degree) participated in the management of the projects through their *Samities*. Presently, with the decline in catalytic supervision and collective spirit (as discussed earlier), farmers have become individualistic, and collective participation in the project cycle has eroded. Thus, from a state of strict bureaucratic control, Betagi-Pomora projects have moved to complete individualistic practice. This individualistic allocation of power is applauded by some studies and is seen as synonymous to "empowerment" (for example Rahman, 1991, p.15). But, in fact, such individualistic pursuits have thwarted the growth of the culture of collective decisions and joint-implementation. Farmers are driven largely by personal interests now. They show little interest in the projects or *Samity* affairs. Farmers' participation in *Samity* is confined to occasional presence in meetings, limited discussion on general forums, and irregular attendance in voting sessions. Therefore, individualistic power does not necessarily reflect collective empowerment of farmers as a group.

The monitoring and evaluation of the three projects are solely the government's responsibility. Farmers have virtually no voice in the performance appraisal of the projects. No institutional form of accountability (to farmers) exists.

As regards benefit distribution, Chandra farmers are supposed to get 50% of SF products and the government is to get the rest. But farmers, like in many other instances (see Chapter 6), have not yet received formal contract from the government regarding benefit-sharing, which has caused dissatisfaction and suspicion among farmers. In Betagi-Pomora, farmers officially get all the products. However, as already discussed (in Chapters 5 and 6), despite the formal rights, farmers have only marginally benefited from SF products. The products are largely absorbed by the commercial network, over which farmers do not have any significant control.

In summary, farmers' participation in decision making and evaluation of the projects is negligible. They have participated in the implementation of the projects within strict bureaucratic limits. However, even when farmers have enjoyed sufficient freedom, they tended to assume individualistic attitudes and to drift away from collective action. Therefore, it seems that neither strict external control nor individualistic autonomy could ensure meaningful participation. Collective participation, through an organized cooperative structure, has not occurred to any significant extent. The benefit sharing arrangements are ambiguous and make room for suspicion. Besides, the unconducive variables of the frame (for example, the unequal power structures, exploitative social relations, unfavourable roles of government and local government offices, and the dominance of market forces), do not allow farmers to fully enjoy SF's benefits.

Conclusion: Marginal Achievements and the Subterranean Purposes of SF

The stated goals of SF have only been marginally achieved. The achievements hardly reflect the ambitious initial formulations. The only notable achievements are the marginal increase in farmers' economic status (including better living standard) and social recognition. Other promised goals remain largely unrealized. For example, no major qualitative change has been observed in the general pattern of women's life and status as a result of SF intervention. Owing to the factional conflicts, patronage influences, and lack of catalytic support, the growth of cooperation remains selective and transient. The collective spirit, which is mainly sparked by crises, has not been institutionalized in any proper organizational structure. The rate of literacy and adoption of family planning measures are limited. Farmers' understanding of SF reflects the bureaucratic explanation of participation where people are invited to join and perform within strict limits set by the state. The principal economic products of SF have been absorbed by the commercial market network; and general villagers have virtually no share in them. Participation in decisions and evaluation of the project remains insignificant.

From the above discussion, SF's lofty goals seem too ambitious, almost to the extent of being unrealistic. Are foresters and donors -- the protagonists of SF -- unaware or uninformed about the frame which constrains SF's achievements? Considering the educational background and wealth of experience of foresters, this does not seem logical. Similarly, international donors, with their unrestrained access to systematic and up-to-date data on Bangladesh, wide spectrum of experience, and long involvement with the country, are most unlikely to be numb to the unfavourable surroundings. Therefore, it can be logically argued that foresters and donors are well aware of the constraints and limited achievements of SF. Why then they are pursuing such unpromising goal? My experience suggests that SF is effectively achieving some crucial purposes for donors and the national government. These purposes are neither explicitly stated in SF documents, nor discussed publicly in the official quarters. They remain as a kind of tacit agenda for SF. In what follows, I shall shed light on these purposes of SF.

As Chapter 3 has suggested, the enthusiasm of south Asian governments in pursuing participatory projects (including SF) can be largely explained by their concern for attracting greater external fund. Oakley et al. (1991, p.273) explain the rationale for the government's interest in foreign-aided projects very simply: "projects mean money". SF is one of the most prospective avenues for attracting greater international fund for the Bangladesh government. Chapter 6 (especially, Table 6.16) has shown the magnitude of external dependence of SF projects. In fact, SF in Bangladesh has a legacy of external dependence. The first formal Community Forestry Project (1981-87) in the northern part of the country was implemented by a Asian Development Bank loan of US$ 11 million, and the technical guidance of the United Nations Development Programme and the Food and Agriculture Organization (Manandhar, 1987, p.5). The trend of external influence continues with equal vigour until today. As a follow-up of the first project, a country wide programme has been carried out in the name of *Thana* Afforestation and Nursery Development Project since 1989 (GOB, 1992, p.2). Presently, this is the largest SF project in Bangladesh with about 88% external funding towards its total cost. Fetching international fund, as my informal interviews with senior officials and politicians suggest, is probably the most active reason for the government's ever-widening interests in SF. A retired chief of the country's premier forestry research institute told me:

> In the international arena, you will see that, from time to time some fashions emerge and we play the "yes-man" part. The fancy of "forestry extension" is over; and now, this is the age of SF....Donors are mainly interested in SF; the equation is simple: we need fund because we are poor and we go for SF because they [donors] are ready to fund it....Our best officers have been placed in formulating SF proposals, because we cannot afford to lose donors because of a bad plan-paper.

Indeed, SF has become a major ground for attracting external fund. Even the country's prime minister seems to be personally involved in the campaign for foreign fund, when she makes the following appeal (GOB, 1992, p.1, emphasis added):

> *The success of various initiatives of our government* for sustainable development including afforestation, reforestation, tree plantation on marginal and vacant lands *is...contingent on external assistance* in the form of resource and technology transfer. *We look forward to generous assistance from development partners.*

Besides acquiring foreign fund, SF also helps the government in fortifying its control and authority in the administration of forests in a number ways. SF is serving as a "buffer" to the endangered state forests. It is reducing pressure on government forests, and thereby, protecting the state's domain which has been under continuous threat of encroachment. Shortly after the launch of SF, the government has been able to regain control over much encroached lands. In Chandra, for example, local foresters reported that 310.25 acres of land was previously under encroachment; but with the introduction of SF project, they have been able to recover 100 acres of land. The following excerpts from the interviews of two centrally placed foresters manifest the forest department's interests in pursuing SF:

> SF is needed to save the core forests from pilferage and encroachment....Reserved forests have to be maintained at any cost in broader national interest.
>
> Till recently, we worked alone for the benefit of the people. But they misunderstood our sincere efforts and damaged state plantations by all conceivable means. Now we are calling them to work in SF plantations and providing them with all necessary benefits and supports....We hope that they will restrain themselves from the misdeeds they used to do before, stop tampering with state plantations and help us in protecting national forests in broader national interests.

During the course of informal discussions, many officials have asserted that SF is "no substitute" for traditional government forestry activities (consisting of plantation, revenue and policing tasks). SF's purpose is to lessen the pressure on government "reserves". Therefore, SF, as seen by the government, is not about widening local people's access to and share in forests, rather it is an attempt to divert pressure on government territories, and thereby, safeguarding the state's authority and control over local communities. On rare occasions, these otherwise reclusive goals of SF surface. For example, some recent studies have noted that the "encroachers" are the primary target of agroforestry and woodlot plantations, so as to restrict further encroachment and recover government lands (FMP, 1992, p.2; Karim, 1990, p.61). Quddus et al. (1992, p.2) have also noted that SF efforts have primarily been directed to protect the existing government forests from further encroachment and to expand public forest areas by utilizing the marginal and fallow lands. Similarly, foresters at Chandra and Pomora have told me that SF has enabled them to regain control over the land they "lost to encroachers", and also to revitalize their decayed authority over local population. As Chapter 6 has argued, the government's interest in SF is also a part of its "strategic compromise" with the people. It stems from the government's realization that it can no longer govern forests singlehandedly; and has to make some room for people to save its endangered authority over forests from total collapse (through such processes as encroachment and public opposition).

For the international aid agencies, SF projects, with their overwhelming dependence on foreign funds, are a way to exert control over decisions and actions of the national government, and a testing ground for promoting their ideas and innovations. The western democratic and populist values of participation and empowerment, commonly incorporated in the SF agenda, are largely the results of overwhelming donor influence on SF projects (Chapter 3). Many senior public officials have told me that international aid agencies are often the "predominant voice" in setting the objectives of SF; and the input from field offices are not given due attention. Consequently, as Locke (1995) and Oakley et al. (1991) have observed in the general context of developing countries, a "propagandist language" (Locke, 1995, p.20) has been prominent in setting SF's objectives, which are often remote from field realities. Cernea (1985) also noted that a "cloud of rhetoric" frequently looms over the (externally-funded) public participatory schemes including SF. For donors, SF in Bangladesh, like in other states of south Asia, serves as "a method of influencing departmental behaviour and [of ensuring] that the unequal partner, the line department, agrees to...a project whose objectives it may not share" (Locke, 1995, pp.22-23).

SF also spreads a philanthropic message for both the government and donors across the country-side. Even the most illiterate farmer in Chandra, for example, is well aware of (and grateful for) the "generosity" of the Ford Foundation. Frequently, farmers used expressions such as *sohay* (support), *upokar* (help), *daan* (gift, donation) or *vala kam* (righteous deed) while referring to the activities of the donor agency. Similarly, SF, to some extent, helps to portray and promote the image of public officials as the "noble providers" of services to the community. Many farmers saw the activities of the project as almost a personal favour to them from local officials and donors. With the introduction of SF, there has been some improvement in the government's relationship with local people in Chandra. Considering the age-old enmity between the forest department and local communities, such a popularity is obviously very comforting to the central authorities in the government. However, with the passage of time, as the general experience suggests, the government's services to farmers gradually decline; and the improved relationship does not generally last long (as happened in the case of Betagi-Pomora).

In summary, although the stated goals remain largely unrealized, SF is furthering some strategic purposes of the government and international donors. Despite its marginal achievements of formal goals, SF will prosper as long as it fulfils these interests of its protagonists. Considering the effectiveness of SF in fostering these interests, one may reasonably predict a more pompous future for SF, although as far as people's participation and empowerment are concerned, SF largely remains a rhetoric.

Notes

1. See for example Proceedings of Betagi-Pomora *Thana* Committee meetings: 25.5.1980-22.2.1982; Minutes of the Landless Farmer Rehabilitation Projects (Betagi-Pomora): 1978-1982; Selected official correspondences of the Divisional Forest Officer (Chittagong), Deputy Chief Conservator of Forests (Extension Circle) and Officer in Charge (Pomora) regarding Betagi-Pomora projects: 1980-1991; ARDP Project Proposal; Consultant Sociologist's and Research Officers' reports on ARDP: 1994. For general references to selected major goals of the Betagi, Pomora and Chandra projects, see Rahman (1992; 1991); Alim (1988); Ahmed and Azad (1987); ARDP (1989).
2. Literally *purdah* means "veil". It is a complex institution. It connotes a physical "system of isolation of women from outsiders" (Begum, 1988, p.11) and also a mental process of internalization of values of modesty, shyness, honour and shame (Feldman and McCarthy, 1981, p.26) and "imposition of high standards of female modesty" (Chaudhury and Ahmed, 1980, p.5). Chaudhury and Ahmed (1980, p.6) argue that, although it mainly draws on Muslim tradition, the Hindus also maintain *purdah*.
3. I discussed with forestry researchers at the Bangladesh Forest Research Institute and Institute of Forestry about some of the knowledge and perceptions of women in the study-areas; and I was told that the information women possessed were, on most counts, scientific and accurate.
4. Khan (1997) partially draws on this section.

8 Conclusions

Introduction

This concluding chapter recapitulates and summarizes the core themes and findings of the previous chapters, and sheds some light on the road ahead. The next section reviews the political-economic trends in forest commercialization which have alienated local communities from forests. Then, the performance of SF as a strategy for redressing the process of alienation is discussed. Subsequently, the tacit but strategic purposes of SF are summarized. The later sections provide some food for thought for prospective future research and action on SF development in Bangladesh.

Commercialization and Alienation

I have dwelt on forest resource use from a historical perspective, supported by corresponding evidences (Chapters 2 and 5). The enquiry revealed the commercial processes which have systematically alienated local people from the use and management of neighbouring forests. The commercial forces were primarily set in motion by the colonial intrusion on the Bengal forests; and after the political independence of the nation, the trends in commercial exploitation and the consequent alienation of people continued to propel somewhat independently. These processes were generally manifested by the reservation and state appropriation of local forests, primarily, for revenue and commercial exploitation (a significant portion of which hitherto used to meet the local community needs); the introduction of bureaucratic forest management; and the penetration of commercial forces into local economy and forests. These processes contributed to the growth of an alliance of vested interests among commercial forces (which generally included urban loggers, local elites, forest based industries and local businesses, and a section of state officials). There was a radical change in the very philosophy of forest management. Under the changed situation, forest management centred around predominantly commercial considerations of profit and revenue maximization. These commercial forces and philosophies, acting collectively, caused public alienation by way of gradual diminution of local rights in forests; pushing local people into severe competition for forest resources under unequal and exploitative terms; and displacing them from their land and habitat. The trends in commercialization and alienation also contributed to the collapse of community restrictions and customary prohibitions, which regulated forest use in the pre-commercial period.

In the field study-areas (in Bangladesh), forest use today is controlled by an alliance of urban traders and their local agents, local elites, industrial and commercial consumers, and a section of public officials. Commercial use by this "alliance", mainly in the form of illegal logging, dominates forest exploitation in the case-study localities. The commercial interests have been the "winners" in the competition for forest resource use and the "losers" have been the poor and weaker households, who depend on neighbouring forests for subsistence needs. They have been deprived even of their minimal share in forests. A section of poor villagers are used by local elites in illegal commercial logging. Currently, as observed in Chapter 5, majority of the poor local people share a feeling of deprivation and alienation.

Social Forestry's Role in Redressing Alienation

Since the end of the 1970s, apparently in an attempt to redress the sores of public alienation, SF was launched with quite ambitious goals. The goals typically included the promotion of empowerment and social equity by ensuring wider participation of local people in forest management; the upliftment of their income and living standard; and the creation of local institutions for cooperation. SF projects also aimed at equitable distribution of benefits, and meeting the basic needs of local people on a priority basis. However, my investigation into the performance of SF showed that the stated goals remained largely unrealized and they hardly reflected the grandiose initial formulations. Furthermore, I observed that the working-ground or the "frame" of SF was not generally conducive to fulfilling its ambitious stated goals.

The review of the case-studies suggests that SF's achievements have been very limited. Except the marginal rise in income and some degrees of social recognition for the participating farmers, SF has not contributed to any significant large-scale empowerment or qualitative change in the life of farmers. Moreover, it has hardly any major influence or impact on the causes of alienation (as discussed in Chapters 2 and 5). I will now summarize the impacts of SF on the major trends which have alienating affects on local communities. First, SF has not enabled the farmers to establish control over market forces. Farmers are still exposed to varied manipulations by middlemen, local elites, urban traders, loggers, and other commercial forces. Farmers have failed to tap the potential sources of market, to resist artificial price fluctuations (caused mainly by middlemen), and to ensure just price for their (SF) products. Second, though some farmers were allotted land, their decisions about land-use remained conditional upon the government's choice and control. Besides, the participating farmers are a very small section of the local community. The grant of land to a particular (privileged) section has caused discontent among other poor and weak sections of the villagers, who are also equally deserving. In this sense, SF has rather added fuel to inter-group enmity among local people. Third, SF farmers have no control on the working of the commercial logging network (Chapter 5). They are still vulnerable to various manipulations and extortions by the network as before (for example, some farmers have continued to work as low-paid contractual labourer in timber-logging. Trees from their farms are plundered by *mastans*, assigned by the loggers). Fourth, the factional conflicts and fragmentation (among patronage groups and other interest-clusters) and similar dispositions of community conflicts in the case-study localities remain as acute as ever. SF does not seem to have any major bearing on mitigating such conflicts, even within the project-areas. Fifth, the development of a cooperative institution was expected to foster collectivism (vis-à-vis individualism) among farmers. But as discussed in Chapters 6 and 7, such collective spirit was transient and could not be sustained beyond a short time. Cooperation and collaboration among farmers have not been institutionalized by a permanent organizational structure. In summary, SF's performance as a "compensatory dividend" for the process of alienation has only been marginal and superficial.

Social Forestry's Subterranean Agenda

On the face of its limited achievements, the propagandist goals of SF turn out to be largely unrealistic. Why then the protagonists of SF pursue such unpromising goals? In response, Chapter 7 has argued that SF is serving some tacit but strategic purposes for its protagonists -- the national government and international donors. For example, it is enabling the government to regain control over its "lost frontiers" (the state lands which were lost by encroachment and agricultural expansion). The government is reassuming its lost authority over local communities through SF interventions. SF is a part of the government's "strategic compromise" with people whereby the government is trying to save its endangered land and authority. More importantly, SF has proved to be a fertile ground for fetching international funds. Similarly, for the donors SF serves as a flag-carrier, a conveyer of missionary image in the country side, and a way to police the working of the national government. As long as SF fulfils these purposes for its protagonists, it will most likely to prosper, even though the stated goals may remain largely unrealized.

The Road Ahead: Some Points to Ponder

The performance of SF does not present a heartening picture. The social milieu within which SF and other rural development programmes (for example agricultural or health schemes) perform, constrain the growth of participatory and collective organisations. The "inequitable social order" promotes diverse structures of extortion and manipulation, where "personal relations with patrons determine access to resources and power" (Hobley, 1991, p.116), and benefits are siphoned-off to the elites (Chapter 6). These realities reiterate the most common conclusion (on the prospect of collective and participatory initiatives), which maintains that "without revolutionary changes in the socio-economic structure, adequate and effective public participation is not feasible" (Zaman, 1984, p.256), because "ultimately, these structures...determine the success or failure of participatory forms of forest management" (Hobley, 1991, p.117). Such recommendations of basic qualitative change in the social order or comprehensive institutional reform are widely found in major literature on Bangladesh (for example Jansen, 1986, pp.306-318; Sobhan, 1982, pp.17-61; Hartmann and Boyce, 1983, pp.207-209); and one typical example (De Vylder, 1982, p.156) is:

> Government policies are doomed to failure unless and until the exploited majority manages to change the social structure within which they are living and within which these schemes get implemented.

Although I quite appreciate the good intentions of such arguments, these ambitious prescriptions contain vast elements of naivety and romanticism, and therefore, remain largely impracticable. In search of more plausible answers, I followed a rather modest way of asking my respondents (mainly, SF farmers and foresters; also academics, experts, common villagers, local elites, and political leaders) for their ideas and opinions on how to increase the chances of success of participatory programmes, such as SF. In the course of interviews and/or discussions, a wide of range of issues and ideas surfaced. I have organized the responses systematically and noted that many of these ideas conform to the experiences reported in literature around the world. Enumerated below some of the principal suggestions:

Sustaining Innovations through Institutionalization

One chronic problem of development programmes in Bangladesh is that they do not survive after the withdrawal of catalytic support and external fund. The initial enthusiasm evaporates shortly and the success does not last long. The Betagi-Pomora projects are cases in point. As I showed in the earlier chapters, the initial successes (in terms of collective spirit, group-enthusiasm, efficient catalytic services and support, increased production and technological innovations) of the projects had not been institutionalized for perpetuation. The strong catalytic support and patronage by the pioneers, which the projects initially enjoyed, are virtually absent now. Consequently, these projects have become, as the farmers astutely put it, "the dead sons of the government". The common trend in the country's development pursuit is to experiment with a project, and then switch over to a new one without sustaining the former.

One clear recommendation which emerged from my discussion with different quarters is that SF programmes need to be "institutionalized", which "in operational sense refers to the long-term process whereby project activities are continued after external funding (and/or support) have been phased out" (Sadik, 1991, p.246). A host of factors may be needed to foster the process of institutionalization. In context of my case-studies, the following may be considered.

The common tendency of whimsical switching from one project to another, without institutionalizing the projects, contributes to sudden demise of many prospective initiatives. This tendency serves as a major barrier to institutionalization. Ahmed (1986, cited in Task Force, 1987, p.49) also notes that, "innovations are apt to die away with the demise of personalities behind such innovations unless these are institutionalised at the instance of the government". Hence, some forms of long-term catalytic support are required for sustenance of SF projects. Zaman (1984, p.255) also concluded from his extensive survey of rural development programmes in Bangladesh that, "public involvement in rural development programs cannot be attained without the active support of the government". This is neither to suggest that SF projects always have to be spoon-fed, nor that they need to be indefinitely funded by the government. But, as I have repeatedly argued, the surrounding political-economic structures (for example social relations and power structures, market, public polices, government and local government offices, land) which regulate the performance of SF are hostile and unfavourable.

Farmers will require substantial support and services from the government to face these hostilities. In citing examples of such areas where SF farmers may require support, the Task Force (1987, p.92) conclude that the "landless and poor groups require assistance in mediation, conflict resolution or legal action to defend their rights to participate in, and eventually reap the benefits of forestry activities".

For institutionalizing SF projects, their production and activities need to be economically rewarding enough to convince the farmers to solely devote themselves to farm-activities. Fernandes (1983, p.20) argues that in order to create any significant impact on rural poverty alleviation, "it is important to turn afforestation into an income generating enterprise". From the experience of Joint Forestry Management in India, it is contended that, "these programmes can succeed only if the benefit or compensation package offered to the community is financially realistic and appealing to them" (SPWD, 1992, p.10). Similarly, reviewing the Nepalese experience, Malla (1993, p.31) concludes that, "community forestry can no longer remain totally oriented to meeting subsistence needs....The needs of rapidly developing market will also have to be addressed". However, making SF an economically viable production system is a difficult task, which, in turn, will require the following. First, technological innovation through operational research, and proper dissemination of information to farmers. Second, logistic support and services by the government (especially for the marketing and storage of SF products). Third, resolution of the problems of tenurial insecurity and ownership pattern of SF lands. No universal mode of land ownership can be prescribed, simply because it will depend on the specific contextual realities. Participatory research, involving farmers and different groups of villagers, appears to be effective in finding ways of resolving these issues. As regards Betagi-Pomora, for example, a strong suggestion is that, instead of the "privatized" allotment of land to individual farmers, the land ownership may be accorded to the farmers' *Samity* (so that, the farmers as "a group" and, not as "individuals", become the owners); or to the farmers, on leasehold basis, with the whole group (*Samity*) acting as the "collateral". Ahmed and Azad (1987, pp.50-51) also recommend such collective responsibility of land for SF, with the contention that, "in case of an individual member's failure to adhere to the agreed goals and disciplines, the group is made accountable...to see that the land...is used in accordance with the core goals of SF". In the same vein, Shepherd (1992a, p.13) suggests that, "what perhaps needs trying is some arrangement which gives villages adjacent to forest strong but not ultimate rights -- a form of leasehold renewed in return for good stewardship". However, it may be reiterated that, these alternatives are suggestive only, and by no means universal. These options need further exploration.

Creating a Participatory Environment

Although widely used, there is no universally accepted definition of the term "participatory environment" (see for example Gronow and Shrestha, 1991; Chambers, 1983). For my analysis, a conducive context for public involvement in SF (that is, participatory environment) is determined by the performance in two operational areas: the facilitative role of the government, and the performance of local people's ("grassroots") organizations.

At several points in the book, I have shown the limits and failure of the government, particularly of the forest department, in fostering a participatory environment for SF. However, as Zaman (1984, p.255) rightly notes, this does not suggest that "bureaucracy is unessential". Rather, as the situation stands now, the forest department is most likely to continue to occupy the "key position" in SF for many years to come (Task Force, 1987, p.107). Therefore, it is practically irrelevant, if not useless, to suggest that the forest department should be relegated and substituted by an idyllic people's organisation, as an alternative institution for SF management. Given these realities, first, I admitted the central role of the state foresters in SF management; and then, tried to concentrate on the ways to make the forest department more responsive to the needs and challenge of SF in the field. I asked my respondents for suggestions and a variety of responses were collected. A selected list is as follows. First, the forest headquarters may delegate sufficient authority to field offices, empowering field staffs to identify problems and to take initiatives. A system of reward and incentive for innovation and commitment also needs to be devised. Second, there seems to be an immediate need for an institutional arrangement to provide regular training and to coordinate the wide array of actors engaged in SF activities. The need for coordinated action in natural resource management in general and SF in particular has been repeatedly argued in Bangladesh context (see Hasan and Mulamoottil, 1994; Task Force, 1987; BARC, 1991; S. Rahman, 1994). Third, farmers require practical field-oriented ("problem-solving") research and an institutional network for prompt dissemination of the innovations and technologies developed in forest research organizations. Fourth, a departmental system needs to be developed for periodic monitoring and evaluation of SF activities, and for ensuring a system of feedback to accommodate corrective measures. Fifth, the policy proposals are required to be initiated from the field offices through mutual exchange of information (between foresters and farmers) and sending them "up the system" (Detailed accounts of such bottom-up policy-planning and farmer-focused action-research for rural development strategies including SF can be found in Shepherd, 1992a, pp.21-25; Task force, 1987, pp.12-16; Chambers et al., 1989a, Parts 3 and 4; Chambers, 1983, Chapters 7 and 8; Bhuiyan, 1994; Burkey, 1993, pp.56-70; Oakley et al., 1991, Chapter 5). Sixth, foresters need to have an appreciation of farmers' daily life and subsistence strategies. One good way to acquire this insight may be for the forest department to ally with voluntary organizations; because such organizations, as Shepherd (1992a, p.22) argues, "in many countries are currently ahead of forestry departments in their understanding of village life". The possible role of voluntary organizations in SF and the prospective areas of their cooperation with public forestry organizations have been discussed by Magno et al. (1992), Alim (1993), Hoq (n.d.), S. Rahman (1994), BARC (1991), and Ahmed (1994). Seventh, the avenues and incentives for public involvement in forestry need to be enshrined and documented in formal policies and legislations. There is a need for a major reshuffling of policies for SF in Bangladesh. Such a change can also lead to a change in foresters' orientation and values (Gronow and Shrestha, 1991, p.10).

At the macro level, as has been cogently argued by Chowdhry (1992), Gronow and Shrestha (1991), Fernandes and Kulkarni (1983), Shepherd (1992a), and Lai and Khan (1992), a shift of emphasis of forest management from "custodial policing" to villagers' needs and participation is imperative. However, as Chowdhry (1992, p.180) notes, "to change policies and mind-set of over one hundred years does not happen overnight". It will require strong commitment from the country's political leadership, wide interaction between villagers and foresters, de-bureaucratization within government departments, and support of the donors. There can be many ways of achieving these purposes. The "participatory workshops" and other forms of reorientation training have been fruitfully used in some parts of south Asia to dilute the stifling bureaucratic atmosphere within the forestry organisations (for details on these techniques, see Gronow and Shrestha, 1988; 1991). Similarly, a variety of participatory tools of management and research have been developed to promote increased interaction between villagers and government extension agents (see Chambers et al., 1989a; Chambers, 1983; Scoones and Thompson, 1994). Particular training needs of the actors in SF (farmers, foresters, researchers, managers) have already been chalked out for Bangladesh (Task Force, 1987, pp.104-105). More importantly, any policy reform will require coordinated support from donor agencies, which, as Lai and Khan (1992, p.42) suggest, can be achieved by creating a "regular forum for government-donor consultation". Therefore, as the above points show, much of the "ground work" and guide-lines for improving the government's role in SF and creating a participatory environment, already exist, on the basis of which the future efforts (suited to specific needs and context) may be launched.

Once the foresters embark upon an "extension-advisory role", the role of farmers' local organization will become gradually significant. Creating the much needed "vested interest" (Fernandes, 1983; Prabhu, 1983; Gadgil, 1983) among SF farmers will, among other things, require a local organization which can bargain with the central authorities to ensure that farmers have rights to the resources and benefits; they have a voice in the decisions which affect them; and that they receive logistics and services from the government (and other external agencies) on a regular basis.

As previously discussed, the role of farmers' cooperative organization (*Samity*) in my case-studies has been insignificant. Farmers themselves came up with a number of practical suggestions for increasing the effectiveness of their *Samities*. First, some (forest) protection, regeneration and management functions may be transferred to the *Samities*. Such transfer, of course, needs to be backed up by delegating some legal powers as well. In fact, this proposition of transferring selected responsibilities of forest management to local organizations has gained wide currency and has been tested in some countries (see Shepherd, 1992a; Lai and Khan, 1992; Gadgil, 1983). The rationale is that local management can reduce cost and create public-enthusiasm; and also that "in view of the limited resources of the Forest Service and the problematic situation concerning recurrent costs and sustainability, much more responsibility must be shared by local population" (Lai and Khan, 1992, p.41). Shepherd (1992a, p.23) also opines that, "it may make good sense...to deploy scarce foresters as trainers of village-level extension workers" and to let village workers play the major role in forestry development in the country-side. Second, farmers have suggested that there needs to be regular election for the *Samity* leadership, and that the leadership is required to represent the different classes of farmers. Hobley (1991, pp.116-117), from her survey of Nepal SF, has also concluded that, the forestry "committees" are not likely to be effective without broad-based representation of all groups of farmers. Third, SF farmers will require catalytic and advisory supports from the government and other external agencies to strengthen their *Samities*.

The above suggestions for improvement need further exploration and have to be applied in consideration of the specific context. Because, as Runge (1986, p.49) argues, "if a variety of responses to problem of resource management are possible, the incentive leading to a particular institutional choice must result from the physical and social environment in which the choice is made".

Some Research Implications

There is a sizeable number of (both investigated and prospective) research areas of this study. For convenience of discussion, three broad areas of enquiry may be traced, based on the arrangement of chapters of this book. First, the "study of alienation" which concerns itself with the historical and contemporary alienating processes, and their implications for forest use (Chapters 2 and 5). Second, the "analysis of the frame of SF" which deals with the examination of the surrounding environment (and the political-economic variables therein) which regulates the performance of a particular SF project (Chapters 3 and 6). Third, the "impact and performance appraisal of SF intervention" which evaluates the level and significance of SF's achievements in light of its stated objectives (Chapters 3 and 7).

The Study of Alienation in Forest Resource Use

The study of alienation remains a relatively less-explored area in the analysis of forest resource use. To my knowledge, very little work have been done from this particular perspective in explaining the problems of forest use. Occasional indication and discussion on the process of alienation and its implications are available in a number of studies (see for example Ribot, 1995; Guha, 1989; Gadgil, 1983; 1989; Fernandes et al., 1988; Nadkarni et al., 1989; Fernandes, 1983; Shiva et al., 1983; Joshi, 1983; Haimendorf, 1982). However, we still await a substantial text which pulls together the many strands of the process of alienation, both as a historical process, and as an academic perspective of research on forest resource use. For the purpose of this book, I found this perspective very helpful. The study attempted to analyze forestry and SF from a historical perspective. The history of the case-study localities illuminated the process of public alienation from forest management. The study of alienation has been vital for the analyses of SF (and forest resource use in general). It helped to shed light on the background of the emergence of SF by focusing on, first, the progressive commercialization of forests and the diminution of local rights; second, the consolidation of enmity between local people and the state; and third, the failure of state efforts in forest conservation and the gradual realization among the governmental quarters that forests could not be managed singlehandedly. The study of alienation, thus, focused on the stage which led to the rise of SF and the political-economic trends which, over the years, have caused people's exclusion from forests.

The Study of the Frame and Performance of Social Forestry

The possible enquiries in this research area are particularly diverse and wide-ranging. It is nearly impossible to bring these prospective issues of enquiry under a single umbrella of discussion. However, during the course of field work, I felt that some areas are more clearly noticeable than others, and offer immediate opportunities for further exploration. The implications of patronage network and caste arrangements, for example, are crucial for SF and there are many probable research questions on these social aspects (see Appendix 3). Similarly, the different land tenurial and market arrangements in context of a particular SF project need careful research. Given the widespread inequity in the social system, SF's role in reaching the disadvantaged sections of rural communities or promoting social equity, can be a subject of rigorous investigation. As this study suggests, a summary of some prospective research areas on Bangladesh SF can be found in Appendix 3.

Going for the Blind Uncle

SF, as it is manifested through the evidences and case-studies of the book, portrays a bureaucratic version of public participation. It is yet another technical programme, planned and administered by the professionals, with the stated aim of "benefiting the people". Here, "participation" is an "invitation" to people to take part in a technically-oriented bureaucratic afforestation programme, where even the so-called "benefits for people" are also conceived and designed by the professionals. The government's flirtation with SF seems to be a part of the "strategic compromise" with local communities to vanguard its endangered territories and to reclaim the lost authority (Chapters 6 and 7).

However, having said that, SF has certainly offered some benefits (mainly in the forms of cash income and some degree of social recognition) to the participating farmers, who previously had nothing but a long trail of deprivation and alienation. Moreover, besides the financial achievements, SF has got a number of attractive potentials. First, in recent years, SF has successfully attracted the much-needed commitment from the country's political leadership. The Bangladesh prime minister herself, has called the nation "to turn afforestation into a social movement"; and it is hoped that her commitment in SF campaign "will pave the way for a wide social mobilization for afforestation and formulation of appropriate government policies" (GOB, 1992, p.12). Second, even in a land-scarce country like Bangladesh, a sizeable land area may still be available for SF purposes, as estimated in the following table:

Table 8.1 Marginal or strip lands potentially available for social forestry (in million hectares)

Marginal land beside roads and embankments	0.08
Homestead lands	0.27
Privately owned hillocks	0.05
(Fallow) land under the tea-estates	0.11
Degraded public forests and other lands	1.00
Total	1.51

Source: Bhuiyan, 1994, p.35 (Translation).

In view of the country's total land and forest areas, the prospective land seems significant; and if fruitfully brought under SF, it is expected to produce "dramatic results" (Bhuiyan, 1994, p.35). Third, as discussed in the previous section, the needs, constraints and suggested remedies have been widely analyzed for SF development in Bangladesh. Much of the "ground-work" already exists. The base is there, on which possible action can be initiated.

Thus SF, even with all its limitations, still remains a prospective alternative development strategy for the country's forestry sector in crisis. However, as Banerjee (n.d., p.25) rightly notes,

> SF is not to be projected as a panacea for all...economic and environmental ills in the rural area, but to the extent it helps the impoverished rural people to lead a somewhat less strenuous and more dignified life, the overall objective would well have been fulfilled.

Instead of the pomp and grandeur which often surround SF projects, a modest down-to-earth approach is desirable today. Interestingly, farmers seem to have grasped this sense of practicality and judiciousness quite well. They many a times reminded me of the oft-quoted Bengali proverb which connotes: "if you don't have a caring uncle, it is preferable to go even for a blind uncle than to have none". For the farmers, as it became clear to me in the course of intimate interviews, participating in a SF project, is a choice between "getting something" and "getting nothing" (or as the farmers quaintly put it, a choice between having a "blind uncle" and having "no uncle" at all). To this group of historically deprived and desperately needy people, SF at least serves the purpose of the "blind uncle", and which, by itself, is no little contribution.

Appendices

Appendix 1. Bangladesh: A Profile of Basic Information

Geographic location:	Lies in the north eastern part of South Asia between 20 34' and 26 38' north latitude and 88 01' and 92 41' east longitude. A deltaic country, formed at the confluence of three mightiest river systems of the world -- the Ganges, Bramaputra and Meghna.
Land area (in sq.miles):	55598
Climate:	Sub-tropical monsoon climate. While there are six seasons, Winter, Summer and Monsoon are prominent. Average annual rainfall varies from 1429 to 4338 ml. metre. Winter temperature ranges from minimum of 7.22-12.77 [degrees] celsius to maximum of 23.88-31.11 degrees. Summer temperature ranges from maximum 34 degrees celsius to minimum 21 degrees.
Population:	Population: 120 million (approx.). Growth rate: 2.17 per annum. Density of population per sq. mile: 1917. Literacy rate: 24.8% (for population 5 years and above). There are 15.1m Households in the country spread over 60315 *mouzas* (revenue villages)
Per capita income (at factor cost; base:1984-85=100):	Taka 4333 (in 1991-1992) (1 GBP= 60 Taka, approx.)

Sectoral share of forestry to GDP (gross domestic products): GDP (in million Taka): Sectoral share of forestry (in million Taka):	2.46% (1991-1992) 534820 (1991-1992) 13147 (1991-1992)
Total cultivable land: Per capita cultivated land holding:	9.03 ha (hectare). 0.08 ha.
Total annual wood production: Annual per capita consumption of timber and fuel wood:	9.6m cu m (cubic meter) 0.1 and 0.05 cu m (respectively)
Labour force employed in the forestry sector:	2% of the total labour force
Administrative set-up:	Parliamentary form of government, with Prime Minister as the chief executive. The country is divided into 6 administrative Divisions, 64 Districts and 489 *Thanas*. A Division consists of a number of Districts. Division and District are primarily under the authority of career bureaucrats known as Divisional Commissioner and Deputy Commissioner respectively. A District is sub-divided into *Thanas*. A *Thana* (known as *Upazila* before 1991) consists of some Unions. *Thana* and Union are headed by respective Chairmen, elected by universal adult franchise; and these are the major local government institutions in the country. *Thana* Chairman is assisted by a cadre of public officials from various departments of the bureaucracy.

Source: BBS, 1992; Hussain, 1992; Magno and Ali, n.d.; GOB, 1990.

Appendix 2. Sample Size Calculation for the Field-Survey

Chandra

Required sample size: $n = (ts/d)^2$

P (margin of error) = 0.15
Y (mean)= 16270 (calculated from pilot survey)
s (standard deviation) = 4393.44 (-ditto-)

t (40, .05) = 2.02
d = P/100 x Y
= .15 x 16270
= 2440.5
Therefore, $n = (2.02 \times 4393.44/2440.5)^2$
= 13.22 (13)

Income distribution in Taka	Population size (Ni)	Sample size (ni)
9000 - 12000	10 (N1)	3
12001 - 15000	12 (N2)	4
15001 - 18000	10 (N3)	3
18001 - 25000	8 (N4)	3
Total:	40	13

n1 = N1/N x n = 10/40 x 13 = 3.25 (3)

similarly, n2 = 3.9 (4)
n3 = 3.25 (3)
n4 = 2.6 (3)

Pomora

Y = 22100; s = 10293.45; d at 15% = 3315 (d = P/100 x Y)
t = 2.04
$n = (ts/d)^2$
$= (2.04 \times 10293.45/3315)^2$
= 40

Income distribution in Taka	Population size (Ni)	Sample size (ni)
4000 - 15000	59 (N1)	17 (n1)
15001 - 30000	62 (N2)	18 (n2)
30001 - 40000	10 (N3)	3 (n3)
40001 - 60000	7 (N4)	2 (n4)
Total :	138	40

n1 = N1/N x n = 59/138 x 40 = 17.10 (17)
similarly, n2 = 17.97 (18)
n3 = 2.90 (3)
n4 = 2.02 (2)

Betagi

Y = 39160; s = 14339; d at 15% = 5874 (d = P/100 x Y)
t = 1.96 (2)
n = $(ts/d)^2$
= $(2 \times 14339/5872)^2$
= 23.84 (24)

Income distribution in Taka	Population size (Ni)	Sample size (ni)
7000 - 30000	21 (N1)	7 (n1)
30001 - 40000	20 (N2)	6 (n2)
40001 - 50000	19 (N3)	6 (n3)
50001 - 60000	15 (N4)	4 (n4)
60001 - 80000	2 (N5)	1 (n5)
Total :	77	24

n1 = N1/N x n = 21/77 x 24 = 6.55 (7)
similarly, n2 = 6.23 (6)
n3 = 5.92 (6)
n4 = 4.36 (4)
n5 = 0.63 (1)

Appendix 3. A Selection of Possible Areas of Research on Bangladesh Social Forestry (as Suggested by the Book)

This table summarizes some of the prospective areas of research on Bangladesh SF. Some of the questions are partially covered by this book and the rest provides clues for future investigation. However, the following list of research-areas and primary references is by no means exhaustive and may kindly be treated as merely indicative.

Suggested areas	Probable nature of enquiry: Few examples
1. Patronage network in rural Bangladesh and its implication for SF projects	* How does the patron-client network relates to SF? * What are its socio-economic and political implications for the performance of SF? * Does these ties delimit the prospect of cooperative or collective action for SF management? (see for example Scott, 1972; 1985; Khan, 1996; Hobley, 1991; Zaman, 1984; Alavi, 1973; Sarkar, 1980; Mashreque and Amin, 1993)
2. Social Caste system and its implication for SF	* Is caste a barrier to social and public involvement in participatory forestry projects? * Does caste contribute to non-egalitarian access to SF production and distribution? * Is there a correlation between caste-status and productive performance in SF? (see for example Saxena, 1990; Mukherjee, 1971; Karim, 1956; 1981; Chowdhury, 1978; Ali, 1985)
3. Factionalism, party-politics and group-alignment in rural Bangladesh and their implications for SF	* Do "factional politics", "interest-alignment" and "confrontation" affect the performance of SF in its different work-areas (such as selection, recruitment and training of the farmers; decision-making and benefit-sharing arrangements; marketing of produces)? * Can SF "keep clear" of the factional politics? (see for example Zaman, 1984; Jahangir, 1982; Mashreque, 1982; Islam, 1974; Zaman, 1977; Blair, 1978)

4. Social inequality and its implication for SF	* Can there be "distributive justice" in a SF project, or can SF reach the poorest stratum of the communities given the general inequality of Bangladesh society? * Is inequality contributing to the skewed pattern of asset accumulation from SF production? (see for example Barraclough and Ghimire, 1990; Uphoff, 1982; ILO, 1977; Bandhu and Garg, 1986)
5. Land relations in SF	* Is (and to what extent) land a motivational factor in SF? * Is there a correlation between land tenurial security and productive efficiency of SF? * Is "giving away" government land "socially justifiable"? * Is privatization of land (as it relates to SF) desirable? (see for example Lynch, 1992; Januzzi and Peach, 1980; Reddell and Fortmann, 1983; Rahman, 1992; Brokensa and Castro, 1984; Bruce, 1989)
6. Market dynamics in SF	* Are the farmers in control of the market for SF? * Do the markets ensure proper return for SF production? * To what extent the rise of intermediate forces (such as the middlemen) affects production and distribution in SF initiatives? * Is collective marketing more desirable than privatization or state monopolization of market? (see for example Gray, n.d.; Foley and Barnard, 1984; Shah, 1988)

7. Bureaucratic responses to SF	* How do the concerned state officials perceive SF and do their perceptions affect the performance of SF? * How do they respond to the changing demands of participatory approaches to forest and environment management? * How effective is the public officials' role as the "motivator" or "catalyst" in SF development? * Is there a change emerging in the traditional elitist orientation of the officials? (see for example Korten and Uphoff, 1981; Anderson and Huber, 1988; Chambers et al., 1989; Ahmed, 1980; Guha, 1988; Foley and Barnard, 1984; Poffenberger, 1990; Korten and Alfonso, 1983)
8. Role of public policy (as it relates to SF)	* To what extent the government policies inhibit or facilitate the performance of SF at the micro and macro levels? * What are the policy implications for SF; and the (suggestive) remedies for mitigating the discrepancies between policy and field-realities? (see for example Poffenberger, 1990; 1990a; Ali, 1991; 1991a; Fernandes and Kulkarni, 1983; Hossain, 1992; Repetto and Gillis, 1988)
9. Prospect of cooperative and collective action in SF management	* What political-economic factors contribute to cooperation or collective action for SF management? * Can these factors be perpetuated or replicated through a proper institutional structure at a general level (vis-a-vis a particular micro location)? (see for example Wade, 1988; Curtis, 1991; Poffenberger, 1990a; ILO, 1988; Korten, 1980)

10. Gender in SF	* Is there a correlation between gender and productive efficiency of SF? * Is SF "gender-neutral"? What are the manifestations of gender inequality in SF? * Comparative performance and contribution of male and female workers in SF. * Is women's participation contingent on the prevailing social surroundings (that is, the "frame") of a SF project? (see for example Wickramasinghe, 1992; Dean, 1986; Ahmed, 1992; Khandakar, 1994)
11. Local institutions and their relation to SF	* What are the attitude and perceptions of local government leaders towards SF (and the implications thereof)? * Can the existing local institutions carry out the tasks of SF? An assessment of the role of local institutions in mass-scale SF interventions in rural Bangladesh. * Are the local institutions suitable for innovation and experimentation in rural social development ? (see for example Blair, 1982; Korten, 1980; Uphoff, 1982; 1986)
12. Farmers' response to risk and uncertainty in SF	* How do farmers allocate scarce resources in the face of disasters and crises? * Changing trends in farmers' behaviour under condition of risk and contingency. (see for example Wade, 1988; Shahabuddin, 1991)
13. Farmers' response to the transferred technologies and innovations	* How do the farmers react to innovations and new technology? * Tracing the social factors which lead to the acceptance or rejection of a particular (new) technology or innovation by the farmers. * Social credibility and acceptability of innovations; or the "social face" of innovation (see for example Cernea, 1985; Chambers et al., 1989a)

Appendix 4. Photographs

Photograph 1. A typical homestead at the Chandra SF project

Photograph 2. Women working in road construction scheme in Chandra

Photograph 3. A poor homestead at the Pomora project
Note: The common cooking fuel sources are barks, leaves, twigs, dried shrubs. Farmers rarely use wood as fuel -- a luxury which they can not afford.

Photograph 4. Illicit supply of fuel-wood from Chittagong Hill Tracts
Note: The wood comes through river up to Rangunia, and then loaded into trucks for its ultimate journey to the illegal timber depots at Lamburhat and Balirhat in the Chittagong city-suburb (Location: Ishakhali Bridge, Rangunia).

Photograph 5. A hill fountain (locally called *Ush*) in Pomora
Note: There are about five Ush in the Pomora locality, which are the common sources of water for farmers. Water scarcity is acute in the project area.

Photograph 6. A typical tea-stall at Betagi
Note: Tea-stalls are one of the major consumers of fuel-wood grown in the SF projects.

Photograph 7. An abandoned site of Forest Patrol Office (Surat Singh Dhala, Pomora, Rangunia)

Note: This was a result of the intense conflict between local people and the foresters. After the retreat of foresters, two families have built new houses on this public land under the patronage of local elites.

Photograph 8. The aftermath of an "eviction operation"

Note: A family (in Surat Singh Dhala, Pomora, Rangunia) was evicted by the government for "encroaching" on public land. The photo was taken immediately after the "eviction operation" whereby the whole homestead (including house, trees and gardens) was levelled down. The family tried in vain to build this temporary structure in an attempt to signal their claim on the plot where they had lived for about 13 years.

Appendix 5. Organizational Chart of the Bangladesh Forest Department

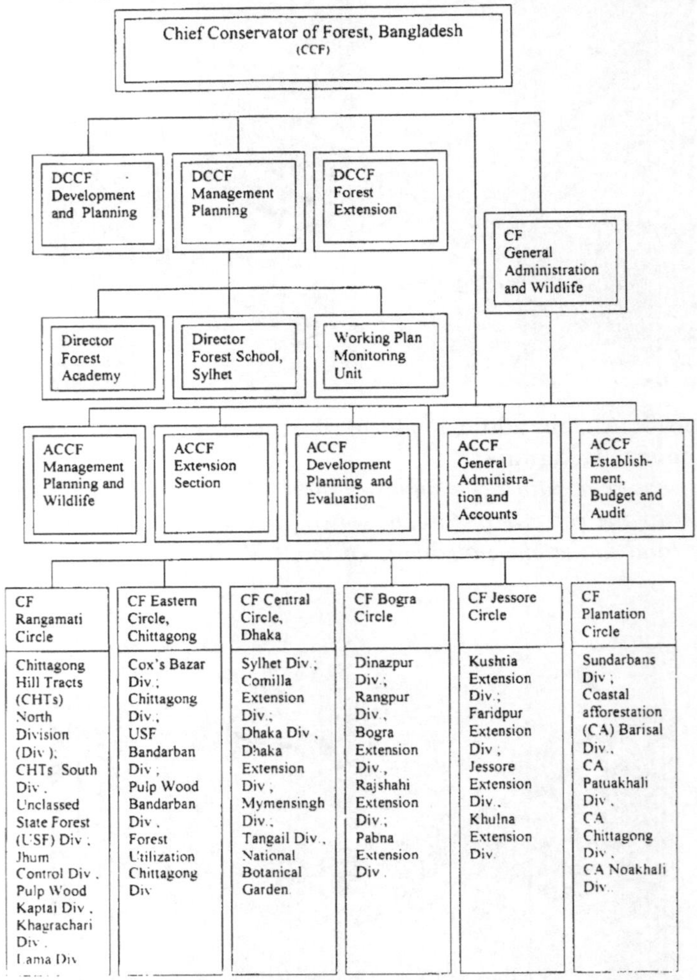

*Notes: * DCCF = Deputy Chief Conservator of Forests; CF = Conservator of Forests; ACCF = Assistant Chief Conservator of Forests. * A Forest Circle, headed by a CF, consists of a number of Forest Divisions. Each Division is headed by a Divisional Forest Officer (DFO), and is administratively divided into some Forest Sub-divisions, Ranges and Beats. The DFO is assisted by a number of Assistant Conservators of Forests (ACF), Forest Range Officers (FRO), and Beat Officers (BO) in the administration of Sub-divisions, Ranges, and Beats respectively.*

Bibliography

Books, Theses, Journals, Government and Archival Publications

Abdullah, T.A. and Zeidenstein, S.A. (1982) *Village Women of Bangladesh: Prospects for Change*, Pergamon Press, Oxford.

Abedin, M.Z. and Quddus, M.A. (1991) "Agroforestry Systems in Bangladesh with Particular Reference to Economics and Tenurial Issues, In: Mellink et al., (eds.), *op.cit.*, pp.13-33.

Abedin, M.Z.; Lai, C.K. and Ali, M.O. (eds.) (1990) *Homestead Plantation and Agroforestry in Bangladesh*, (proceedings of a National Workshop held in July 17-19, 1988, Joydebpur), Bangladesh Agricultural Research Council, and Winrock International, Dhaka.

ADAB (Association of Development Agencies in Bangladesh) (1992) Environment and Development: Bangladesh NGOs' Perspective on Policy and Action, A Position Paper for UNCED (June 1992, Rio), Dhaka.

ADB (Asian Development Bank) (1989) *Appraisal of the Upazila Afforestation and Nursery Development Project in Bangladesh*, Internal Report, ADB Project Lap:BAN 16087, Dhaka.

Adhikari, J. (1990) "Is Community Forestry a New Concept? An Analysis of the Past and Present Policies Affecting Forest Management in Nepal", *Society and Natural Resources*, 3:257-266.

Agarwal, A. and Narain, S. (1985) *The State of India's Environment 1984-85: The Second Citizen's Report*, Centre for Science and Environment, New Delhi.

Agarwal, B. (1986) *Cold Hearths and Barren Slopes: The woodfuel Crisis in the Third World*, Zed Books, London.

Ahamed, F.U. (1993) *Responding to Environmental Degradation: Some Implications of a Social Forestry Project in Bangladesh*, M.Phil. Dissertation, Selwyn College, University of Cambridge, Cambridge.

Ahmad, A.N.M.M. (1991) "Women's Participation in Politics: Bangladesh Perspective", *Chittagong University Studies*, 12(1):1-18.

Ahmad, K.M. (1990) *Agroforestry, Fuelwood and Rural Development in Bangladesh*, M.Sc. Dissertation, School of Agricultural and Forest Sciences, University College of North Wales, Bangor.

Ahmad, Q.H. (1985) "Female Employment in Bangladesh: A Review of Status and Policy", In: Q.H. Ahmad, M.A.Khan, S. Khan, and J.A. Rahman, (eds.), *Situation of Women in Bangladesh, Ministry of Social Welfare and Women's Affairs*, Dhaka, pp.19- 64.

Ahmad Y.S. (1938) *Working Plan for the Dacca-Mymensingh Forest Division (1937-38 to 1957-58)*, Government of Bengal, Bengal Government Press, Alipore.

Ahmed A.U. (1986) "Issues in Social Forestry: The Betagi-Pomora Case, A Personal Account", Paper presented at *Workshop on Social Forestry*, Bangladesh Centre for Advanced Studies, (held in Ford Foundation Guest House), September 12-14, 1986, Dhaka.

Ahmed, A.U. and Azad, A.K. (1987) *Social Forestry in Bangladesh: A Case Study of Betagi and Pomora Experiments*, University of Chittagong, Chittagong.

Ahmed, B. (1989) "Sociological Aspects of the Study Villages", In: M.G. Sattar (ed.) *Six Villages of Bangladesh: Benchmark Situation of Six Villages in Different Agro-ecological Zones of Bangladesh*, Bangladesh Academy for Rural Development, Comilla, pp.17-41.

Ahmed, B. (1994) *Benchmark and Other Issues*, Agroforestry Research and Demonstration Project, Forest Department, Dhaka.

Ahmed, E. (1980) *Bureaucratic Elites in Segmented Economic Growth: Bangladesh and Pakistan*, University Press Limited, Dhaka.

Ahmed, M.R. (1987) "Social Forestry: The Key to Socio-economic Development of the Rural People", (in Bengali) *The Daily Sangbad*, December 28, 1987, Dhaka.

Ahmed, M.R. (1991) "Social Forestry: Theories, Concepts and Applications", Paper presented at seminar on *Management of Social Forestry Projects*, February 17-20, 1991, World Food Programme, Dhaka.

Ahmed, M.R. (1992) "The Unseen Workers: Socio-cultural Profile of Women in Bangladesh Agriculture", *Society and Natural Resources*, 5(4):375-390.

Ahmed, M. R. (1993) "Community Forestry Development in Bangladesh: Constraints, Priorities and Strategies", Paper presented at the *Forests, Trees and People Programme Review and Planning Meeting*, December 14-17, 1993, Regional Community Forestry Training Centre, Bangkok.

Ahmed, M.R. (1994a) "Approaches and Roles of NGO's in Social Forestry Development in Bangladesh", In: M.R Ahmed, (ed.), *ibid*, pp.77-82.

Ahmed, M.R. (n.d.) *Social Forestry: Some Issues of Concepts,Theories and Applications*, mimeo, Institute of Forestry, Chittagong.

Ahmed, M.R. (ed.) (1994) *Social Forestry and Community Development*,

Proceedings of a national workshop, held in October 5-10, 1991, Institute of Forestry, Chittagong.

Ahmed, M.R. and Bhuiyan, M.K. (1993) "Seeking New Dimension in Policy and Legislation for Community Forestry Development in Bangladesh", Paper presented at the Regional Workshop on *Policy and Legislation on Community Forestry*, January 27-29, 1993, Regional Community Forestry Training Centre, Bangkok.

Ahmed, R.U. (1988) *A Review of The Management of the Forests of Bangladesh*, B.Sc (Honours) Review Paper 2, Institute of Forestry, Chittagong.

Ahmed, T. (1990) "The Third World Dilemma of Social Work and Social Services: A case Study on Rural Social Service Model from Deedar", Paper presented at the *Second Workshop of the European Network of Bangladesh Studies*, September 1990, Bath.

Ahmed, T. (1991) *Decentralisation and The Local State under Peripheral Capitalism: A Study in the Political Economy of Local Government in Bangladesh*, Ph.D Thesis, Centre for Development Studies, University of Wales, Swansea.

Ahsan, K. (1991) "Social Forestry and its Usefulness" (in Bangali), *Palli Unnayan* (Rural Development Journal), January 1991:90-96.

Akhter, M.Y. (1989) "Patron-Client relations in Bangladesh Organizations", *Administrative Change*, 27(1):28-41.

Alam, A.M.Q. and Rahman, M. (1993) "Foreign Assistance and Industrialization in Bangladesh", *Administrative Change*, 20(1-2):61-82.

Alam, M.S. (1992) *The Forest Act*, Pallab Publishers, Dhaka.

Alavi, H. (1973) "Peasant Classes and Primordial Loyalties", *The Journal of Peasant Studies*, 1.

Ali, A.F.I. (1985) "The Concept of Caste: A New Perspective", *Chittagong University Studies (Social Science)*, 8(1):47-75.

Ali, S.I. (1987) "Trees and Tenure in Bangladesh: Some Legal Aspects", In:

S. Huq, (ed.), *Trees and Tenure in Bangladesh*, Proceedings of a discussion meeting held on December 10, 1987, Bangladesh Centre for Advanced Studies, Dhaka, pp.14-16.

Ali, S.S. (1991) "Policy Issues for Social Forestry in Bangladesh", Paper presented at *Social Forestry and Community Development Workshop*, October 5-10, 1991, Institute of Forestry, Chittagong.

Ali, S.S. (1991a) *General Policy Issues on Social Forestry/Agroforestry Projects*, mimeo, Forest Department, Dhaka.

Alim, A. (1984) *Integrated Rural Development, Protection of Forest, Use of*

Forest Soil Potential and Connection of Rural Poor to Waste and Marginal Land (Social Forestry), unpublished paper, Institute of Forestry, Chittagong.

Alim, A. (1988) *Forestry with the People and for the People*, Institute of Forestry, Chittagong.

Alim, A. (1993) "Afforestation and Environmental Conservation: The Role of Public Media" (in Bengali), *Adhuna*, 2(2):10-19.

Almond, G.A. and Verba, S. (1963) *The Civic Culture: Political Attitudes and Democracy in Five Nations*, Princeton University Press, Princeton, New Jersey.

Anderson, R.S. and Huber, W. (1988) *The Hour of The Fox: Tropical Forests, The World Bank and Indigenous People in Central India*, University of Washington Press, Seattle and London.

Anisuzzaman, M. (1979) *Bangladesh Public Administration and Society*, Bangladesh Books International Limited, Dhaka.

Anonymous (1991) *An Overview of Community Forestry with Special Reference to the Manikganj Community Forestry Project*, B.Sc.(Honours) Review Paper 154, Institute of Forestry, Chittagong.

Anonymous (1991a) *Status of Social Forestry in Bangladesh*, B.Sc.(Honours) Review Paper 138, Institute of Forestry, Chittagong.

Anonymous (1994) *An Investigation into The Ecology of Chandra Sal (Shorea robusta Gaertn) Forest, Gaziur, Bangladesh*, M.Sc. Thesis, Jahangir Nagar University, Dhaka.

Anonymous (n.d.) *Forest Policy of Bangladesh: Seeking For New Directions*, B.Sc. (Honours) Review Paper 179, Institute of Forestry, Chittagong.

APD (Amended Project Document) (1989) *Amended Project Document II: Bihar Social Forestry Project for Chotanagpur and Santhal Parganas, 1988/89 to 1990/91*, March 1989, Bihar Social Forestry Project, Bihar.

Appadurai, A. (1986) "Class, Cooperation and Well-Irrigation in Western India", In: D.W. Attwood, M. Israel, and N. Wagle, (eds.), *Dynamics of an Agrarian Society: Maharashtra, India*, University of Toronto Press, Toronto.

ARDP (Agroforestry Research and Demonstration Project) (1989) *Project Paper*, ARDP, Ministry of Agriculture and Forests, and Ford Foundation, Dhaka.

ARDP (Agroforestry Research and Demonstration Project) (n.d.) *Project Paper*, ARDP, Ministry of Environment and Forests, and Ford Foundation, Dhaka.

ARDP (Agroforestry Research and Demonstration Project) (n.d.[a]) *Technical Assistance Project Proposal*, ARDP, Ministry of Environment and

Forests, and Ford Foundation, Dhaka.

Arnold, J.E.M (1990) *Beyond Community Woodlots: Programmes with Participation*, Social Forestry Network Paper 11(e), Overseas Development Institute, London.

Arnold, J.E.M (1990a) *Social Forestry and Communal Management in India*, Social Forestry Network Paper 11(b), Overseas Development Institute, London.

Arnold, J.E.M. (1993) "Management of Forest Resources as Common Property", *Commonwealth Forestry Review*, 72(3):157-161.

Arnold, J.E.M and Stewart, W.C. (1989) *Common Property Resource Management in India: Report to the Indian Agriculture Division*, World Bank, India.

Arnold, J.E.M and Stewart, W.C. (1991) *Common Property Resource Management in India*, Tropical Forestry Paper 24, Oxford Forestry Institute, Oxford.

Arnold, J.E.M., Bergman, A. and Djurfeldt, G. (1988) *Forestry for the Poor? An Evaluation of the SIDA Supported Social Forestry Project in Tamil Nadu, India*, SIDA Report 1987/8, Swedish International Development Authority, Stockholm.

Arnold, J.E.M., Bergman, A., and Djurfeldt, G. (1990) *Evaluation of the SIDA Supported Bihar Social Forestry Project for Chotanagpur and Santhal Parganas, India*, Swedish International Development Authority, Stockholm.

Asaduzzaman, M. (1989) *Social Forestry in Bangladesh*, Research Report 115, Bangladesh Institute of Development Studies, Dhaka.

Attwood, D.W. (1988) "Social and Political Pre-conditions for Successful Co-operatives: The Co-operative Sugar Factories of Western India", In: D.W. Attwood and B.S. Baviskar, (eds.), *op.cit.*, pp.69-90.

Attwood, D.W. and Baviskar, B.S. (1988a) "Introduction", In: D.W. Attwood and B.S. Baviskar, (eds.), *ibid*, pp.1-20.

Attwood, D.W. and Baviskar, B.S. (eds.) (1988) *Who Shares? Cooperatives and Rural Development*, Oxford University Press, Delhi.

Bachkheti, N.D. (1984) *Social Forestry in India: Problems and Prospects*, Birla Institute of Scientific Research, Economic Research Division, Radiant Publishers, New Delhi.

Bajracharya, D. (1983) "Deforestaion in the Food/Fuelwood Context: Historical and Political Perspective from Nepal", *Mountain Research and Development*, 3:227-240.

Bandhu, D. and Garg, R.K. (1986) *Social Forestry and Tribal Development*, Indian Environmental Society, New Delhi.

Bandyopadhyay, S. and Eschen, D. V. (1988) "Villager Failure to Co-operate: Some Evidence from West Bengal, India", In: D.W. Attwood and B.S. Baviskar, (eds.), *op.cit.*, pp.112-145.

Banerjee, U. (1989) "Participatory Forest Management in West Bengal", In: WBFD, *op.cit.*, pp.1-4.

Banerjee, U. (n.d.) *Social Forestry Management in West Bengal: Experiences and Issues*, West Bengal Forest Department, Calcutta.

Banko Janakari (1993) *Banko Janakari: A Journal of Forestry Information for Nepal*, Vol. 4(1), March 1993, Forestry Research Division, Kathmandu.

Baral, N.R. (1993) "Where Is Our Community Forestry?", *Banko Janakari*, 4(1):12-15.

Baran P.A. (1992) "On The Political Economy of Backwardness", In: C.K. Wilber and K.P. Jameson, (eds.), *op.cit.*, pp.94-106.

BARC (Bangladesh Agricultural Research Council) (1991) *Bangladesh Agroforestry Plan 1990-1995: An Agenda for Policy, Research and Action*, BARC, Dhaka.

Barraclough, S. and Ghimire, K. (1990) *The Social Dynamics of Deforestation in Developing Countries: Principal Issues and Research Priorities*, Discussion Paper 16, United Nations Research Institute for Social Development, Geneva.

Barth, F. (1981) *Selected Essays of Fredrik Barth, Vol. 1-2*, Routledge and Kegan Paul, London, Boston and Henley.

Bartlett, A.G. (1992) "A Review of Community Forestry Advances in Nepal", *Commonwealth Forestry Review*, 71(2):95-100.

Bartlett, A.G. and Malla, Y.B. (1992) "Local Forest Management and Forest Policy in Nepal", *Journal of World Forest Resource Management*, 6:99-116.

Bartlett, A.G., Nurse, M.C., Chhetri, R.B. and Kharel, S. (1992) "Towards Sustainable Community Forestry: An Evaluation of Community Forestry Through User-Groups in Central Nepal", Paper presented at a conference on *Sustainable and Effective Management Systems for Community Forestry*, January 15-17, 1992, Regional Community Forestry Training Centre, Bangkok, pp.55-68.

Baten, S.A. (1969) *Revised Working Plan for the Forests of Chittagong Division (for the period from 1968-69 to 1977-78)*, Forest Department, Dhaka.

Baviskar, B.S. (1988) "Dairy Co-operatives and Rural Development in Gujarat", In: D.W. Attwood and B.S. Baviskar, (eds.), *op.cit.*, pp.345-361.

BBS (Bangladesh Bureau of Statistics) (1989) *Small Area Atlas of Bangladesh*

(Chittagong District), BBS, Ministry of Planning, Dhaka.

BBS (Bangladesh Bureau of Statistics) (1990) *Report on Survey of Farm Forestry, 1988*, BBS, Ministry of Planning, Dhaka.

BBS (Bangladesh Bureau of Statistics) (1992) *Statistical Yearbook of Bangladesh, 1992*, BBS, Ministry of Planning, Dhaka.

BDR (Bengal District Records) (1978), *Bengal District Records, Chittagong, Vol.1:1760-1787*, S. Islam, (ed.), Dhaka University, Dhaka.

Beck, T. (1992) "Survival Strategies and Power Amongst the Poorest in a West Bengal Village", In: C.K. Wilber and K.P. Jameson, (eds.), *op.cit.*, pp.478-497.

Begum, N.N. (1988) *Pay or Purdah: Women and Income Earning in Rural Bangladesh*, Winrock International Institute for Agricultural Development, and Bangladesh Agricultural Research Council, Dhaka.

Beldt, R.V.D., Hatch, C., Sastry, C., Raintree, J., Ghosh, S. and Saxena, N.C. (1994) "Farm Forestry Comes of Age: The Emerging Trend in Asia", *Farm Forestry News*, 6(2):1-7.

Bell, J. (1987) *Doing Your Research Project: A Guide for First-time Researchers in Education and Social Science*, Open University Press, England.

Bensel, T. and Remedio, E. (1994) "Woodfuel Markets in Cebu: Link to Deforestation or Reforestation?", *Farm Forestry News*, 6(2):7-9.

Bentley, W.R., Singh, G.B., and Chatterjee, N. (1987) "Tenure and Agroforestry Potentials in India", In: J.B. Raintree, (ed.), *op.cit.*, pp.231-243.

Bertocci, P.J. (1976) "Rural Development in Bangladesh: An Introduction", In: R.D Stevens, H. Alavi, and P.J. Bertocci, (eds.), *Rural Development in Bangladesh and Pakistan*, East West Centre, Honolulu, pp.1-28.

Bertocci, P.J. (1979) "Structural Fragmentation and Peasant Classes in Bangladesh", *Journal of Social Studies*, 5.

BFRI (Forest Research Institute) (1982) *Forest Statistics of Bangladesh*, Forest Economics and Statistics Division, BFRI, Chittagong.

BFRI (Bangladesh Forest Research Institute) (1992) *Forest Statistics of Bangladesh*, Forest Economics and Statistics Division, Bulletin 3, BFRI, Chittagong.

Bhaduri, A. and Rahman, M.A (eds.) (1982) *Studies in Rural Participation*, International Labour Office, Published by Oxford and IBH Publishing Company, New Delhi.

Bhattarai, T.N. (1990) "Community Forestry and Research Initiatives in the South Asia Sub-Region: Research Policy for Community Forestry in Asia Pacific Region", In: M.E. Stevens et al., (eds.), *op.cit.*, pp.65-71.

Bhuiyan, A.A. (1977) "Some Thoughts on Future Forest Policy and Administration in Bangladesh", In: *Proceedings of the First Bangladesh National Conference on Forestry*, February 11-15, 1977, Forest Department, and Ministry of Agriculture, Dhaka.

Bhuiyan, A.A. (1982) "Development of Community Forestry in the Khas and Scattered Forest of Chittagong District", In: *Proceedings of the Second Bangladesh National Conference on Forestry*, Forest Department, and Ministry of Agriculture, Dhaka, pp.227-237.

Bhuiyan, A.A. (1984) *Social and Community Forestry as an Instrument of Forest Policy and Development in Bangladesh*, M.Sc. Dissertation, Department of Forestry, University of Oxford, Oxford.

Bhuiyan, A.A. (1994) "Social Forestry as a Strategy for Economic Development" (in Bengali), In: M.R. Ahmed, (ed.), *op.cit.*, pp.23-46.

Bhuiyan, A.A. (n.d.) *Programme and Progress of Social Forestry in Bangladesh*, mimeo, Forest Department, Dhaka.

Bhuiyan, M.M. (1994) "Farming System Research and Extension for Social Forestry Development", In: M.R Ahmed, (ed.), *op.cit.*, pp.83-90.

Bista, D.B. (1991) *Fatalism and Development: Nepal's Struggle for Modernisation*, Orient Longman, Calcutta.

Bista, R.B., Sijapati, B.B. and Shrestha, S.M. (1986) "Forest Management and Utilization", In: *Four Special Study Reports by National Consultants for CFDP*, Misc. Document 45, HMG/UNDP/FAO, Community Forestry Development Project, Kathmandu.

Biswas, S.K. (1988) *Forest Administration in India with Special Reference to West Bengal*, Chugh Publications, Allahabad.

Blaikie, P. (1988) "The Explanation of Land Degradation in Nepal", In: J. Ives and D.C. Pitt, (eds.), *op.cit.*, pp.132-158.

Blair, H.W. (1978) "Rural Development, Class Structure and Bureaucracy in Bangladesh", *World Development*, 6(1):65-82.

Blair, H.W. (1982) "The Elusiveness of Equity: Institutional Approaches to Rural Development in Bangladesh", In: N.T. Uphoff, (ed.), *op.cit.*, pp.387-477.

Blair, H.W. (1985) "Participation, Public Policy, Political Economy and Development in Rural Bangladesh, 1958-85", *World Development*, 13(12):1231-1247.

Borg, W.R. (1981) *Applying Educational Research: A Practical Guide for Teachers*, Longman, New York.

Bottomore, T.B. (1964) *Elites and Society*, Pelican Books Limited, Middlesex.

BRAC (Bangladesh Rural Advancement Committee) (1986) *The Net: Power Structure in Ten Villages*, Second Edition, BRAC, Dhaka.

BRAC (Bangladesh Rural Advancement Committee) (1989) *BRAC Social Forestry Programme*, Paper prepared for the Agroforestry Working Group, BRAC, Dhaka.

Brandt, W. (1980) *North-South: A Programme for Survival*, Pan World Affairs, Pan Books, London and Sydney.

Briscoe, J. (1979) *The Political Economy of Energy Use in Rural Bangladesh*, Environmental Systems Program, Harvard University, Cambridge, Massachusetts.

Brokensha, D. and Castro, A.P. (1984) *Fuelwood, Agro-forestry and Natural Resource Management: The Development Significance of Land Tenure and Other Resource Management/Utilization Systems*, Institute for Development Anthropology, New York.

Bruce, J.W. (1989) *Community Forestry: Rapid Appraisal of Tree and Land Tenure*, Food and Agriculture Organisation, Rome.

Bruce, J.W., Fortmann, L. and Riddell, J. (1985) "Trees and Tenure: An Introduction", In: L. Fortmann and J. Riddell, (eds.), *op.cit.*, pp.vii-xvii.

Burch, W.R. (1988) "The Uses of Social Science in Training of Professional Social Foresters", *Journal of World Forest Resource Management*, 3(2):73-109.

Burch, W.R. (1990) "Foreword", In: M. Poffenberger, (ed.), *op.cit.*, p.xv.

Burkey, S. (1993) *People First: A Guide to Self-reliant Participatory Rural Development*, Zed Books Limited, London and New Jersey.

Burkitt, B. (1984) *Radical Political Economy: An Introduction to the Alternative Economics*, Wheatsheaf Books Limited, Sussex.

Byron, R.N. (1984) "Status, Trend and Proposed Direction of Bangladesh Forestry", Paper presented at *Forestry Extension Seminar*, Forestry Development and Training Centre, Kaptai.

Castro, A.P. (1988) "Southern Mount Kenya and Colonial Forest Conflicts", In: J.F. Richards and R.P. Tucker, (eds.), *op.cit.*, pp.33-55.

Cernea, M.M. (1981) *Land Tenure Systems and Social Implications of Forestry Development Programs*, Staff Working Paper 452, World Bank, Washington.

Cernea, M.M. (1988) "Alternative Social Forestry Development Strategies", In: J. Ives and D.C. Pitt, (eds.), *op.cit.*, pp.159-190.

Cernea, M.M. (1989) *User Groups as Producers in Participatory Afforestation Strategies*, World Bank Paper 70, World Bank, Washington.

Cernea, M.M. (1990) *Beyond Community Woodlots: Programmes with Participation*, Social Forestry Network Paper 11(e), Overseas Development Institute, London.

Cernea, M.M. (ed.) (1985) *Putting People First: Sociological Variables in*

Rural Development, World Bank, Washington, Published by Oxford University Press, London and New York.

Cernea, M.M. (ed.) (1991) *Putting People First: Sociological Variables in Rural Development*, Second Edition, World Bank, Washington, Published by Oxford University Press, London and New York.

CFD (Chittagong Forest Division) (1994) *Tree Planting Campaign, 1994 (in Bengali)*, Brochure on the occasion of "Tree Planting Week", CFD, Chittagong.

Chacko, V.J. (1965) *A Manual on Sampling Techniques for Forest Surveys*, Forest Research Institute, Dehra Dun.

Chaffey et al. (1995) *A Forest Inventory of Sundarbans, Bangladesh*, Overseas Development Administration, London.

Chambers, R. (1983) *Rural Development: Putting the Last First*, Longman Scientific and Technical, Essex.

Chambers, R. (1991) "Shortcut and Participatory Methods for Gaining Social Information for Projects", In: M.M. Cernea, (ed.), *op.cit.*, pp.513-537.

Chambers, R., Saxena, N.C. and Shah, T. (1989) *To the Hands of the Poor: Water and Trees*, Oxford and IBH Publishing Company Private Limited, New Delhi.

Chambers, R., Pacey, A. and Thrupp, L.A. (eds.) (1989a) *Farmer First: Farmer Innovation and Agricultural Research*, Intermediate Technology Publications, London.

Chandra, N.S. and Poffenberger, M. (1989) "Community Forest Management in West Bengal: FPC Case Studies", In: WBFD, *op.cit.*, pp.22-47.

Chaudhury, R.H. and Ahmed, N.R. (1980) *Female Status in Bangladesh*, Bangladesh Institute of Development Studies, Dhaka.

Choudhury, A.H. (1977) *Forest Policy of Bangladesh*, mimeo, Bangladesh Forest Industries Development Corporation, Dhaka.

Choudhury, J.H. (1993) *Draft Revised Forest Management Plan for Chittagong Forest Division (1991-92 to 2000-2001)*, Vol.1, Forest Working Plan Division, Chittagong.

Choudhury, R.K. (1993) "The Untold Saga of Women: Challenges and Successes", *Grassroots*, 1-2(v-vi):66-69.

Chowdhry, K. (1992) "Population Growth, Forest Management and Equity Issues", In: F.R. Miller and K.L. Adam, (eds.), *Wise Management of Tropical Forests*, Proceedings of the Oxford Conference on Tropical Forests 1992, Oxford Forestry Institute, Oxford, pp.175-181.

Chowdhry, K. (n.d.) "Social Forestry: Who Benefits?", In: Y.S. Rao et al. (eds.), *op.cit.*, pp.55-66.

Chowdhury, A. (1978) *A Bangladesh Village: A Study of Social*

Stratification, Centre for Social Studies, Dhaka University, Dhaka.

Chowdhury, A.H. (1977) "Forest Policy of Bangladesh", In: *Proceedings of First Bangladesh Conference on Forestry*, February 11-15, 1977, Ministry of Agriculture and Forests, Dhaka.

Chowdhury, J.A. (1993) *Final Report of the Resource Survey Consultant, Thana* Afforestation and Nursery Development Project, BGD/84/054, Forest Department, Dhaka.

Chowdhury, M.R. (1957) *Working Scheme for the Mymensingh Division 1955-56 to 1959-60*, East Pakistan Government Press, East Pakistan.

Chowdhury, R.A. and Hussain, Z. (1990) "Participatory Planning and Extension Strategies", In: M.Z. Abedin et al., (eds.), *op.cit.*, pp.100-105.

Chowdhury, R.I., Miah, M.M.R., Hossain, M., Chowdhury, A.F.H. and Quddus, A.H.G. (1979) *Tribal Leadership and Political Integration: A Study of Chakma and Mong Tribes of Chittagong Hill Tracts*, University of Chittagong, Chittagong.

CHT (Chittagong Hill Tracts) Commission (1991) *"Life is not Ours": Land and Human Rights in The CHTs, Bangladesh*, The Report of the Chittagong Hill Tracts Commission, International Work Group for Indigenous affairs, Copenhagen.

Clapham, C. (ed.) (1982) *Private Patronage and Public Power: Political Clientelism in the Modern State*, Frances Pinter (Publishers) Limited, London.

Clausi, R.E. and Bopp, J.R. (eds.) (1989) *The Social Sciences in Asian Forestry Curricula: Readings from the Literacy of Social Sciences in Forestry*, Tropical Forestry Institute, Yale, and School of Forestry and Environmental Studies, New Haven, Connecticut.

Cohen, M.J. and Uphoff, N.T. (1980) "Participation's Place in Rural Development: Seeking Clarity through Specificity", *World Development*, 8(3):213-235.

Colchester, M. (1989) *Pirates, Squatters and Poachers: The Political Ecology of Dispossession of The Native Peoples of Sarawak*, Survival International and INSAN Report, Kuala Lumpur.

Colchester, M. (1994) "Sustaining the Forests: The Community-based Approach in South and South-East Asia", *Development and Change*, 25(1994):69-100.

Cooper, A. (1983) "Sharecroppers and Landlords in Bengal, 1930-1950: The Dependency Web and its Implication", *The Journal of Peasants Studies*, 10(2-3):227-255.

Cowan, J.M. (1923) *Working Plan for the Forests of the Chittagong Division, Bengal*, Bengal Government Press, Calcutta.

Crittenden, R. and Lea, D.A.W. (1989) *Integrated Rural Development Programmes in Papua New Guinea*, Monograph 28, PNG Institute of Applied Social and Economic Research, Port Moresby, pp.1-132.

Curtis, D. (1991) *Beyond Government: Organisations for Common Benefit*, Macmillan Education Limited, London.

Dalmacio, M.V.(1989) *Agroforestry for Forest Land Management Systems in Bangladesh*, Working Paper 21, FAO/BGD/85/085, Forest Department, Dhaka.

Das, S. (1982) "The Forest Management Practices in Bangladesh", In: Proceedings of the Second Bangladesh National Conference on Forestry, January 21-26, 1982, Forest Department, and Ministry of Agriculture, Dhaka, pp.6-17.

Dean, S. (1986) *Women in Cooperatives*, COPAC, Rome.

De Vylder, S. (1982) *Agriculture in Chains, Bangladesh: A Case Study in Contradictions and Constraints*, Zed Press, London.

Donal O'Brien, B.C. (1979) "Ruling Class and Peasantry in Senegal, 1960-1976: The Politics of a Monocrop Economy", In: R.C. O'Brien,(ed.), *op.cit.*, pp.209-227.

Douglas, J.J. (1982) *Consumption and Supply of Wood and Bamboo in Bangladesh*, Field Document 2, UNDP/FAO Project BGD/78/010, Bangladesh Planning Commission, Dhaka.

Dumont, R. (1984) *Report on Rural Development in Bangladesh*, (Chapter xiv), mimeo, Dhaka.

Dwivedi, A.P. (1980) *Forestry in India*, Jugal Kishore and Company, Dehra Dun.

Edwardes, M. (1967) *British India 1772-1947: A Survey of the Nature and Effects of Alien Rule*, Sidgwick and Jackson Limited, London.

Elwin, V. (ed.) (1963) *A New Deal for Tribal India*, Ministry of Home Affairs, Government of India, New Delhi.

Engels, F. (1951) "Karl Marx: A Contribution to the Critique of Political Economy", In: K. Marx and F. Engels, *op.cit.*, pp.332-341.

Evers, Hans-Dieter (1987) "The Bureaucratization of Southeast Asia", *Comparative Study of Society and History*.

Falconer, J. (1987) *Forestry Extension: A Review of Key Issues*, Social Forestry Network Paper 4(e), Overseas Development Institute, London.

FAO (Food and Agriculture Organisation of the United Nations) (1978) *Forestry for Local Community Development*, Forestry Paper No. 7, FAO, Rome.

FAO (Food and Agriculture Organisation of the United Nations) (1989) *Women in Community Forestry: A Field Guide for Project Design and*

Implementation, FAO, Rome.

Feeny, D. (1988) "Agricultural Expansion and Forest Depletion in Thailand, 1900-1975", In: J.F. Richards and R.P. Tucker, (eds.), *op.cit.*,pp.112-143.

Feldman, S. and McCarthy, F.E. (1981) "Conditions Influencing Rural and Town Women's Participation in the Labour Force", In: C.B. Seely, (ed.), *Women, Politics and Literature in Bengal*, Asian Studies Center, Michigan State University, East Lansing.

Fernandes, W. (1983) "Towards a New Forest Policy: An Introduction", In: W. Fernandes and S. Kulkarni, (eds.), *op.cit.*, pp.1-22.

Fernandes, W. and Kulkarni, S. (1983) *Towards a New Forest Policy: People's Rights and Environmental Needs*, Indian Social Institute, New Delhi.

Fernandes, W., Geeta, M. and Viegas, P. (1988) *Forest, Environment and Tribal Economy: Deforestation, Impoverishment and Marginalisation in Orrissa*, Tribes in India Series 2, Indian Social Institute, New Delhi.

Fisher, R.J. (1989) *Indigenous Systems of Common Property Forest Management in Nepal*, Working Paper 18, East West Environment and Policy Institute, Honolulu.

Fisher, R.J. and Gilmour, D.A. (1990) "Putting the Community Forestry at the Centre of Community Forestry Research", In: M.E. Stevens et al., (eds)., *op.cit.*, pp.73-80.

FMP (Forestry Master Plan) (1992) *Forestry Master Plan: Participatory Forestry*, FMP Project 372001/22, Ministry of Environment and Forests, UNDP/FAO Project BGD/88/025, Dhaka.

FMP (Forestry Master Plan) (1992a) *Forestry Master Plan: Forest Institutions*, FMP Project 372001/23, Ministry of Environment and Forests, UNDP/FAO Project BGD/88/025, Dhaka.

FMP (Forestry Master Plan) (1993) *Forestry Master Plan: Economics and Marketing*, FMP Project 372001/31, Ministry of Environment and Forests, UNDP/FAO Project BGD/88/025, Dhaka.

Foley, G. and Barnard, G. (1984) *Farm and Community Forestry*, Technical Report 3, Earthscan, London.

Fortmann, L. (1987) "Tree Tenure: An Analytical Framework for Agroforestry Projects", In: J.B. Raintree, (ed.), *op.cit.*, pp.17-33.

Fortmann, L. (1987a) "Tenure and Social Forestry: Research Issues and Methods", In: S. Huq, (ed.), *Trees and Tenure in Bangladesh*, Proceedings of a discussion meeting held on December 10, 1987, Bangladesh Centre for Advanced Studies, Dhaka, pp.8-13.

Fortmann, L. and Bruce, J.W. (1988) *Whose Trees? Property Dimensions of*

Forestry, Westview Press, Boulder, Colorado.

Fortmann, L. and Riddell, J. (1985) *Trees and Tenure: An Annotated Bibliography for Agroforesters and Others*, Land Tenure Center, Madison, Wisconsin, and International Centre for Research on Agroforestry, Nairobi.

Fox, J. and Fisher, R. (1990) "Community Organisations and Government Bureaucracies in Social Forestry", In: M.E. Stevens et al., (eds.), *op.cit.*, pp.113-119.

Frank, A.G. (1992) "The Development and Underdevelopment", In: C.K. Wilber and K.P. Jameson, (eds.), *op.cit.*, pp.107-118.

Fox, R.G. (1971) *Kin Clan Raja and Rule: State-Hinterland Relations in Preindustrial India*, University of California Press, Berkeley, Los Angeles and London.

FRI (Forest Research Institute) (1961) *100 Years of Indian Forestry:1861-1961*, Vol.1, FRI, Dehra Dun.

FRMP (Forest Resources Management Project) (1992) *Project Proforma*, FRMP, Ministry of Environment and Forest, Forest Department, Dhaka.

Gadgil, M. (1983) "Forestry with a Social Purpose", In: W. Fernandes and S. Kulkarni, (eds.), *op.cit.*, pp.111-133.

Gadgil, M. (1985) "Social Restraints on Resource Utilization: The Indian Experience", In: J.A. McNeely and D. Pitt, (eds.), *Culture and Conservation: The Human Dimension in Environmental Planning*, Croom Helm, pp. 135-154.

Gadgil, M. (1987) *Culture, Perceptions and Attitudes to the Environment*, Technical Report 30, Centre for Ecological Science, Bangalore.

Gadgil, M. (1989) *Deforestation: Problems and Prospects*, Foundation Day Lecture, May 12, 1989, Society for Promotion of Wastelands Development, New Delhi.

Gadgil, M. and Iyer, P. (1989) "On the Diversification of Common Property Resource Use by Indian Society", In: F. Berkes, (ed.), *Common Property Resources: Ecology and Community-based Sustainable Development*, Belhaven Press, London.

Gadgil, M., Prashad, N. and Ali, R. (1982) *Forest Management in India: A Critical Review*, Indian Institute of Science, Bangalore.

Gain, P. and Moral, S. (1994) "Vanishing Forests, Dying People", *The Dhaka Courier*, 10(39), 29 April 1994:18-20.

Ganjanapan, A. (n.d.) *Community Forestry in Northern Thailand: Learning from Local Practice*, Chiang Mai University, Thailand.

Gardner, K. (1994) "Purdah, Female Power and Cultural Change: A Sylheti Example", *Journal of Social Studies*, 65:1-24.

Gavron, H. (1966) *The Captive Wife*, Routeledge and Kegan Paul, London.

Gerry, C. (1979) "The Crisis of the Self-Employed: Petty Production and Capitalist Production in Dakar", In: R.C. O'Brien, (ed.), *op.cit.*, pp.126-155.

Geyer, R.F. and Schweitzer, D. (1981) *Alienation: Problems of Meaning, Theory and Method*, Routledge and Kegan Paul Limited, London and Boston.

Ghani, Q. (1955) *Working Plan for the Forest of the Chittagong Division (1950-51 to 1969-70)*, East Bengal Government Press, East Bengal.

Gilmour, D.A. (1989) *Forest Resources and Indigenous Management in Nepal*, Working Paper 17, East West Environment and Policy Institute, Honolulu.

Gilmour, D.A., King, G.C. and Hobley, M. (1989) "Management of Forest for Local Use in the Hills of Nepal: Changing Forest Management Paradigms", *Journal of World Forest Resource Management*, 4:93-110.

Gilpin, A. (1966) *Dictionary of Economic Terms*, Butter Worth and Company Limited, London.

Glenn, N.D. (1978) "The General Social Surveys: Editorial Introduction to a Symposium", *Contemporary Sociology*, 7.

GOB (Government of Bangladesh) (1979) *Bangladesh National Forest Policy*, Forest Department, Dhaka.

GOB (Government of Bangladesh) (1979a) *The Second Five Year Plan 1980-1985*, Planning Commission, Ministry of Planning, Dhaka.

GOB (Government of Bangladesh) (1990) *The Fourth Five Year Plan 1990-1995*, Planning Commission, Ministry of Planning, Dhaka.

GOB (Government of Bangladesh) (1992) *A New Pledge for a Greener Bangladesh*, External Publicity Wing, Ministry of Information, Dhaka.

GOB (Government of Bangladesh) (n.d.) *Environment in Bangladesh and the Directorate of Environment* (in Bengali), Directorate of Environment, Bangladesh Government Press, Dhaka.

GOEP (Government of East Pakistan) (1955) *Plantation Techniques in East Pakistan (Tectona Grandis)*, Directorate of Forests, Dhaka.

GOEP (Government of East Pakistan) (1961) *District Census Report, Chittagong, Parts 1-5*, (Compiled by A. Rashid), Pakistan Government Publication, Karachi.

GOEP (Government of East Pakistan) (1964) *Annual Progress Report of Forest Administration in East Pakistan for the Year 1957-58*, Directorate of Forests, Dhaka.

GOEP (Government of East Pakistan) (1964a) *Annual Progress Report on Forest Administration in East Pakistan for the Year 1956-57*,

Directorate of Forests, Dhaka.

GOEP (Government of East Pakistan) (1970) *East Pakistan District Gazetteers (Chittagong)*, S.N.H. Rizvi, (ed.), East Pakistan Government Press, East Pakistan.

GOEP (Government of East Pakistan) (n.d.) *East Pakistan Forest Manual*, Part 2, Directorate of Forests, Dhaka.

GOI (Government of India) (1919) *Indian Industrial Commission Report 1916-18*, Superintendent of Government Printing, Calcutta.

GOK (Government of Karnataka) (1980) *Project for Social Forestry with World Bank Assistance*, Government Secretariat, Bangalore.

GOMP (Government of Madhya Pradesh) (1980) *Impact of Bastar Forestry Project on Tribal Economy: An Economic Study*, Tribal Research and Development Institute, Tribal and Harijan Welfare Department, Bhopal.

Gordon, G. (1979) "Finding Out About Child (0-5 Years) Feeding Practices", Paper presented at the *Conference on Rapid Rural Appraisal*, 4-7 December, 1979, Institute of Development Studies, Sussex.

Goulet, D. (1989) "Participation in Development: New Avenues", *World Development*, 17(2):165-178.

Gow, D.D and VanSant, J. (1981) *Beyond the Rhetoric of Rural Development Participation: How Can it be Done?*, IRD Working Paper 9, Development Alternatives Inc., Washington.

Graham, M. (1893) *Journal of a Residence in India 1809-11*, Second Edition, London.

Gray, A. (n.d.) *Between the Spice of Life and Melting Pot: Biodiversity Conservation and its Impact on Indigenous Peoples*, Document 70, International Work Group for Indigenous Affairs, Copenhagen.

Griffith, A.L. (1941) *Teak Plantation Technique*, Indian Forest Records, Vol. 5(2), Government of India Publications.

Gronow, J. and Shrestha, N.K. (1988) *Manual for a Reorientation Workshop on Community Forestry*, Koshi Hills Community Forestry Project, Dhankuta.

Gronow, J. and Shrestha, N.K. (1991) *From Mistrust to Participation: The Creation of a Participatory Environment for Community Forestry in Nepal*, Social Forestry Network Paper 12(b), Overseas development Institute, London.

Guha, R. (1989) *The Unquiet Woods: Ecological Change and Peasant Resistance in Himalaya*, Oxford University Press, New Delhi.

Gujral, R.S (1988) "Homesteads and Agroforestry Systems in Asia", Paper presented at a workshop on *Homestead Plantation and Agroforestry*, July 17-19, 1988, Dhaka.

Gujral, R.S (1991) "Types of Agroforestry Systems in Asia-Pacific Region", In: Mellink et al., (eds.), *op.cit.*, pp.195-201.

Gupta, A.K. (1986) "Sociology of Stress: Why do Common Property Resource Management Projects Fail?", In: Proceedings of The Conference on *Common Property Resource Management*, April 21-26, 1985, (published in 1986), National Academy Press, Washington, pp.305-321.

Hadden, S.G. (1982) "Rural Electrification and Decentralized Decision-Making in Rajasthan, India", In: N.T. Uphoff, (ed.), *op.cit.*, pp.225-288.

Haimendorf, C. Von-Furer (1955) *Himalayan Barbary*, London.

Haimendorf, C. Von-Furer (1982) *The Tribes of India: The Struggle for Survival*, Oxford University Press, Delhi.

Hannan, F.H. and Islam, N. (1986) *Women in Agriculture: An Annotated Bibliography*, Bangladesh Academy for Rural Development, Comilla.

Hardin, G. (1968) "The Tragedy of the Commons", *Science*, 162:1243-1248.

Harriss, B.(1989) "Agricultural Merchants, Capital and Class Formation in India", *Sociologia Ruralis*, 29(2).

Hartmann, B. and Boyce, J.K. (1983) *A Quiet Violence: Views from a Bangladesh Village*, University Press Limited, Dhaka.

Hasan, S. and Mulamoottil, G. (1994) "Natural Resource Management in Bangladesh", *Ambio*, 23(2):141-145.

Hayami, Y. (1980) "Economic Approach to Village Community and Institution", *Journal of Rural Development*, 3(April):27-49.

Hazell, P.B.R. and Ramasamy, C. (1991) *The Green Revolution Reconsidered: The Impact of High Yielding Rice Varieties in South India*, John Hopkins University Press, Baltimore.

Heaver, R. (1982) *Bureaucratic Politics and Incentives in the Management of Rural Development*, Staff Working Paper No. 538, World Bank, Washington.

Hisham, M.A., Sharma, J. and Ngaiza, A. (1991) *Whose Trees? A People's View of Forestry Aid*, Panos Publications Limited, London.

Hobley, M. (1991) "From Passive to Active Participatory Forestry: Nepal", In: P. Oakley et al., *op.cit.*, pp.107-118.

Hocking, A. (1991) *Village and Farm Forestry in Bangladesh*, Final Report on Participation of Women in the Programme, Prepared for Swiss Development Cooperation, Berne and Dhaka.

Hodges, R. (1988) *Primitive and Peasant Markets*, Blackwells, Oxford.

Hoq, A.K.M.F (n.d.) *Agroforestry Development, Research and Training in Bangladesh*, Mimeo, Forest Department, Dhaka.

Hoque, A.K.F. (n.d.) *A Case Study of Pomora Community Forestry Project*,

B.Sc. (Forestry) Review Paper 16, Institute of Forestry, Chittagong.

Howes, M. and Jabbar, M.A. (1986) *Rural Fuel Shortages in Bangladesh: The Evidence from Four Villages*, Discussion Paper 213, Institute of Development Studies, Sussex.

Huque, A.S. (1985) "The Politics of Local Government in Rural Bangladesh", *Public Administration and Development*, 5(3):205-217.

Hunter, W.W. (1876) *Statistical Account of Bengal (Chittagong Hill Tracts, Chittagong, Noakhali, Tipperah, Hill Tipperah)*, Vol.6, Trubner and Company, London.

Hunter, W.W. (1893) *The Indian Empire: Its Peoples, History, and Products*, Third Edition, W.H. Allen and Company, London.

Hussain, M.M. (1987) *Forestry and Forest Conservation Strategy in Bangladesh*, Seminar Paper, Bangladesh Public Administration Training Centre, Dhaka.

Hussain, M.M. (1992) *Study on National Forest Policy in Bangladesh*, Mimeo, Forest Department, Dhaka.

Hussain, M.M. (1992a) "Country Paper: Bangladesh", Paper prepared for the workshop on *Sustainable and Effective Management Systems for Community Forestry*, January 15-17, 1992, Regional Community Forestry Training Centre, Bangkok.

Hyden, G. (1988) "Approaches to Co-operative Development: Blueprint versus Greenhouse", In: D.W. Attwood and B.S. Baviskar, (eds.), *op.cit.*, pp.149-171.

Hye, H.A. (1991) *Integrated Rural Development: A Review of Major Developments*, National Institute of Local Government, Dhaka.

Hye, H.A. (ed.) (1985) *Decentralisation, Local Government and Resource Mobilisation*, Bangladesh Academy for Rural Development, Comilla.

Hyman, H.H. (1987) *Secondary Analysis of Sample Surveys*, Wesleyan University Press, Middletown, Conn..

IBRAD (Indian Institute of Bio-social Research and Development) (n.d.) *Progress of Joint Forestry Management*, IBRAD, Calcutta.

ILO (International Labour Office) (1977) *Poverty and Landlessness in Rural Asia*, ILO, Geneva.

ILO (International Labour Office) (1988) *Cooperative Management and Administration*, Indian Edition, Oxford & IBH Publishing Company, New Delhi.

Islam, A.K.M.A. (1974) *A Bangladesh Village: Conflict and Cohesion*, Schenkman Publishing Company, Cambridge, Massachusetts.

Islam, S. (1978) "Economic Life and Activities", In: BDR, *op.cit.*, p.314.

Isokun, M.I. (1985) *Administration of Justice: A Study of the Nigerian System of Justice with Special Reference to the Informal Justice System in Bendel State, Nigeria*, Ph.D Thesis, Centre for Development Studies, University of Wales, Swansea.

Ives, J. and Pitt, D.C. (eds.) (1988) *Deforestation: Social Dynamics in Watersheds and Mountain Ecosystems*, Routledge, London.

Jahangir, B.K. (1979) *Differentiation, Polarisation and Confrontation in Rural Bangladesh*, Centre for Social Studies, Dhaka University, Dhaka.

Jahangir, B.K. (1982) *Rural Society, Power Structure and Class Practice*, Centre for Social Studies, Dhaka University, Dhaka.

Jahangir, B.K. (1989) "Political Economy of Local Government in Bangladesh", In: H. W. Blair, (ed.), *Can Rural Development be Financed from Below*? Local Resource Mobilisation in Bangladesh, University Press Limited, Dhaka.

Jannuzi, F.T. and Peach, J.T. (1980) *The Agrarian Structure of Bangladesh: An Impediment to Development*, Westview Press, Boulder, Colorado.

Jansen, E.G. (1986) *Rural Bangladesh: Competition for Scarce Resources*, Norwegian University Press, Oslo.

Jansen, E.G. (1990) "Process of Polarization and the Breaking up of Patron-Client Relationships in Rural Bangladesh", In: O.D.K. Norbye, (ed.), *Bangladesh Faces the Future*, University Press Limited, Dhaka.

Jayaraman, K.S. (n.d.) *Trees for the Poor*, The Panos Institute, London.

Jenkins, R. (1992) "Industrialization and the Global Economy", In: T. Hewitt, H. Johnson and D. Wield, (eds.), *Industrialization and Development*, Oxford University Press (in association with the Open University), Oxford.

Jhonson, K., Olson, E.A. and Manandhar, S. (1982) "Environmental Knowledge and Respond to Natural Hazards in Mountainous Nepal", *Mountain Research and Development*, 2(2):175-188.

Jodha, N.S. (1988) "Population Growth and Common Property Resources: Micro Level Evidence from India", Paper presented at the *Meeting on the Consequences of Rapid Population Growth in Less Developed Countries*, August 23-26, 1988, (unpublished manuscript).

Jodha, N.S. (1990) "Rural Common Property Resources: Contributions and Crisis", *Economic and Political Weekly Quarterly Review of Agriculture*, 25(26).

Johl, S.S. and Mudahar, M.S. (1982) "The Dynamics of Institutional Change and Rural Development in Punjab, India", In: N.T. Uphoff, (ed.), *op.cit.*, pp.134-224.

Jones, S. (1978) "Tribal Underdevelopment in India", *Development and Change*, 9(1):41-70.

Joshi, G. (1983) "Forest Policy and Tribal Development: Problems of Implementation, Ecology and Exploitation" In: W. Fernandes and S. Kulkarni, (eds.), *op.cit.*, pp.25-47.

Kader, M.B. (1989) *A Study on Community Forestry Project within the Jurisdiction of Bogra Forest Extension Division*, M.Sc. Review Paper, Bangladesh Forest Academy, Chittagong.

Kanel, K. (1993) "Community Forestry and the 1993 Forestry Legislation: Implications for Policy and Implementation", *Banko Janakari*, 4(1):2-5.

Karim, A.H.M.M. (1990) "Agroforestry on Forest Land: Experiences in the Dinajpur Forest Division", In: M.Z. Abedin et al., (eds.), *op.cit.*, pp.54-62.

Karim, A.K.N. (1956) *Changing Society in India and Pakistan*, Ideal Publications, Dhaka.

Karim, A.K.N. (1981) *The Dynamics of Bangladesh Society*, Vikash Publishing House Private Limited, New Delhi.

Kartasubrata, J. (1991) "Planning and Implementation Aspects based on Some Successful Agroforestry Projects in Indonesia", In: W. Mellink et al., (eds.), *op.cit.*, pp.232-250.

Kaufman, H. (1960) *The Forest Ranger: A Study in Administrative Behaviour*, The John Hopkins Press, Baltimore.

Kayastha, B.P. (1991) *Elements of Community Forestry in Nepal*, Published by Sabriti Devi, Kathmandu.

Kerkvliet, B.J. (1990) *Everyday Politics in the Philippines: Class and Status Relations in a Central Luzon Village*, University of California Press, Berkeley, Los Angeles and London.

Khaleque, K. (1984) *Prospects of Social Forestry in the Garo Villages of Madhupur Garh*, Department of Sociology, Dhaka University, Dhaka.

Khan, A.I. and Khan, M.A. (1988) "Agroforestry for the Poor: The Proshika Experience", Paper presented in *National Workshop on Homestead Plantations and Agroforestry in Bangladesh*, July 17-19, 1988, Bangladesh Agricultural Research Institute, Gazipur.

Khan, M.A. (1985) "U.N. Decade for Women: Background, Objectives and National Notion", In: Q.H. Ahmad, M.A. Khan, S. Khan, and J.A. Rahman, (eds.), *Situation of Women in Bangladesh*, Ministry of Social Welfare and Women's Affairs, Dhaka, pp.1-17.

Khan, M.M. (1980) *Bureaucratic Self-Preservation*, Dhaka University, Dhaka.

Khan, M.M. (1985) "Process of Decentralisation in Bangladesh", In: H.A. Hye, ed., *Decentralisation, Local Government Institutions and Resource Mobilisation*, Bangladesh Academy for Rural Development, Comilla.

Khan, N.A. (1993) "Towards an Understanding of Participation: The

Conceptual Labyrinth Revisited", *Administrative Change*, 20(1-2):106-120.

Khan, N.A. (1994) "Participation in South Asian Social Forestry: Chasing a Mirage?", *(Indian) Journal of Rural Development*, 13(3):343-355.

Khan, N.A. (1994a) "In Chase of Cooperation: Conditions for Collective Action Reconsidered", *Social Science Review*, 11(2):97-110.

Khan, N.A. (1994b) "Coordination in the Field Administration: The Case of Forest Department", (in Bengali), *Chittagong University Studies (Social Science)*, 24(1):137-145.

Khan, N.A. (1996) *Of Murubbi and Kamla: Patronage and Social Forestry in Bangladesh*, Papers in International Development 19, Centre for Development Studies, University of Wales, Swansea.

Khan, N.A. (1996a) "Revisiting Community Forestry Developments in Nepal: A Selected Review of Performance", *Asian Journal of Environmental Management*, 4(2):95-101.

Khan, N.A. (1996b) "Of Sirs and *Sahibs*: Bureaucratic Constraints on Community Forestry Programmes in Bangladesh", In: *Proceedings of the Symposium on Resource Management: Vision and Issues*, July 5-8, 1996, Centre for Resource Management, Lincoln University, Canterbury, New Zealand, pp.110-121.

Khan, N.A. (1997) "Leisure and Recreation among Women of Selected Hill-Farming Families in Bangladesh", *Journal of Leisure Research*, 29(1):5-20.

Khan, N.A. (1997[a]) "Social Research Approaches to Agroforestry Management in Asia: A Primer on the Sociology of Agroforestry", *Asian Journal of Environmental Management*, 5(2):103-108.

Khan, N.A. (forthcoming) "Land Tenurial Dynamics and Participatory Forestry Management in Bangladesh", *Public Administration and Development*.

Khan, N.A. (forthcoming[a]) "Interview with the *Sahibs*: Bureaucratic Constraints on Community Forestry Programmes in Bangladesh", *Social Development Review*.

Khan, N.A. and Begum, S.A. (1997) "Participation in Social Forestry Re-examined: A Bangladesh Case Study", *Development in Practice*, 7(3):260-266.

Khan, N.A. and Hasan, S. (1995) "Forest Management through People's Participation: Practices and Problems in India and Nepal", *Asian Journal of Environmental Management*, 3(1):51-60.

Khan, N.A. and Hasan, S. (1995a) "Social Forestry in India: Widening the Gap?", *Asian Profile*, 23(6):504-510.

Khan, N.A. and Khan, A.A. (1993) "Community Forestry and Public Participation: A Case Study form Rural Bangladesh", *Grassroots*, 2(7).

Khan, N.A., Rahman, A and Ali, I.A.F. (1996) "Implications of Land Tenure at the Betagi Community Forestry Project: Some Emerging Issues", *Chittagong University Studies (SocialScience)*, 25(1).

Khan, S.A. (1980) *Revised Working Plan for the Forest of Chittagong Division (for the years 1978-79 to 1987-88)*, Vol. 1, Forest Working Plan Division, Chittagong.

Khan, S.A. and Khan, N.A. (1994) "Non-Wood Forest Products of Bangladesh: An Overview", *Bangladesh Journal of Forest Science*, 23(1):67-73.

Khandakar, K. (1994) "Women in Social Forestry", In: M.R. Ahmed, (ed.), *op.cit.*, pp.103-107.

Kharel, S. (1993) "Women's Participation in Community Forestry: The Nepal-Australia Community Forestry Project's Experience", *Banko Janakari*, 4(1):73-75.

Khattak, G.M. (1979) "History of Forest Management in Bangladesh", *Pakistan Journal of Forestry*, July 1979.

King, G.C., Hobley, M. and Gilmour, D.A. (1990) "Management of Forests for Local Use in the Hills of Nepal: Towards the Development of Participatory Forest Management", *Journal of World Forest Resource Management*, 5:1-13.

Kirchhofer, J. and Mercer, E. (1984) *Putting Social and Community Forestry in Perspective in the Asia-Pacific Region*, Working Paper, Environment and Policy institute, East-West Center, Honolulu.

Klein, M.A. (1979) "Colonial Rule and Structural Change : The Case of Sine-Saloum", In: R.C. O'Brien, (ed.), *op.cit.*, pp.65-99.

Koirala, B.P. (1985) "People's Participation in Community Forestry: A Must for Nepal", In: *Community Forestry: Socio-Economic Aspects*, Regional Office for Asia and the Pacific, Food and Agriculture Organisation, Bangkok.

Korten, D.C. (1980) "Community Organisation and Rural Development: A Learning Process Approach", *Public Administration Review*, September-October 1980:490-511.

Korten, D.C. (ed.) (1987) *Community Management: Asian Experience and Perspectives*, Kumarian Press, West Hartford, Connecticut.

Korten, D.C. and Uphoff, N. (1981) *Bureaucratic Reorientation for Participatory Rural Development*, NASPAA Working Papers, NASPAA, Washington.

Korten, D.C. and Alfonso, F.B. (eds.) (1983) *Bureaucracy and the Poor:*

Closing the Gap, Kumarian Press, West Hartford, Connecticut.

La Palombara, J. (ed.) (1963) *Bureaucracy and Political Development*, Princeton University Press, Princeton, New Jersey.

Lai, C.K. (1990) "Design of Agroforestry Systems with Special Reference to Bangladesh", In: M.Z. Abedin et al., (eds.), *op.cit.*, pp.90-99.

Lai, C.K. and Khan, A. (1990) "Participatory Forestry in Bangladesh: First Step Towards Sustainability", In: L.L. Rebugio, *Readings in Rural Development*, Vol. 3, Institute of Forestry, Chittagong.

Lai, C.K. and Khan, A. (1992) "Forest Policy in the Sahel: Institutional Constraints on Social Forestry in Mali", In: G. Shepherd, (ed.), *op.cit.*, pp.29-44.

Lee, R.G., Field, D.R. and Burch, W.R. (eds.) (1990) *Community and Forestry: Continuities in the Sociology of Natural Resources*, International Book Distributors, Dehra Dun.

Lin, C. (1976) *Foundation of Social Research*, McGraw-Hill Book Company.

Locke, C. (1992) *West Bengal Joint Forestry Management Project*, A report to ODA on an orientation visit to West Bengal, August 24-27, 1992, Overseas Development Administration, London.

Locke, C. (1995) *Planning for the Participation of Vulnerable Groups in Communal Management of Forest Resources: The Case of the Western Ghats Forestry Project*, Ph.D Thesis, Centre for Development Studies, University of Wales, Swansea.

Ludz, P.C. (1981) "A Forgotten Intellectual Tradition of the Alienation Concept", In: R.F. Geyer and D. Schweitzer, (eds.), *op.cit.*, pp.21-35.

Ludden, D. (1985) *Peasant History in South India*, Princeton University Press, New Jersey and Surrey.

Lynch, O.J (1992) *Securing Community based Tenurial Right in the Tropical Forests of Asia: An Overview of Current and Prospective Strategies*, World Resources Institute, Washington.

Magno, V.C. (1986) *Community Forestry Handbook*, FAO/UNDP BGD/81/028, Field Document 1, Forest Department, and Food and Agriculture Organisation, Dhaka.

Magno, V.C. and Ali, S.S. (n.d.) *Handbook on Social Forestry*, *Upazila (Thana)* Afforestation and Nursery Development Project, Forest Department, Dhaka.

Magno, V.C., Ali, S.S. and Castro, C.P (1992) *Introduction to Social Forestry*, *Upazila (Thana)* Afforestation and Nursery Development Project, UNDP/FAO Project BGD/84/054, Forest Department, Dhaka.

Magubane, B.M. (1979) *The Political Economy of Race and Class in South Africa*, Monthly Review Press, New York and London.

Maine, H. (1905) *Ancient Law: Its Connection with the Early History of Society and its Relation to Modern Ideas*, Routledge, London.

Malhotra, K.C., Deb, D., Dutta, M., Vasulu, T.S., Yadav, G. and Adhikari M. (n.d.) *Role of Non-Timber Forest Products in Village Economy: A Household Survey in Jamboni Range, Midnapore District, West Bengal*, Indian Institute of Bio-Social Research and Development, Calcutta.

Malla, Y.B. (1993) "Changing Role of the Forest Resource Market: An Ignored Dimension of Community Forestry", *Banko Janakari*, 4(1):28-36.

Manandhar, P.K. (1985) *Introduction to Agroforestry Programme in Bangladesh*, Community Forestry Project BGD/81/028, Forest Department, Bogra.

Manandhar, P.K. (1986) *Interim Report on Agroforestry Farms initiated in Rangpur Dinajpur Forest Divisions in 1984/85*, Community Forestry Project BGD/81/028, Forest Department, Bogra.

Manandhar, P.K. (1987) *Agroforestry Programme in Bangladesh (Final Report)*, Community Forestry Project BGD/81/028,Forest Department, Bogra.

Marx, K. (1951) "Preface to a Contribution to the Critique of Political Economy", In: K. Marx and F. Engels, *op.cit.*, pp.327-331.

Marx, K. (1951a) "The British Rule in India", In: K. Marx and F. Engels, *op.cit.*, pp.312-318.

Marx, K. (1975) "Debates on the Law on Thefts of Wood", In: K. Marx and F. Engels, 1975, *op.cit.*, pp.224-263.

Marx, K. (n.d.) *Notes on Indian History (664-1858)*, Second Impression, Foreign Languages Publishing House, Moscow.

Marx, K. and Engels, F. (1951) *Selected Works in Two Volumes*, Vol. 1, Lawrence and Wishart Limited, London.

Marx, K. and Engels, F. (1975) *Collected Works*, Vol. 1, Lawrence and Wishart Limited, London.

Mashreque, S.M. (1982) "Kinship and Rural Factionalism: A Study of a Village in Bangladesh", *Chittagong University Studies*, 7(2).

Mashreque, S.M. and Amin, R.M. (1993) "Patron-Client Relations and Fictive Kinship in Bangladesh", *The Eastern Anthropologist*, 46(1):19-30.

Maslekar, A.R. (1983) *Managing the Forests*, Jugal Kishore and Company, Dehra Dun.

Mathur, H.M. (1985) "Participatory Development and Public Administration", *Administrative Change*, 12(2):126-162.

Mellink, W., Rao, Y.S. and MacDicken, K.G. (eds.) (1991) *Agroforestry in the Asia and the Pacific*, Regional Office for Asia and the Pacific,

Food and Agriculture Organisation, Bangkok.

Messerschmidt, D.A. and Hammett, A.L. (1993) "Indigenous Knowledge of Alternative Forest Resources Extraction and Marketing: Significance for Community Forestry planning and Management", *Banko Janakari*, 4(1)32-36.

Miah, A.L. (1989) *Community Forestry and Development with Special Reference to Betagi*, M.Sc. Review Paper, Bangladesh Forest Academy, Chittagong.

Midgeley, J., Hall, A., Hardiman, M. and Narine, D. (1986) *Community Participation, Social Development and the State*, Methuen and Company, London and New York.

Mies, M. (1986) *Indian Women in Subsistence and Agricultural Labor*, International Labour Office, Geneva.

Mills, C.W. (1959) *The Sociological Imagination*, Oxford University Press, New York.

Momen, M.A. (1993) "Land Tenure in Bangladesh", In: *Our Common Estate*, The Royal Institute of Chartered Surveyors, London, pp.59-77.

MPFDC (Madhya Pradesh State Forestry Development Corporation) (1978) *Introducing M.P. Forestry Technical Assistance Project, Bastar, India*, General Bulletin, MPFDC, Bhopal.

Mukherjee, R. (1957) *The Dynamics of Rural Society: A Study of the Economic Structure of Bengal Village*, Akademic Verlag, Berlin.

Musgrave, P. (1990) *Effective Social Forestry: A Fact Pact: Suggestions for Practice*, Church of Bangladesh Social Development Programme, Dhaka.

Myrdal, G. (1971) *Asian Drama: An Inquiry into the Poverty of Nations*, The Twentieth Century Fund, Allen Lane, the Penguin Press, London.

Nachmias, C.F. and Nachmias, D. (1992) *Research Methods in Social Sciences*, Fourth Edition, Edward Arnold (a division of Hodder and Stoughton), London, Melbourne and Auckland.

Nadkarni, M.V., Pasha, S.A. and Prabhakar, L.S. (1989) *The Political Economy of Forest Use and Management*, Sage Publications, New Delhi.

Nanayakkara, V.R. (1991) "Agroforestry Systems in Sri Lanka", In: W. Mellink et al., (eds.), *op.cit.*, pp.148-167.

Ngaiza, A. (1991) "Learning from The Past? The East Usambara Catchment Forest Project, Tanzania", In: M.A. Hisham et al., *op.cit.*, pp.58-98.

Nikitin, P.I. (1983) *The Fundamentals of Political Economy*, Progress Publishers, Moscow.

Noronha, R. and Spears, J.S. (1985) "Sociological Variables in Forestry

Project Design", In: M.M. Cernea, (ed.), *op. cit.*, pp.227-266.

Nurse, M.C., Singh, H.B., Paudyal, B.R. and Bonjan, S. (1993) "Beat-level Planning: Development of a Management Information System for Community Forestry", *Banko Janakari*, 4(1):91-95.

Oakley, P. and Marsden, D. (1984) *Approaches to Participation in Rural Development*, International Labour Office, Geneva.

Oakley P. et al. (1991) *Projects with People: The Practice of Participation in Rural Development*, International Labour Office, Geneva.

O'Brien, R.C. (ed.) (1979) *The Political Economy of Underdevelopment: Dependence in Senegal*, Vol. 3, Sage Series on African Modernization and Development, Sage Publications Inc., Beverly Hills and London.

Ollman, B. (1971) *Alienation: Marx's Conception of Man in Capitalist Society*, Cambridge University Press, London and New York.

Olson, M. (1971) *The Logic of Collective Action: Public Goods and the Theory of Groups*, Harvard University Press, Harvard, Massachusetts.

Om Consultants (1986) *Management Study*, Summary Report, SIDA-aided Tamil Nadu Social Forestry Project, Om Consultants (India) Private Limited, Bangalore.

Padhi, G.S. (1982) *Forestry in India: A Critical Study*, International Book Distributors, Dehra Dun.

Palit, S. (1989) "Present Status of Forest Protection Committees", In: WBFD, *op.cit.*, pp.5-8.

Palit, S. (1991) "Participatory Management of Forests", *Indian Forester*, 117(5):342-349.

Palit, S. (1992) "People's Participation in Forest Management", *Indian Forester*, July 1992:447-455.

Palit, S. (1992a) "Implementing Joint Forest Management", Paper presented at the *National Workshop on Joint Forest Management*, August 17-19, 1992, organized by the Society for Promotion of Wastelands Development, Surajkund.

Parker, K.J., Clausi, R.E. and Burch, W.R. (1989) *The Social Sciences in Asian Forestry Curricula*, Report on the workshop held in November 27-December 2, 1988 in Khon Kach, Thailand, F/FRED of United States Agency for International Development, USA.

Picciotto, R. (1992) *Participatory Development: Myths and Dilemmas*, Policy Research Working Paper, WPS 930, World Bank, Washington.

Poffenberger, M. (1990a) *Joint Management of Forest Lands: Experiences from South Asia*, A Ford Foundation Program Statement, Ford Foundation, New Delhi.

Poffenberger, M. (ed.) (1990b) *Forest Management Partnerships:*

Regenerating India's Forests, Executive Summary of the Workshop on Sustainable Forestry, September 10-12, 1990, Ford Foundation, New Delhi.

Poffenberger, M. (ed.) (1990) *Keepers of the Forest: Land Management Alternatives in Southeast Asia*, Kumarian Press, West Hartford, Connecticut.

Poffenberger, M. and Singh, S. (1992) "Forest Management Partnerships: Regenerating India's Forests", *Unasylva*, 170(43):45-52.

Prabhu, P. (1983) "Social Forestry: An Adivasi Viewpoint", In: W. Fernandes and S. Kulkarni, (eds.), *op.cit.*, pp.135-143.

Price, N. (1995) *The Value of Children amongst the Kikuyu of Central Province, Kenya: An Anthropological Perspective on Fertility*, Papers in International Development 13, Centre for Development Studies, University of Wales, Swansea.

Quddus, A.H.G. (1978) *Perceptions and Attitude of National Level Elite Members of India Toward Social, Economic and Political Issues*, M.A.(Sociology) Thesis, University of Northern Iowa, Iowa.

Quddus, A.H.G., Ali, S.M.I., Bhuiyan, A.M.A. and Hossain, M. (1992) *Greening the Hills: The Betagi-Pomora Agroforestry Experience*, Bangladesh Agricultural Research Council, and Winrock International, Dhaka.

Quddus, M.A., Solaiman, M. and Karim, M.R. (1985) *Rural Women in Households in Bangladesh: With a Case Study of Three Villages in Comilla*, Bangladesh Academy for Rural Development, Comilla.

Rahman, A. (1987) "Socio-Economic Impact of Pomora Community Forestry Project", *Bano Biggyan Patrika* (Bangladesh Journal of Forest Science), 16(1-2):10-24.

Rahman, A. (1991) "Social Forestry in Bangladesh: The Betagi-Pomora Experience", Paper presented at the Seminar on *Betagi-Pomora Community Forestry Projects*, December 15, 1991, Bangladesh Agricultural Research Council, Dhaka.

Rahman, A. (1992) "Participatory Approach to Forest Resource Management in Bangladesh: Some Major Initiatives and Lessons Learned", Paper prepared for International seminar on *Forest Administration and Management*, September 1992, Michigan School of Natural Resources, Michigan.

Rahman, A. (1994) "Betagi Community Forestry Project as a Social Forestry Model in Bangladesh", In: M.R. Ahmed, (ed.), *op.cit.*, pp.91-102.

Rahman, A. (n.d.) *Agroforestry System under the Betagi and Pomora Community Forestry Projects of Bangladesh*, mimeo, Bangladesh Forest

Research Institute, Chittagong.

Rahman, A. and Khan, N.A. (1994) "The Economic Potentials and Problems of Lemon Farming at the Betagi Social Forestry Project", *Bangladesh Journal of Forest Science*, 23(2):67-73.

Rahman, A. (1981) *Rural Power Structure: A Study of the Local Level Leaders in Bangladesh*, Bangladesh Books International, Dhaka.

Rahman, A. (1988) "The State, Local Power Brokers and Rural Development in Bangladesh: A study of the selected Upazila Chairmen", Paper presented at *The Second National Conference of Bangladesh Sociological Association*, April 1988, Dhaka.

Rahman, L.M. (1993) "History of Forest Conservation in Indo-Bangladesh", *Aranya: A Forestry Periodical*, 2(2):21-24.

Rahman, M.H. (1994) *Decentralisation and Rural Society in Bangladesh: A Study of Bureaucratic Constraints on Access in the Upazila Structure*, Ph.D Thesis, Centre for Development Studies, University of Wales, Swansea.

Rahman, S. (1994) "Agroforestry in Bangladesh: Need for a Collaborative Effort", *Grassroots*, 3(xi):5-15.

Raintree, J.B. (ed.) (1987) *Land, Trees And Tenure: Proceedings of an International Workshop on Tenure Issues in Agroforestry*, Held in May 27-31, 1985, International Centre for Research on Agroforestry, Nairobi, and Land Tenure Centre, Madison.

Randhawa, R.K. (1976) "The Forests in the life of the Tribals", *The Administrator*, 21(b.1, Spring).

Rao, Y.S. (1989) "Forestry Research in the Asia-Pacific Region: A Review of Recent Developments", Paper prepared for the *Expert Consultation for the Asia Pacific Network on Forestry Research*, August 22-25, 1989, Bangkok.

Rao, Y.S., Vergara, N.T. and Lovelace, G.W. (eds.) (n.d) *Community Forestry: Socio-Economic Aspects*, Regional Office for Asia and the Pacific, Food and Agriculture Organisation, Bangkok.

Rashiduzzaman, M. (1968) *Politics and Administration in Local Council: A Study of Union and District Councils in East Pakistan*, Oxford University Press, Dhaka.

Rebugio, L.L. (1985) *Social Forestry as a Resource System*, Occasional Paper 4, College of Forestry, University of Philippines at Los Banos, Laguna.

Reddy, G.R. (1982) "Panchayati Raj and Rural Development in Andhra Pradesh, India", In: N.T. Uphoff, (ed.), *op.cit.*, pp.45-133.

Rehman, M.S. (1983) "Background and Attitudes of Thana Officials in Bangladesh: A Pilot Study", *Administrative Change*, 10(1-2):52-68.

Rehman, M.S. (1986) "Administrative Elite in Bangladesh: Their Attitudinal Profile Towards Socio-Cultural Development", *Administrative Change*, 13(1-2):55-82.

Repetto, R. and Gillis, M. (eds.) (1988) *Public Policies and Misuse of Forest Resources*, A World Resources Institute Book, Cambridge University Press, Cambridge, New York, and Melbourne.

Rew, A. (1986) "Institutional Culture and Persistent Structures in Papua New Guinea", *Public Administration and Development*, 6(4):391-400.

Rew, A. (1996) *Development Management and Ethnic Identity in New Britain, Papua New Guinea*, Papers in International Development 18, Centre for Development Studies, University of Wales, Swansea.

Ribot, J.C. (1995) "From Exclusion to Participation: Turning Senegal's Forestry Policy Around?", *World Development*, 23(9):1587-1599.

Richards, J.F. and Tucker, R.P. (eds.) (1988) *World Deforestation in the Twentieth Century*, Duke University Press, Durham and London.

Riddell, J.C. (1987) "Land Tenure and Agroforestry: A Regional Overview", In: J.B. Raintree, (ed.), *op.cit.*, pp.1-16.

Riddell, J.C. and Fortmann, L. (1983) *Land Tenure Issues in Agroforestry*, Land Tenure Centre, Madison.

Roberts, G.K. (1971) *A Dictionary of Political Analysis*, Longman Group Limited, London.

Rocheleau, D.E. (1987) "Women, Trees and Tenure: Implications for Agroforestry Research and Development", In: J.B. Raintree, (ed.), *op.cit.*, pp.79-120.

Romm, J. (1981) "The Uncultivated Half of India", *Indian Forester*, 107(1-2).

Rotenstreich, N. (1989) *Alienation: The Concept and its Reception*, E.J. Brill Publishers, The Netherlands.

Roy, A. (1983) "Panchayat Elections: Class Mobilisation without Class Struggle", *Economic and Political Weekly*, 18(26):1141.

Roy Burman, B.K. (1987) "Historical Process in Respect of Communal Land System and Poverty Alleviation among Tribals", *Social Action*, 37(4):391-419.

Roy, M.K. (1987) *Forest Sector Planning and Development in Bangladesh*, Master of Forestry Dissertation, Department of Forestry, Australian National University, Canberra.

Runge, C.F. (1986) "Common Property and Collective Action in Economic Development", In: Proceedings of the conference on *Common Property Resource Management*, April 21-26,1985, (published in 1986), National Academy Press, Washington, pp.31-59.

Sadik N. (ed.) (1991) *Population Policies and Programmes: Lessons Learned from Two Decades of Experience*, Published for United Nations Population Fund, New York University Press, New York and London.

Saha, S.K. (1986) "Historical Premises of India's Tribal Problem", *Journal of Contemporary Asia*, 16(3):274-319.

Saha, S.K. (1988) "The Territorial Dimension of India's Tribal Problem", In: M. Shepperdson and C. Simmons, (eds.), *The Indian National Congress and the Political Economy of India 1885-1985*, Avebury, Aldershot, pp.253-269.

Saint, K. (n.d.) *Community and the Common Lands: The Banished Factor in Land Reforms*, unpublished manuscript, India.

Salehuddin, A.B.M. (n.d.) *Social Forestry in Bangladesh*, mimeo, Bangladesh Forest Research Institute, Chittagong.

Salehuddin, A.B.M. and Rebugio, L.L. (1990) "Social Forestry in Bangladesh", In: Abedin et al., (eds.), *op.cit.*, pp.65-74.

SALT (Sloping Agricultural Land Technology Project) (1993) *Technical Assistance Project Proposal Format*, Chittagong Hill Tracts Development Board, Chittagong Hill Tracts.

Sarkar, P. (1980) "Darma Atmiyo: Fictive Kin Relationship in Rural Bangladesh", *The Eastern Anthropologist*, 33(1).

Sarker, A.E. (1990) *State Intervention in Rural Development: A Case Study of Bangladesh*, Ph.D Thesis, University of Liverpool, Liverpool.

Sarma, M.T.R. (n.d.) "Employment and Income Aspects of Social Forestry Programmes in India and the Philippines", In: Y.S. Rao et al., (eds.), pp.131-145.

Saxena, N.C. (1990) *Trees on Farm Lands in North-West India: Field Data from Six Villages*, Social Forestry Network Paper 11(d), Overseas Development Institute, London.

Saxena, N.C. (1991) "Village Tree planting in North-West India", *The Commonwealth Forestry Review*, 70(3), No.223:153-168.

Saxena, N.C. (1992) "Eucalyptus on Farmlands in India: What went Wrong?", *Unasylva*, 170(43):53-58.

Schacht, R. (1981) "Economic Alienation: With or without Tears", In: R.F. Geyer and D. Schweitzer, (eds.), *op.cit.*, pp.36-67.

Scoones, T. and Thompson, J. (ed.) (1994) *Beyond Farmer First: Rural People's Knowledge, Agricultural Research and Extension Practice*, Intermediate Technology Publications, London.

Scott, J.C. (1972) "Patron-Client Politics and Political Change in Southeast Asia", *The American Political Science Review*, 66 (1):91-113.

Scott, J.C. (1985) *Weapons of the Weak: Everyday Forms of Peasant Resistance*, Yale University Press, New Haven and London.

Selltiz, C., Jahoda, M., Deutsch, M. and Cook, S.W. (1962) *Research Methods in Social Relations*, Second Edition, Holt, Rinehart and Winston, New York.

Sen, D. and Das, P.K. (1987) *The Management of People's Participation in Community Forestry: Some Issues*, Social Forestry Network Paper 4(d), Overseas Development Institute, London.

Shah, T. (1988) *Gains from Social Forestry: Lessons from West Bengal*, Discussion Paper 243, Institute of Development Studies, Sussex.

Shahabuddin, Q. (1991) *Peasant Behaviour under Uncertainty: Decision-making among Low-income Farmers in Bangladesh*, Bangladesh Agricultural Research Council, and Winrock International, Dhaka.

Sharma, B.D. (1978) *Tribal Development: The Concept and the Frame*, Prachi Prakasan, Delhi.

Sharma, J. (1991) "Grassroots and Green Hills: The Hill Forest Development Project, Nepal", In: M.A. Hisham et al., (op.cit.), pp.100-138.

Sharma, R.A. (1993) "The Socioeconomic Evaluation of Social Forestry Policy in India", *Ambio*, 22(4):219-224.

Sharma, R.S. (1965) *Indian Feudalism: c.300-1200*, Centre for Advanced Study in Ancient Indian History and Culture, University of Calcutta, Calcutta.

Sharma, U.R. (1993) "Community Forestry: Some Conceptual Issues", *Banko Janakari*, 4(1):9-11.

Shepherd, G. (1990) *Forestry, Social Forestry, Fuelwood and the Environment: A Tour of the Horizon*, Social Forestry Network Paper 11(a), Overseas Development Institute, London.

Shepherd, G. (1992a) "Forest Policies and Forest Politics", In: G. Shepherd, (ed.), *ibid.*, pp.5-27.

Shepherd, G. (1993) "Local and National Level Forest Management Strategies -- Competing priorities at the Forest Boundary: The Case of Madagascar and Cameroon", *Commonwealth Forestry Review*, 72(4):316-320.

Shepherd, G. (ed.) (1992) *Forest Policies, Forest Politics*, Agricultural Occasional Paper 13, Overseas Development Institute, London.

Shrestha, K.B. and Budhathoki, P. (1993) "Problems and Prospects of Community Forestry in the Terai Region of Nepal", *Banko Janakari*, 4(1):24-27.

Shrestha, M.L. (1993) "Community Forestry in Nepal: Need for Uniformity", *Banko Janakari*, 4(1):6-8.

Shiva, V., Sharatchandra, H.C. and Bandyopadhyay, J. (1983) "The Challenge of Social Forestry" In: W. Fernandes and S. Kulkarni, (eds.), *op.cit.*, pp.48-72.

Shiva, V., Sharatchandra, H.C. and Bandyopadhyay, J. (1987) "Social Forestry for Whom?", In: D.C. Korten, (ed.), *op.cit.*, pp.238-246.

Siddiquee, M.N.A. (1993) "Local Governance and Agricultural Development in Bangladesh: The Experience of Upazila System", *The Journal of Rural Development*, 23(1):1-26.

Singh, C. (1986) *Common Property and Common Poverty: India's Forests, Forest Dwellers and the Law*, Oxford University Press, New Delhi.

Singh, C. (1988) Contribution in *National Social Forestry Project: Mid-Term Review*, World Bank, United States Agency for International Development, and Government of India, New Delhi.

Singh, S. and Khare, A. (1993) "People's Participation in Forest Management", *Commonwealth Forestry Review*, 72(4):279-283.

Sobhan, R. (1968) *Basic Democracies, Works Programme and Rural Development in East Pakistan*, Bureau of Economic Research, Dhaka University, Dhaka.

Sobhan, R. (1982) "The Crisis of External Dependence: Some Policy Options for Bangladesh", *Journal of Social Studies*, 18(October 1982).

Spears, J. (1978) "The Changing Emphasis in the World Bank Forestry Lending: A Summary of Recent Experiences and Problems", Paper presented in the *Eight World Forestry Congress*, Jakarta.

SPWD (Society for Promotion of Wastelands Development) (1992) *Joint Forest Management: Concept and Opportunities*, Proceedings of a national workshop held at Surajkund (India), August 1992, SPWD, New Delhi.

Srivastava, B.P. and Pant, M.M. (1979) "Social Forestry on a Cost-Benefit Analysis Framework", *Indian Forester*, 105(1).

Stebbing, E.P. (1923) *The Forests of India*, Vol.1, The Bodley Head Limited, London.

Stein, B. (1980) *Peasant State and Society in Medieval South India*, Oxford University Press, Delhi, Oxford and New York.

Stevens, M.E., Bhumibhamon, S. and Wood, H. (eds.) (1990) *Research Policy for Community Forestry in Asia Pacific Region*, Proceedings of a seminar, January 8-11, 1990, Report 5, Regional Community Forestry Training Center, Bangkok.

Tamrakar, S.M. and Nelson, D. (1991) *Potential Community Forestry Land in Nepal*, HMGN/FAO/UNDP, Community Forestry Development Project, Field Document 16, NEP/85/017, Ministry of Forest and Soil Conservation, Kathmandu.

TANDP (*Thana* Afforestation and Nursery Development Project) (1993) *A*

Guide to Social Forestry, Nursery Development, and Plantation Techniques, (in Bengali), TANDP, Ministry for Environment and Forest, Forest Department, Dhaka.

TANDP (*Thana* Afforestation and Nursery Development Project) (1994) *Project Proforma*, TANDP, Ministry for Environment and Forest, Forest Department, Dhaka.

TANDP (*Thana* Afforestation and Nursery Development Project) (n.d.) *Thana Afforestation and Nursery Development Project: Background, Technology and Application* (in Bengali), TANDP, Ministry for Environment and Forest, Forest Department, Dhaka.

Task Force (1987) *Participatory Forestry in Bangladesh: Concepts, Experiences and Recommendations*, A Task Force Report to the Ministry of Agriculture, Dhaka.

Task Force (1991) *Report of the Task Forces on Bangladesh Development Strategies for the 1990s: Environment Policy*, Vol. 4, University Press Limited, Dhaka.

Tepper, E.L. (1976) "The Administration of Rural Reform: Structural Constraints and Political Dilemmas", In: R.D Stevens, H. Alavi and P.J. Bertocci, (eds.), *Rural Development in Bangladesh and Pakistan*, East West Centre, Honolulu, pp.29-59.

Tiwari, K.M. (1983) *Social Forestry for Rural Development*, International Book distributors, Dehra Dun.

Tiwari, K.M. (1986) "Sir Dietrich Brandis and Social Forestry", In: *Tributes to Sir Dietrich Brandis*, Forest Research Institute Press, Dehra Dun.

Torrance, J. (1981) "Alienation and Estrangement as Elements of Social Structure", In: R.F. Geyer and D. Schweitzer, (eds.), *op.cit.*, pp.68-96.

Trainer, T. (1989) *Developed to Death: Rethinking Third World Development*, Green Print, London.

Tucker, R.P. (1988) "The British Empire and India's Forest Resources: The Timberlands of Assam and Kumaon, 1914-1950", In: J.F. Richards and R.P. Tucker, (eds.), *op.cit.*, pp.91-111.

Tuladhar, A.R. (1991) "Agroforestry in Nepal", In: Mellink et al., (eds.), *op.cit.*, pp.106-109.

United Nations Statistical Commission (1979) *National Household Survey Capability Programme*, Draft outline of the proposal, Tabled at UN Statistical Commission, New York.

Uphoff, N.T. (1982a) "Introduction to South Asian Cases", In: N.T. Uphoff, (ed.), *ibid.*, pp.33-44.

Uphoff, N.T. (1986) *Local Institutional Development: An Analytical Source Book with Cases*, Kumarian Press, West Hartford, Connecticut.

Uphoff, N.T. (1989) *Approaches to Community Participation in Agriculture and Rural Development*, EDI Catalogue No.076/005, Economic Development Institute, World Bank, Washington.

Uphoff, N.T. (ed.) (1982) *Rural Development and Local Organization in Asia*, Vol:1 (Introduction and South Asia), Macmillan India Limited, New Delhi.

Uphoff, N.T., Cohen, J. and Goldsmith, A. (1979) *Feasibility and Application of Rural Development Participation: A State-of-the-Art Paper*, Rural Development Committee, Cornell University, Ithaca.

Vergara, N. et al. (1986) *Social Forestry Research Issues*, Overseas Development Institute, London.

Wade, R. (1988) *Village Republics: Economic Condition for Collective Action in South India*, Cambridge University Press, Cambridge.

Wadud, A.A.N.M. (1989) *Need for Change in Forest Act*, B.Sc. (Honours) Review Paper 111, Institute of Forestry, Chittagong.

Wallace, B.J., Ahsan, R.M., Hussain, S.H. and Ahsan, E. (1987) *The Invisible Resource: Women and Work in Rural Bangladesh*, Westview Press, Boulber, Colorado.

Wanigasundara, M. (n.d.) *Farming with Ahimsa*, The Panos Institute, London.

Watson, D.J. (1989) "The Evolution of Appropriate Resource Management Systems", In: F. Berkes, (ed.), *Common Property Resources: Ecology and Community-Based Sustainable Development*, Belhaven Press, London.

WBFD (West Bengal Forest Department) (1986) *Progress Report on Social Forestry*, Social Forestry Wing, Forest Department, Calcutta.

WBFD (West Bengal Forest Department) (1989) *Forest Regeneration through Community Protection: The West Bengal Experience*, Proceedings of the working group meeting on Forest Protection Committees, Calcutta, June 21-22, 1989, K.C. Malhotra and M. Poffenberger, (eds.), Forest Department, Calcutta.

Westergaard, K.(1985) *State and Rural Society in Bangladesh: A Study in Relationships*, Curzon, London.

White, S. (1992) *Arguing with the Crocodile: Gender and Class in Bangladesh*, Zed Books Limited, London and New Jersey, and University Press Limited, Dhaka.

Wickramasinghe, A. (1992) "Women and Equity in Forestry: A Case Study in Sri Lanka" , Paper presented at the workshop on *Sustainable and Effective Management Systems for Community Forestry*, January 15-17, 1992, Regional Community Forestry Training Centre, Bangkok.

Wignaraja, P. (1984) "Towards a Theory and Practice of Rural Development",

Development, 2.

Wilber, C.K. and Jameson, K.P. (1992) *The Political Economy of Development and Underdevelopment*, Fifth Edition, McGraw-Hill Inc., Singapore.

Wilkins, E.J. (1970) *An Introduction to Sociology*, Macdonald and Evans Limited, London.

Wood, G.D. (1994) *Bangladesh: Whose Ideas, Whose Interests?*, Intermediate Technology-Bangladesh, University Press Limited, Dhaka.

World Bank (1994) "The World Bank in Forest Sector, India: Policies and Issues in Forest Sector Development", *Indian Forester*, 120(4):291-313.

World Bank (1979) *Bangladesh: Food Policy Issues*, (December 19, 1979), World Bank, Washington and Dhaka.

Worsley, P. (ed.) (1977) *Introducing Sociology*, Penguin Books, Harmondsworth, Middlesex.

WRI (World Resources Institute) (1987) *The Tropical Forestry Action Plan*, New York.

Zabala, N.Q. (1990) *Principles of Silviculture*, UNDP/FAO/85/011, Field Document 7, Institute of Forestry, Chittagong.

Zabala, N.Q. (1990a) *Agroforestry*, UNDP/FAO/85/011, Field Document 1, Institute of Forestry, Chittagong.

Zaman, M.Q. (1977) *Social Conflict and Political Process in Rural Bangladesh: A Case Study*, M.Phil. Dissertation, Institute of Bangladesh Studies, Rajshahi University, Rajshahi.

Zaman, W.A (1984) *Public Participation in Development and Health Programs: Lessons from Rural Bangladesh*, University Press of America, Lanham and New York.

Newspapers, Magazines and Television Programmes

The Bangladesh Observer: 19 February 1991;

The Bangladesh Times: 14 August 1991;

The Daily Azadi: 22 September 1992; 9 May 1994; 17 June 1994;

The Daily Ittefaq: 7 July 1994;

The Daily Janakantha: 3 Ashar 1401 Bengali Year; 16 February 1994; 28 Februray 1994; 28 March 1994; 3 May 1994; 20 September 1994;

The Daily Purbakon: 22 April 1994;

The Daily Sangbad: 28 December 1987;

The Dainik Bangla: 26 July 1994; 7 August 1994;

The Dhaka Courier: 29 April 1994; 31 May - 6 June 1991;

The Friday: 20 February 1992;
The Janamat 26 May - 1 June 1995;
The New Nation: 10 February 1990;
Avimot, Bangladesh Television: 9 April 1994;
Samoya, Bangladesh Television: 23 May 1994.

Records, Files and Miscellaneous Documents of Public, Local Government and *Samity* Offices

Agroforestry Research and Demonstration Project (ARDP) Office, Dhaka

ARDP Project Proposal;
ARDP (amended) Project Proposal;
Participants' Profile;
Plantation Manual;
Reports of Research Officers on the Project sites (5 sites);
Staffing Pattern file;
Tour Report of the Consultant Sociologist.

Bangladesh Forest Academy, Chittagong

Draft Forestry Master Plan Documents (8 files);
Forest Policies Manual;
Government of Bangladesh Gazettes, 1993-94;
Gradation List of Bangladesh Civil Service: Forests;
Progress Reports on East Pakistan Forest Directorate, 1961-1967.

Bangladesh Forest Research Institute (BFRI), Chittagong

Bangladesh Forestry Statistics Records, 1982-1992;
BFRI Research Progress Report, 1994;
Draft expansion plan of the BFRI under the Forest Resource Management Project 1994-95;
Draft proposal file on Strengthening and Reorganisation of the BFRI;
Forest Economics and Statistics Division's Time Series Data file on Betagi Pomora Projects;
Profile of Common Tree and Fruit Species of Rural Bangladesh (with their botanical names);
Profile of Non-Wood Forest Products.

Betagi Union Council Office, Chittagong

Election Record files;
Population Survey Records, 1991;
Primary Data: Betagi Union;
Revenue Sources and Earnings, 1989/90 - 1993/94;
Union Expenditure (estimated and actual) file, 1991/92.

Chittagong Forest Division (CFD), Chittagong

Betagi-Pomora Projects' Records, 1978-1994;
Correspondences between Divisional Forest Officer (DFO), Comilla Extension and DFO, Chittagong regarding Betagi-Pomora Projects, 1980-1983;
Correspondences between DFO, Chittagong and Conservator of Forests (CF), Eastern Circle and Deputy Chief Conservator of Forests (DCCF), Extension Circle, regarding Betagi-Pomora Projects 1979-1982;
DFO's Correspondence file on Betagi-Pomora Projects, 1978-1994;
DFO's Enquiry and Investigation Reports on Various Allegations and Problems about Betagi-Pomora Projects, 1987-1990;
DFO's Progress Report on Betagi-Pomora Projects, 1981, 1982;
Encroachment Records;
Khazna (Revenue) Register on Betagi-Pomora Projects;
Land Records file, CFD;
Notification Register (No. 42T. For.) 1935;
Plantation Records, 1969-1994;
Records of Assault on Forest Staffs and Forest Crimes;
Temporary Lease Contract Deed (between Pomora farmers and the forest department).

Dhaka Forest Division, Dhaka

Agreement form concerning benefit distribution under the *Thana* Afforestation and Nursery Development Project (TANDP) (issued by the Ministry of Environment and Forests), 1994;
DFO's (selected) Correspondences with Kaliakair Range Office 1992-1994;
Guideline for Incorporating Non-Governmental Organisations in the TANDP, 1994;
Tour Report of the Land Minister following his visit to Manikganj Community Forestry Project, 18.7.1987;
Working Plans, Dhaka-Mymensingh Division 1938.

Forest Headquarters, Dhaka

Correspondences between DCCF, Extension Circle and DFO, Chittagong, and DFO, Comilla, concerning Betagi-Pomora projects, 1982-85;
Correspondences between DCCF, Extension Circle and Manager, Bangladesh *Krishi* (Agriculture) Bank, Pomora, 1982-83;
Correspondences concerning Land Possession for Betagi-Pomora Projects (between Chief Conservator of Forests (CCF) and CF, Eastern Circle, 1980-1982);
Land under the purview of the forest department (CF, General Administration's office records);
Revenue Earning and Expenditure of the forest department, 1980/81-1988/89 (CF, General Administration's office records);
Work Schedule (and General Guideline for Forest Circles) for Implementing Projects under the ADP (Annual Development Programmes), 1993-94 (CCF's office letter, dt.5.7.93);
World Forestry Statistics (DCCF, Environment Management Wing office).

Forest Utilization Division, Chittagong

Indian Forester (archival collections, 1936-1941);
Railway Sleeper Distribution Records;
Timber Export Statistics.

Forest Working Plan Division, Chittagong

Draft Land Survey and Territorial Jurisdiction Maps of CFD;
Working Plans for the Chittagong Forest Division (CFD), 1923, 1969, 1980, 1993.

Kaliakair Forest Range Office, Gazipur

Chandra Gazette Notifications;
Forest Range Officer (FRO)'s correspondence file (with DFO, Dhaka and Beat Officer (BO), Chandra);
Land Acquisition and Notification file, Kaliakair;
Land Encroachment Records as in 1993, Kaliakair Range;
Mouza Maps;
Plantation Journal of Chandra ARDP, 1993;
Staff Statistics, Kaliakair Range;

Visitor's Book of Chandra *Bungalow*.

Kaliakair Thana Administration, Gazipur

Revenue Earning Statistics of Kaliakair;
Thana Profile Records, 1993-94.

Mouchak Union Council Office, Kaliakair, Gazipur

Expenditure and Revenue file;
Union Statistics file.

Pomora Forest Extension Centre, Rangunia, Chittagong

Correspondences between Officer-in-Charge and Chairmen Betagi-Pomora Unions;
Correspondences between Officer-in-Charge and DFO, Chittagong, regarding Betagi-Pomora Projects, 1980-1991;
Disciplinary Action and Notice (against SF farmers) file, 1979-1978;
Litigation file on Betagi-Pomora, 1979-1985;
Proceedings of *Thana* Selection Committee, 1979-1992;
Proceedings of the Landless Farmer Rehabilitation Project, 1978-1982;
Profile of Officers-in-Charge, 1982-1994;
Samity Election Schedules files, 1980, 1982, 1993.

Pomora Union Council Office, Rangunia, Chittagong

Members' (description and honorarium) file;
Proceedings of meetings on Betagi-Pomora Projects, 25.5.1980-22.2.1982;
Union Election Records;
Union Expenditure (estimated and actual) and Revenue file, 1991/92-1993/94;
Union Statistics file, Pomora.

Rangunia Forest Range Office, Chittagong

Beat Officers' (Pomora, Kodala, Chiringa Beats) Reports to FRO, 1987-1994;
Classification of Forest Lands Register, Rangunia Range;
"Disforested" Area Register, Pomora Beat;
Encroachment Register, Rangunia Range, 1994;
Forest Offence (POR/UDOR/COR) Report file, Rangina Range;

FRO's Correspondence file with the DFO, Chittagong, 1988-1994;
Information file on Brick-kilns and Saw Mills;
Land Area Register, Kodala Beat;
Land Register and Schedule file, Rangunia Range;
List of Notorious Loggers, 1992-1994, Rangunia Range;
List of Saw-Mills situated within 3 Miles of Public Forests;
Manpower Statistics, Rangunia Range;
Mouza, Beat and Range Territorial Demarcation and Plantation Maps, Rangunia Range;
Plantation Chart of Rangunia Range, 1968-1992;
Plantation Records, Pomora Beat, 1968-1989;
Plantation Register, Kodala.

Rangunia Thana Administration, Chittagong

Archival Records on *Khas mahal Tahsilder*;
Archival Records on Revenue Survey (RS);
CO (Revenue)'s corresponces concerning Community Forestry, 1977-1981;
Draft *Thana* Profile records;
Land Revenue statistics;
Mouza wise list of Civil Suits concerning Betagi-Pomora Projects, Circle Officer, Revenue (CO:Revenue)'s Office, 1979-1983;
Proceedings Register of the Rangunia *Thana* Central Cooperative Association, 1977/78-1982/83;
Rangunia *Thana* at a glance: *Thana Nirbahi* Officer (TNO)'s records;
Statistics on Brick-kilns.

Social Forestry Farmers' Samity Offices, Betagi and Pomora

Bank (*Grameen* and *Krishi* banks) Loan Register 1980-1983, Betagi-Pomora;
Betagi Primary School records;
Budget and Expenditure register; Betagi-Pomora;
Chairmen's External Correspondence files, Betagi-Pomora, 1983-1994;
Election (1993) file, Pomora;
Permanent Land Contract Deeds (to Betagi farmers from the *Upazila* Revenue Officer, Rangunia), 1987;
Plot Survey and Demarcation Maps, Pomora;
Proceedings of Group-Meetings, Betagi, 1979-1993;
Proceedings of Group-Meetings, Pomora, 1980-1994;
Savings Register, Betagi-Pomora;

Seedlings Distribution register, Betagi-Pomora;
Visitor's Note Book.